Lifeng Fan
College of Architecture and Civil Engineering
Beijing University of Technology
Beijing, China

Xiuli Du
College of Architecture and Civil Engineering
Beijing University of Technology
Beijing, China

Meng Wang
School of Civil Engineering
Shijiazhuang Tiedao University
Shijiazhuang, China

College of Architecture and Civil Engineering
Beijing University of Technology
Beijing, China

ISBN 978-981-97-1439-1 ISBN 978-981-97-1440-7 (eBook)
https://doi.org/10.1007/978-981-97-1440-7

Jointly published with Science Press
The print edition is not for sale in China (Mainland). Customers from China (Mainland) please order the print book from: Science Press.

This Springer imprint is published by the registered company Springer Nature Singapore Pte Ltd.
The registered company address is: 152 Beach Road, #21-01/04 Gateway East, Singapore 189721, Singapore

Preface

Dynamic loads are usually generated in the construction of rock mass engineering. Dynamic loads propagate in the form of stress waves in the rock mass and affect the stability of the rock mass. The dynamic stability analysis of rock engineering is a difficult problem that needs to be solved urgently in engineering construction. The study of stress wave propagation in jointed and fractured rock bodies is one of the key scientific issues to solve this engineering challenge. Therefore, the analytical methods of stress wave propagation in rock mass have been widely emphasized. This book is based on extensive research and in-depth analysis of domestic and international studies on stress wave propagation in rock mass. It focuses on the analysis method of characteristics for the stress wave propagation in the rock mass. This book can provide a reference for the study of rock mass dynamics and the analysis and control of dynamic stability of rock mass engineering.

The book is divided into nine chapters. Chapter 1 is the introduction, which introduces the background of the study and summarizes the current state of research on stress wave propagation in rock mass. Chapter 2 introduces the basic theory of stress wave propagation. Chapters 3–6 introduce the two-characteristic method for stress wave propagation in intact rock mass, jointed rock mass, complex rock mass and layered complex rock mass, respectively. Chapter 7 describes the three-characteristic method for wave propagation through the micro-defected rock mass. Chapter 8 describes a split three-characteristic method for stress wave propagation through rock mass with double-scale discontinuities. Chapter 9 describes the dual-mesh three-characteristic method for stress wave propagation through the micro-defected rock mass with a thin layer-filled macro-joint. The book is structured according to the characteristics of the content system of the discipline, and strives to be systematic, complete and cutting-edge, guiding readers to be coherent and laying a foundation for engaging in rock dynamics-related research.

The aim of this book is to provide a brief introduction to the analysis of the method of characteristics. The application of the characteristic lines to the study of stress waves through rock mass with discontinuity is discussed, and then the research work carried out in recent years is summarized. This book involves some mathematical analysis, but it has been kept as simple as possible so that readers with a foundation in advanced mathematics and elastic-plastic mechanics can successfully read it. The readers will gain a comprehensive understanding of the characteristic method for studying stress wave propagation by learning from this book.

Beijing, China

Lifeng Fan
Meng Wang
Xiuli Du

Contents

List of Figures

List of Tables

Chapter 1
Introduction

1.1 Background

The study of stress wave propagation in rock mass is important for rock engineering stability analysis. Natural rock mass is a special material composed of a variety of minerals. After various geological tectonic movements, discontinuous ranging from micro-defect to macro-joint is formed inside it. The primary discontinuity set with wider spacing relative to the stress wavelength is briefly called "macro-joint", and the secondary discontinuity set is characterized by large density and small spacing relative to the stress wavelength, briefly called "micro-defect" in this book. The mechanical behavior of the rock mass is characterized by the physical properties and geometrical dimensions of rock blocks as well as by characteristics of discontinuities. The presence of the discontinuities results in increasing overall mechanical compliance and decreasing overall strength of the rock mass.

Due to the presence of discontinuities, modeling the dynamic behavior of the rock mass becomes a difficult task and covers a wide variety of aspects such as mechanics, geology, hydrology and rheology. The stress wave propagation method is always employed for its easy operation, good repeatability and high accuracy. It is commonly recognized that when a stress wave propagates through the rock mass, the discontinuities result in the attenuation of stress wave amplitude, slowing of wave propagation velocity and transformation of waveforms.

The discontinuities are sometimes taken into account implicitly when engineering problems take material in quantities large enough so that the rock mass can be described by the averaged material properties, which are assumed to be continuous. Therefore, the rock mass properties are captured by the effective moduli of the equivalent continuum medium [1–4]. These treatments can be used to avoid the difficulty in determining the complex multiple reflections between individual joints, so they are widely used for analyzing the wave propagation across multiple parallel joints in a long wavelength limit, extremely effective for the micro-cracks or

L. Fan et al., *The Method of Characteristics for Stress Wave Propagation in the Rock Mass*, https://doi.org/10.1007/978-981-97-1440-7_1

highly dense macro-joint cases [5–10]. The effect of the rock and contained discontinuities (both macro-joint and micro-defect) are lumped into the effective elastic modulus and quality factor of the equivalent medium. However, the rock mass is always discontinuous, inhomogeneous, anisotropic, non-uniform and nonlinear in engineering problems. The traditional equivalent simplification inherently results in some limitations. For example, the discreteness of wave attenuation at individual macro-joints is always ignored; the intrinsic frequency-dependent property caused by the micro-defect cannot be considered; the wavelength should be larger than the joint spacing, etc. Viscoelastic equivalence has been introduced to investigate the mechanical behavior of the rock mass and its effect on wave propagation to avoid the limitations induced by traditional elastic equivalence.

1.2 Stare of the Art of Wave Propagation in the Rock Mass

1.2.1 Wave Propagation in the Micro-Defected Rock Mass

Efforts have been made to evaluate the effect of micro-scale discontinuities on the elastic wave propagation through the rock mass. The wave scattering methods have been adopted to study the elastic wave propagation velocity and amplitude attenuation [11–14]. In these methods, the propagating elastic waves are considered to be uniformly scattered by the micro-cracks. These methods are suitable for analyzing elastic wave propagation provided that the crack size is small compared to the wavelength and the crack sparsely distributes. Besides the wave scattering methods, equivalent continuum methods have been developed to investigate the mechanical behavior of micro-defected rock from a homogenized view, and to analyze the effect of pores, impurities, grains or micro-cracks on elastic wave propagation through micro-defected rock [14–17]. In these methods, the background material and the micro-defect together are approximated by an equivalent continuum and the effect of the micro-defect on the rock properties is described by the effective moduli of the equivalent medium, which can be utilized to account for the elastic wave propagation. Viscoelastic medium has also been used as an equivalent continuum to describe the dynamic mechanical behavior of micro-defected rock [6, 18]. Barenblatt et al. [19] presented the basic physical concepts of the motion of liquids in fissured rocks and introduced their mean characteristics. The physical meaning of the viscous component of rock under transmitted waves was discussed, and a possible mechanism for viscoelastic behavior was suggested by Yoshioka and Kikuchi [20]. The inherent existence of micro-defects in rock including grain-boundaries, micro-cracks, micro-cavities and mineral cleavages leads to a viscoelastic behavior in rock. Ichikawa et al. [6] observed a three-dimensional texture of granitic rock by X-ray tomography microscopy (XTM) images using an ultra-bright synchrotron radiation (SR)-CT system. The molecular dynamic computer simulation method was introduced to study the properties of each constituent mineral and its interfaces. A micro–macro model of granite by a homogenization analysis method was established to simulate the viscoelastic behavior. An experimental study on the viscoelastic property of micro-defected sedimentary rock under dynamic loading using transient elastic wave propagation method was performed by Fan and Wong [21], in which a frequency-dependent viscoelastic model was proposed to describe the mechanical behavior of micro-defected sedimentary rock.

1.2.2 Wave Propagation in Jointed Rock Mass

Efforts have also been made to evaluate the effect of macro-scale discontinuities on the elastic wave propagation through the rock mass. Macro-scale discontinuities can be regarded as structural features of thin layers with mechanical behavior different from that of the adjacent intact rock, i.e., fractures, faults and macro-veins, referred to as "macro-joint" for brevity [22–24]. The equivalent medium methods are extended to investigate the dynamic behavior of the rock mass containing macro-joint [25, 26]. Similar to the consideration of micro-defects, the additional compliance of a dilute population of macro-joint is incorporated into an average strain, resulting in effective moduli for the rock mass as homogenized ones, so as to allow a continuum analysis. A transversely isotropic medium with five elastic constants was used as an equivalent medium of the rock mass containing a single set of plane parallel joints based on an average strain consideration. Ma et al. [27] evaluated the difference between the static and dynamic equivalent medium methods for further practical application. Their research showed that the results obtained by the traditional static and dynamic equivalences were not granted for the macro-jointed rock mass. A viscoelastic equivalent medium was used to consider the dynamic behavior of the rock mass containing macro-joint, which predicted the results reasonably well. As an alternative to the continuum method, the displacement discontinuity methods, which treat the joint as an individual boundary in the discrete medium, are applicable to analyze the effect of macro-joint on the wave propagation [28]. The displacement discontinuity methods were originally developed by Mindlin [29] to study the coupling between compressive and shear waves at the boundary of elastically restrained plates. In such methods, joints are assumed large in extent and small in thickness relative to the incident wavelength. Therefore, when elastic waves propagate across such a joint, the stress field is continuous but the displacement field is discontinuous due to the joint deformation. These two assumptions can be mathematically represented as two boundary conditions in the wave equations. Miller [30] established the displacement discontinuity boundary conditions by considering the slip rate-dependent shear behavior of a fracture. In their studies, the shear stress was assumed to be a nonlinear function of frictional slip and slip rate of the fraction. Zhao and Cai [31] presented a theoretical study on normally incident elastic P-wave transmission across single dry joints with a nonlinearly normal deformational behavior. The effect of nonlinear fracture normal behavior on P-wave transmission was examined without the mixture of fracture shear behavior. Successful predictions have been made by using the displacement discontinuity method to analyze the seismic wave propagation through the macro-jointed rock mass.

1.2.3 *Wave Propagation Through the Rock Mass with Double-Scale Discontinuities*

All the previous methods can be used to model the dynamic behavior of rock containing single-scale discontinuities theoretically. In order to investigate the mechanical behavior of the rock mass containing double-scale discontinuities, Auriault and Boutin [32] proposed a homogenization technique which described the macro-scope modeling for a deformable porous medium with double porosity constituted by pores and fractures; their models were obtained under quasi-static loading condition. The homogenization principle was extended as an effective field method to study the wave propagation in an elastic double-porosity system [16], in a poroelastic composite material with analogous micro-structure and in a viscoelastic double-porosity system [33]. The effective field methods were successfully used to predict the velocities and quality factors of elastic waves propagating in a double-porosity medium. However, the constitution of the macro-scale discontinuities cannot be considered explicitly in such methods and the discreteness caused by the macro-scale discontinuities also cannot be described [34]. The manifold method (MM) was used to numerically model the rock mass based on a discrete-continuum approach [35, 36]. The micro-scale discontinuities were evaluated by an equivalent continuum medium method while the macro-scale discontinuities were described by an explicit physical model. A framework was proposed based upon a mesh-based partition of unity method, which was illustrated in two example applications. However, their studies were explored in a numerical investigation, while a complete theoretical derivation of the effective mechanical behavior of the rock mass containing double-scale discontinuities is not available.

References

1. Sheshenin SV, Kalinin EV, Bujakov MI. Equivalent properties of rock strata: Static and dynamic analysis. Int J Numer Anal Met. 1997;21(8):569–79.
2. Ma G, Li M, Wang H, et al. Equivalent discrete fracture network method for numerical estimation of deformability in complexly fractured rock masses. Eng Geol, 2020; 277.
3. Li JC, Ma GW, Zhao J. An equivalent viscoelastic model for rock mass with parallel joints. J Geophys Res-Solid Ea. 2010;115(B3):B03305.
4. Li JC, Li HB, Zhao J. An improved equivalent viscoelastic medium method for wave propagation across layered rock masses. Int J Rock Mech Min. 2015;73:62–9.
5. Nguyen ST, To QD, Vu MN. Extended analytical solutions for effective elastic moduli of cracked porous media. J Appl Geophys. 2017;140:34–41.

6. Ichikawa Y, Kawamura K, Uesugi K, et al. Micro- and macrobehavior of granitic rock: Observations and viscoelastic homogenization analysis. Comput Method Appl M. 2001;191(1–2):47–72.
7. Wu Z, Fan L, Liu Q, et al. Micro-mechanical modeling of the macro-mechanical response and fracture behavior of rock using the numerical manifold method. Eng Geol. 2017;225:49–60.
8. Nguyen ST, Dormieux L. Propagation of micro-cracks in viscoelastic materials: Analytical and numerical methods. Int J Damage Mech. 2014;24(4):562–81.
9. Gao J, Fan L, Wan Z. Thermal cycling effects on the dynamic behavior of granite and microstructural observations. B Eng Geol Environ. 2021;80:8711–23.
10. Jansen DP, Carlson SR, Young RP, et al. Ultrasonic imaging and acoustic emission monitoring of thermally induced microcracks in lac du bonnet granite. J Geophys Res: Solid Earth. 1993;98(B12):22231–43.
11. Iturraran-Viveros U, Sanchez-Sesma FJ, Luzon F. Boundary element simulation of scattering of elastic waves by 3-d cracks. J Appl Geophys. 2008;64(3–4):70–82.
12. Sánchez-Sesma FJ, Iturrarán-Viveros U. Scattering and diffraction of sh waves by a finite crack: An analytical solution. Geophys J Int. 2001;145(3):749–58.
13. Hudson JA. A higher order approximation to the wave propagation constants for a cracked solid. Geophys J Int. 1986;87(1):265–74.
14. Hudson JA, Pointer T, Liu E. Effective-medium theories for fluid-saturated materials with aligned cracks. Geophys Prospect. 2001;49(5):509–22.
15. O'Connell RJ, Budiansky B. Seismic velocities in dry and saturated cracked solids. J Geophys Res. 1974;79(35):5412–26.
16. Markov M, Levine V, Mousatov A, et al. Elastic properties of double-porosity rocks using the differential effective medium model. Geophys Prospect. 2005;53(5):733–54.
17. Müller TM, Gurevich B, Lebedev M. Seismic wave attenuation and dispersion resulting from wave-induced flow in porous rocks a review. Geophysics. 2010;75(5):75A147–64.
18. Fakhimi AA, Fairhurst C. A model for the time-dependent behavior of rock. Int J Rock Mech Min Sci & Geomech Abstr. 1994;31(2):117–26.
19. Barenblatt GI, Zheltov IP, Kochina IN. Basic concepts in the theory of seepage of homogeneous liquids in fissured rocks [strata]. J Appl Math Mech. 1960;24(5):1286–303.
20. Yoshioka N, Kikuchi M. Visco-elastic response of joints to transmission waves. Geophys Res Lett. 1993;20(12):1143–6.
21. Fan LF, Wong LNY. Stress wave transmission across a filled joint with different loading/unloading behavior. Int J Rock Mech Min. 2013;60:227–34.
22. Pariseau WG. An equivalent plasticity theory for jointed rock masses. Int J Rock Mech Min. 1999;36(7):907–18.
23. Liu E, Hudson JA, Pointer T. Equivalent medium representation of fractured rock. J Geophys Res-Solid Ea. 2000;105(B2):2981–3000.
24. Tod SR. The effects of stress and fluid pressure on the anisotropy of interconnected cracks. Geophys J Int. 2002;149(1):149–56.
25. Pyrak-Nolte LJ, Myer LR, Cook NGW. Anisotropy in seismic velocities and amplitudes from multiple parallel fractures. J Geophys Res. 1990;95(B7):11345–58.
26. Pyrak-Nolte LJ, Myer LR, Cook NGW. Transmission of seismic waves across single natural fractures. J Geophys Res. 1990;95(B6):8617–38.
27. Ma GW, Fan LF, Li JC. Evaluation of equivalent medium methods for stress wave propagation in jointed rock mass. Int J Numer Anal Met. 2013;37(7):701–15.
28. Schoenberg M. Elastic wave behavior across linear slip interfaces. J Acoust Soc Am. 1980;68(5):1516–21.
29. Mindlin RD. Waves and vibrations in isotropic elastic plates[M]. New York: Pergamon Press; 1960.
30. Miller RK. An approximate method of analysis of the transmission of elastic waves through a frictional boundary. J Appl Mech-T Asme. 1977;44(4):652–6.
31. Zhao J, Cai JG. Transmission of elastic p-waves across single fractures with a nonlinear normal deformational behavior. Rock Mech Rock Eng. 2001;34(1):3–22.

32. Auriault JL, Boutin C. Deformable porous media with double porosity. Quasi-statics. I: Coupling effects. Transp Porous Media, 1992; 7(1):63–82.
33. Levin VM, Markov MG. Propagation of low-frequency elastic waves in double-porositymedia containing viscoelastic component in the pores. Int J Eng Sci. 2011;49(1):140–54.
34. Zhao J, Zhao XB, Cai JG. A further study of p-wave attenuation across parallel fractures with linear deformational behaviour. Int J Rock Mech Min. 2006;43(5):776–88.
35. Jaeger JC, Cook NGW, Zimmerman R. Fundamentals of rock mechanics. 4th ed. Malden: Blackwell Publishing; 2007.
36. Bandis SC, Lumsden AC, Barton NR. Fundamentals of rock joint deformation. Int J Rock Mech Min. 1983;20(6):249–68.

Chapter 2
Basic Theory of Stress Wave

2.1 Basic Concept of Stress Wave

When the dynamic load is applied to the local surface of the deformable solid, the surface particle directly subjected to the dynamic load leaves the initial equilibrium position due to deformation and moves relative to the adjacent particle. The surface particle is subjected to the force given by the neighboring particle, while the neighboring particle is subjected to the reaction force of the surface particle and leaves the initial equilibrium position. Due to the inertia of the particle, the motion of the neighboring particle lags behind the motion of the surface particle. By analogy, the perturbation caused by the dynamic load on the surface propagates in the medium, and this perturbation from near to far is called the stress wave in the medium. The interface between the perturbed and undisturbed is called the wavefront surface, and the propagation velocity of the perturbation is called the wave velocity.

Stress waves are classified into the following types: (1) Stress waves can be divided into longitudinal waves (P-waves) and transverse waves (S-waves) according to the relationship between the direction of particle vibration and the direction of wave propagation, in which the vibration direction of the particle in the longitudinal wave is parallel to the wave propagation direction, and the vibration direction of the particle in the transverse wave is perpendicular to the wave propagation direction. (2) According to the interaction between the stress wave and the interface, the stress wave can be divided into Rayleigh wave, Leff wave and Stonley wave. Among them, the Rayleigh wave is the interface wave that appears near the elastic half-space or elastic delamination half-space, the Leff wave is the interface wave formed by the superposition of SH waves in the elastic delamination half-space, and the Stonley wave is the interface wave that propagates along the intersection of two different media. (3) According to the size of the stress and the deformation characteristics of the medium, stress waves can be divided into elastic and elastic–plastic waves. Among them, the deformation of the medium after the action of elastic waves can be completely recovered; the deformation of the medium after the action of elasto-plastic waves cannot be

L. Fan et al., *The Method of Characteristics for Stress Wave Propagation in the Rock Mass*, https://doi.org/10.1007/978-981-97-1440-7_2

completely recovered, and there is residual deformation. (4) According to the characteristics of the wave source and the shape of the wavefront, stress waves can be divided into plane waves, column waves and spherical waves. Among them, the plane wave source is plane load, the column wave source is line load, and the spherical wave source is point load. (5) According to the stress characteristics, stress waves can be divided into tensile waves, compressional waves, bending waves, shear waves and torsional waves. (6) According to the geometric characteristics of the propagation medium, stress waves can be divided into one-dimensional waves, two-dimensional waves (plane waves), three-dimensional waves (space waves), etc.

2.2 Reflection and Transmission of Stress Wave

2.2.1 Reflection and Transmission of Stress Waves at Interfaces

When the elastic wave is incident vertically from medium 1 (propagation direction is perpendicular to the interface) to medium 2, the elastic wave is reflected and transmitted at the interface due to the different wave impedances of the two mediums. The stress wave returning to medium 1 is the reflected wave. The stress wave entering medium 2 is the transmitted wave. The reflection and transmission of the stress wave are related to the wave impedance of the medium.

It is assumed that when the elastic wave propagation through the interface of different media, the interface always maintains contact (both pressure and tension and not separated). According to the continuity condition and Newton's third law, the velocity and stress of the particle on two sides of the interface are equal, which can be expressed as

$$v_{\mathrm{i}} + v_{\mathrm{r}} = v_{\mathrm{t}} \tag{2.1}$$

and

$$\sigma_{\mathrm{i}} + \sigma_{\mathrm{r}} = \sigma_{\mathrm{t}} \tag{2.2}$$

where the subscript i is the incident wave, r is the reflected wave, and t is the transmitted wave.

At first, the particle velocity under the action of the stress wave is solved according to the condition of momentum conservation. The increment of momentum generated by the micro-element dx at dt time is equal to the impulse generated by the internal and external forces at dt time. The expression of momentum conservation for the micro-element dx is

$$\rho_0 A_0 \mathrm{d}x \mathrm{d}v = [P(x + \mathrm{d}x, t) - P(x, t)]\mathrm{d}t \tag{2.3}$$

According to Newton's second law, it can be obtained as

$$\rho_0 A_0 \mathrm{d}x \frac{\partial v}{\partial t} = P(x + \mathrm{d}x, t) - P(x, t) = \frac{\partial P}{\partial x}\mathrm{d}x \tag{2.4}$$

Substituting Eq. (2.4) into Eq. (2.3) yields

$$\mathrm{d}\sigma = \pm \rho_0 C_\mathrm{P} \mathrm{d}v \tag{2.5}$$

where " + " is the stress wave propagating along the x-positive direction, and " − " is the stress wave propagating along the x-negative direction.

When the elastic wave propagates to the left side of the interface, according to Eq. (2.5), the particle velocity can be obtained as

$$v_\mathrm{i} = \frac{\sigma_\mathrm{i}}{(\rho_0 C_\mathrm{P})_1} \tag{2.6}$$

where $(\rho_0 C_\mathrm{P})_1$ is the wave impedance of medium 1.

When the elastic wave propagates to the interface, it produces a reflected wave. According to Eq. (2.5), the particle velocity of the reflected wave is

$$v_\mathrm{r} = -\frac{\sigma_\mathrm{r}}{(\rho_0 C_\mathrm{P})_1} \tag{2.7}$$

When the elastic wave propagates through the interface, it produces a transmitted wave. According to Eq. (2.5), the particle velocity of the transmitted wave is

$$v_\mathrm{t} = \frac{\sigma_\mathrm{t}}{(\rho_0 C_\mathrm{P})_2} \tag{2.8}$$

where $(\rho_0 C_\mathrm{P})_2$ is the wave impedance of medium 2.

Substituting Eqs. (2.6) to (2.8) into Eq. (2.1) yields

$$\frac{\sigma_\mathrm{i}}{(\rho_0 C_\mathrm{P})_1} - \frac{\sigma_\mathrm{r}}{(\rho_0 C_\mathrm{P})_1} = \frac{\sigma_\mathrm{t}}{(\rho_0 C_\mathrm{P})_2} \tag{2.9}$$

Equation (2.9) can be combined with Eq. (2.2) to obtain the reflected stress σ_r, reflected velocity v_r, transmitted stress σ_t and transmitted velocity v_t, respectively:

$$\begin{cases} \sigma_\mathrm{r} = \dfrac{(\rho_0 C_\mathrm{P})_2 - (\rho_0 C_\mathrm{P})_1}{(\rho_0 C_\mathrm{P})_2 + (\rho_0 C_\mathrm{P})_1}\sigma_\mathrm{i} \\ v_\mathrm{r} = -\dfrac{(\rho_0 C_\mathrm{P})_2 - (\rho_0 C_\mathrm{P})_1}{(\rho_0 C_\mathrm{P})_2 + (\rho_0 C_\mathrm{P})_1}v_\mathrm{i} \end{cases} \tag{2.10}$$

and

$$\begin{cases} \sigma_t = \dfrac{2(\rho_0 C_P)_2}{(\rho_0 C_P)_1 + (\rho_0 C_P)_2}\sigma_i \\ v_t = \dfrac{2(\rho_0 C_P)_1}{(\rho_0 C_P)_1 + (\rho_0 C_P)_2}v_i \end{cases} \tag{2.11}$$

Assume the following equation holds:

$$\begin{cases} F = \dfrac{(\rho_0 C_P)_2 - (\rho_0 C_P)_1}{(\rho_0 C_P)_1 + (\rho_0 C_P)_2} \\ T = \dfrac{2(\rho_0 C_P)_2}{(\rho_0 C_P)_1 + (\rho_0 C_P)_2} \end{cases} \tag{2.12}$$

where F is the reflection coefficient, and T is the transmission coefficient.

The relationship between F and T is

$$T = 1 + F \tag{2.13}$$

Equation (2.10) can be expressed as

$$\begin{cases} \sigma_r = F\sigma_i \\ v_r = -Fv_i \end{cases} \tag{2.14}$$

Equation (2.11) can be expressed as

$$\begin{cases} \sigma_t = T\sigma_i \\ v_t = \frac{(\rho_0 C_P)_1}{(\rho_0 C_P)_2}Tv_i \end{cases} \tag{2.15}$$

According to Eq. (2.12), it is known that T is positive. The stress (velocity) of the transmitted wave and the stress (velocity) of the incident wave have the same sign. The positive or negative of F depends on the relative magnitudes of the wave impedance of the two media, and the positive or negative of F is discussed in two cases.

(1) When $(\rho_0 C_P)_1 < (\rho_0 C_P)_2$, $F > 0$. The stress (velocity) of the reflected wave and the stress (velocity) of the incident wave have the same sign. In this case, $T > 1$, indicating that the stress wave from a medium with small wave impedance into a medium with large wave impedance will occur when the stress and mass velocity amplification.

(2) When $(\rho_0 C_P)_1 > (\rho_0 C_P)_2$, $F < 0$. The stress (velocity) of the reflected wave and the stress (velocity) of the incident wave have different signs. In this case, $T > 1$, indicating that the stress and mass velocity decrease when the stress wave propagates from a medium with large wave impedance to a medium with small wave impedance.

In summary, even if the two different media ρ_0 and C_P are different, as long as their wave impedance is the same, that is, $(\rho_0 C_P)_1 = (\rho_0 C_P)_2$, the elastic wave will not produce reflections when propagating at the interface of different media.

2.2.2 Reflection of Stress Waves at the Fixed and Free Ends

The reflection of a one-dimensional stress wave at the fixed and free ends can be regarded as a special case of reflection at different media interfaces. When the stress wave is reflected at the free end surface, i.e., $(\rho_0 C_P)_2 \to 0$, $T = 0$ and $F = -1$. At this time, Eq. (2.10) can be expressed as

$$\begin{cases} \sigma_r = -\sigma_i \\ v_r = v_i \end{cases} \tag{2.16}$$

Therefore, the stress wave is reflected at the free end surface when $\sigma_i = -\sigma_r$, and the particle velocity $v = v_i + v_r$, i.e., at the free end surface, the total stress is zero and the particle velocity is doubled. The compressional wave is reflected as a tensile wave, and the tensile wave is reflected as a compressional wave.

When the stress wave is reflected at the fixed end face, i.e., $(\rho_0 C_P)_2 \to \infty$, $T = 2$ and $F = 1$. Equation (2.10) can be expressed as

$$\begin{cases} \sigma_r = \sigma_i \\ v_r = -v_i \end{cases} \tag{2.17}$$

Therefore, the stress wave is reflected at the fixed end face with $v_i = -v_r$ and the total stress $\sigma = \sigma_i + \sigma_r$, i.e., at the fixed end face, the particle velocity is zero and the stress is doubled. The compressional wave is reflected as a compressional wave and the tensile wave is reflected as a tensile wave.

Chapter 3
Two-Characteristic Method for the Wave Propagation Through Intact Rock

3.1 Establishment of the Characteristic Method

3.1.1 Governing Equations for Stress Wave Propagation

Wave propagation in the rock mass has been widely studied. The governing equations for stress wave propagation are usually established based on the motion equation, the continuity equation and the material constitutive equation. Based on the control equation, the characteristic method of stress wave propagation through the rock mass can be solved. The constitutive equations are determined by the material properties, and the equations of motion and continuity can be obtained by the following methods.

The initial cross-sectional area of the bar is set to be A_0 and the initial density is ρ_0. The force analysis of any micro-element dx before deformation (at $t = 0$) is selected, as shown in Fig. 3.1. For the micro-element $\mathrm{d}x$, assuming that the total force on the cross-section R is $P(x, t)$, the total force on the cross-section S can be expressed as

$$P(x + \mathrm{d}x, t) = P(x, t) + \frac{\partial P(x, t)}{\partial x}\mathrm{d}x \tag{3.1}$$

where t is the time, and x is the coordinate of the mass point.

According to Newton's second law, it can be obtained as

$$\rho_0 A_0 \mathrm{d}x\frac{\partial v}{\partial t} = P(x + \mathrm{d}x, t) - P(x, t) = \frac{\partial P}{\partial x}\mathrm{d}x \tag{3.2}$$

Substituting the stress equation $\sigma = P/A_0$ into Eq. (3.2), the equation of motion is obtained as

$$\rho_0\frac{\partial v}{\partial t} = \frac{\partial \sigma}{\partial x} \tag{3.3}$$

L. Fan et al., *The Method of Characteristics for Stress Wave Propagation in the Rock Mass*, https://doi.org/10.1007/978-981-97-1440-7_3

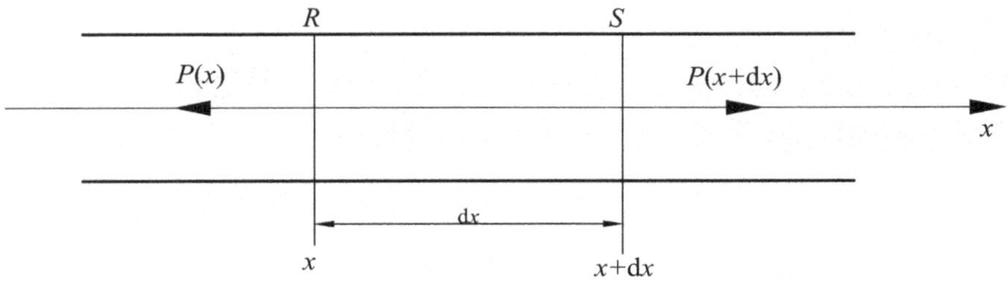

Fig. 3.1 Stress wave propagation in the rock mass

where v is the velocity, and σ is the stress.

The strain ε and velocity v are the first-order derivatives of the displacement u with respect to x and t, respectively. The continuity equations on ε and v can be obtained according to the single-valued continuity condition of the displacement u as

$$\frac{\partial v}{\partial x} = \frac{\partial \varepsilon}{\partial t} \tag{3.4}$$

Assuming that the stress is only a function of the strain, the constitutive equation of the material can be expressed as

$$\sigma = \sigma(\varepsilon) \tag{3.5}$$

The one-dimensional stress wave propagation problem can be solved by establishing the governing equations based on the motion equation, the continuity equation and the constitutive equation, and by combining the initial and boundary conditions to obtain the three unknown quantities σ (x, t), ε (x, t) and v (x, t) given by the initial and boundary conditions.

$\sigma(\varepsilon)$ is a continuous differentiable function, and the partial derivative with respect to x can be expressed as

$$\frac{\partial \sigma}{\partial x} = \frac{\mathrm{d}\sigma}{\mathrm{d}\varepsilon}\frac{\partial \varepsilon}{\partial x} = \frac{\mathrm{d}\sigma}{\mathrm{d}\varepsilon}\frac{\partial^2 u}{\partial x^2} \tag{3.6}$$

Substituting Eq. (3.6) into Eq. (3.3) yields

$$\frac{1}{\rho_0}\frac{\mathrm{d}\sigma}{\mathrm{d}\varepsilon}\frac{\partial^2 u}{\partial x^2} = \frac{\partial^2 u}{\partial t^2} \tag{3.7}$$

Assume the following equation holds:

$$C_{\mathrm{P}}^2 = \frac{1}{\rho_0}\frac{\mathrm{d}\sigma}{\mathrm{d}\varepsilon} \tag{3.8}$$

where C_P is the wave velocity.

Substituting Eq. (3.8) into Eq. (3.7) yields the fluctuation equation as

$$\frac{\partial^2 u}{\partial t^2} - C^2\frac{\partial^2 u}{\partial x^2} = 0 \tag{3.9}$$

Equation (3.9) combined with the initial and boundary conditions can be solved for the stress, strain and velocity.

3.1.2 Two-Characteristic Method and Compatibility Relationship of Two-Characteristic Method

The fluctuation equation is a second-order partial differential equation with the displacement u as the unknown quantity, which can be solved by the characteristic method, where the characteristic line equation in the characteristic method can be found according to the directional derivative method. The directional derivative method can be understood as a linear combination of a second-order partial differential equation or an equivalent set of first-order partial differential equations into a form that contains only the directional derivative of a certain curve L along the plane of the dependent variable (x, t), if the curve is the characteristic line of the corresponding partial differential equation. It is assumed that the differentials of the first-order partial derivatives ε and v of the displacement u along the direction of the curve $L(x, t)$ can be expressed respectively as

$$\mathrm{d}\varepsilon = \frac{\partial \varepsilon}{\partial x}\mathrm{d}x + \frac{\partial \varepsilon}{\partial t}\mathrm{d}t = \frac{\partial^2 u}{\partial x^2}\mathrm{d}x + \frac{\partial^2 u}{\partial t \partial x}\mathrm{d}t \tag{3.10}$$

$$\mathrm{d}v = \frac{\partial v}{\partial x}\mathrm{d}x + \frac{\partial v}{\partial t}\mathrm{d}t = \frac{\partial^2 u}{\partial x \partial t}\mathrm{d}x + \frac{\partial^2 u}{\partial t^2}\mathrm{d}t \tag{3.11}$$

where $\mathrm{d}t$ is the component of the micro-element $\mathrm{d}L$ on the curve $L(x, t)$ on the t-axis; $\mathrm{d}x$ is the component of the micro-element $\mathrm{d}L$ on the curve $L(x, t)$ on the x-axis; and $\mathrm{d}x/\mathrm{d}t$ is the slope of the curve L at the point (x, t).

If the curve $L(x, t)$ is the characteristic line of the fluctuation equation, then there exists a constant λ such that the following equation holds:

$$\mathrm{d}v + \lambda\mathrm{d}\varepsilon = \frac{\partial^2 u}{\partial t^2}\mathrm{d}t + (\lambda\mathrm{d}t + \mathrm{d}x)\frac{\partial^2 u}{\partial x \partial t} + \lambda\frac{\partial^2 u}{\partial x^2}\mathrm{d}x = 0 \tag{3.12}$$

According to Eqs. (3.12) and (3.9), it can be obtained as

$$\frac{1}{\mathrm{d}t} = \frac{0}{\lambda\mathrm{d}t + \mathrm{d}x} = -\frac{C_P^2}{\lambda\mathrm{d}x} \tag{3.13}$$

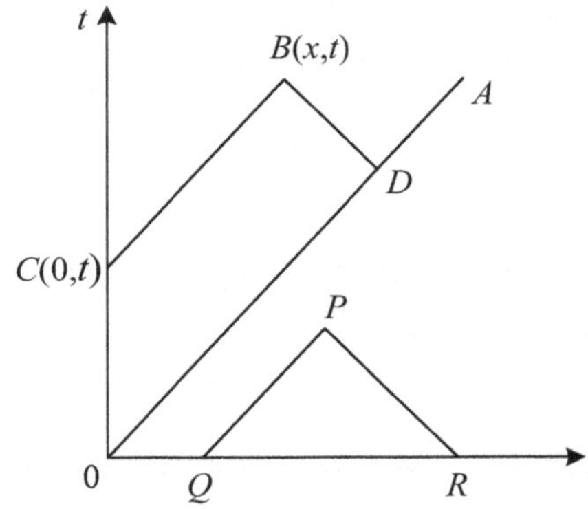

Fig. 3.2 Two-characteristic method

According to Eq. (3.13), the first equation can be expressed as λ = -dx/dt. The second equation can be expressed as dx/dt = $\pm C_P$. In addition, the second equation can also be expressed as

$$dx = \pm C_P dt \tag{3.14}$$

Equation (3.14) is the characteristic line differential equation, which is integrated to obtain the characteristic line. The two characteristic line method is shown in Fig. 3.2.

Substituting Eq. (3.14) into Eq. (3.13) yields $\lambda = \mp C_P$; Eqs. (3.9) and (3.12) can be simplified to the ordinary differential equations along the characteristic line direction

$$dv = \pm C_P d\varepsilon \tag{3.15}$$

3.2 Propagation of Stress Wave in Intact Rock

The stress wave propagation through the intact rock mass can be solved according to the two-characteristic method. Assuming that the intact rock bar is initially stationary and the impact load is given at the rock bar end $x = 0$ when $t = 0$, the initial condition of the rock bar is

$$v(x, 0) = 0 \tag{3.16}$$

and

$$\varepsilon(x, 0) = 0 \tag{3.17}$$

In addition, the boundary condition is

$$v(0, t) = v_0 \tag{3.18}$$

According to the initial and boundary conditions, the stress wave propagation can be solved for any point in Fig. 3.2. According to the initial conditions, it is obtained that along the characteristic line *PQ* and the characteristic line *PR* in the *AOx* region there are

$$v - C_0\varepsilon = v(Q) - C_0\varepsilon(Q) \tag{3.19}$$

and

$$v - C_0\varepsilon = v(R) - C_0\varepsilon(R) \tag{3.20}$$

The velocity and strain at the intersection *P* of the characteristic line *PQ* and the characteristic line *PR* are

$$v(P) = \frac{1}{2}\{[v(R) + v(Q)] + C_0[\varepsilon(R) - \varepsilon(Q)]\} \tag{3.21}$$

and

$$\varepsilon(P) = \frac{1}{2C_0}\{[v(R) - v(Q)] + C_0[\varepsilon(R) + \varepsilon(Q)]\} \tag{3.22}$$

According to the initial condition, it can be known that

$$v(Q) = \varepsilon(Q) = v(R) = \varepsilon(R) = 0 \tag{3.23}$$

Substituting Eq. (3.23) into Eqs. (3.21) and (3.22) yields

$$v(P) = 0 \tag{3.24}$$

and

$$\varepsilon(P) = 0 \tag{3.25}$$

The velocity and strain at any point in the *AOx* region can be known to be zero according to Eqs. (3.24) and (3.25). Therefore, the velocity and strain along the characteristic line *AO* are zero.

The strain and velocity of point *D* on the characteristic line *OA* is zero, then along the characteristic line *BD* in the *AOt* region yields

$$v = -C_0\varepsilon = -\frac{\sigma}{\rho_0 C_0} \tag{3.26}$$

According to the boundary conditions and Eq. (3.26), along the characteristic line *CB* can be obtained as

$$v - C_0\varepsilon = 2v = -2C_0\varepsilon = 2v_0(t_1) \tag{3.27}$$

In addition, along the characteristic line *CB* can be obtained as

$$x = C_0(t - t_1) \tag{3.28}$$

Equation (3.28) is substituted into Eq. (3.27) to obtain

$$v = -C_0\varepsilon = v_0\left(t - \frac{x}{C_0}\right) \tag{3.29}$$

In the present study, the incident wave is assumed to be half-sinusoidal as

$$\sigma^*(0, t) = \begin{cases} A_0 \sin(2\pi f_0 t) & \text{when} \quad 0 \le t \le 1/(2f_0) \\ 0 & \qquad\quad \text{others} \end{cases} \tag{3.30}$$

where A_0 is the amplitude of the incident wave and f_0 is the frequency of the incident wave.

The stress wave propagation through the intact rock can be obtained according to Eq. (3.29).

Figure 3.3 shows the stress wave propagation through the intact rock. It can be seen from Fig. 3.3 that the half-sine incident wave propagates in the rock mass, and the initial time of the transmitted wave is shifted later than the initial time of the incident wave when the stress wave propagates in the intact rock mass. In addition, the amplitude and waveform of the stress wave propagating in the intact rock mass remain constant.

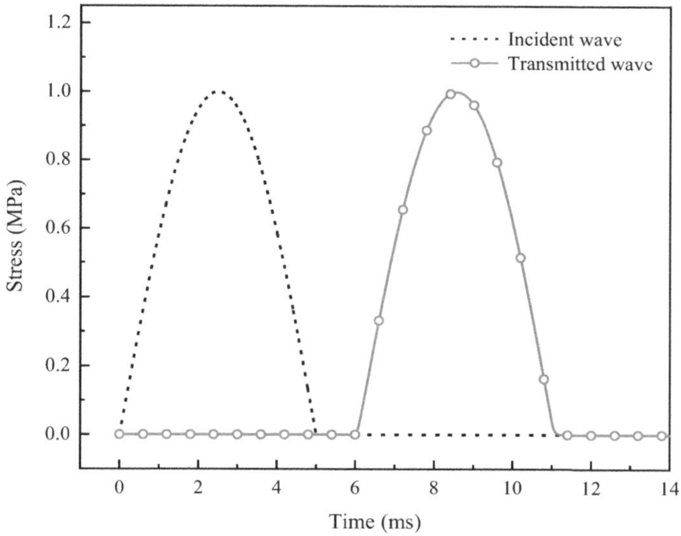

Fig. 3.3 Stress wave propagation through the intact rock

Chapter 4
Two-Characteristic Method for the Wave Propagation Through Joint

4.1 Stress Wave Propagation Through Linearly Deformational Joints

A large number of joints exist in the natural rock mass, which often govern the dynamic mechanical behavior of the rock mass. When joints are subjected to dynamic loading, the stress field on two sides of the joint is usually considered to be continuous, while the displacement field is discontinuous when the joints are with a long length but a small thickness. The difference between the displacement field in front of and behind the joint is the amount of closure (tension) of the joint. When the stress wave with a small amplitude passes through the joints, the joints produce linear deformation, and the linear displacement discontinuity method can be used to study the effect of linearly deformational joints on stress wave propagation, and the linear displacement discontinuity condition at the joints is

$$\sigma^{-} = \sigma^{+} = \sigma \tag{4.1}$$

and

$$u^{-} - u^{+} = \frac{\sigma}{k_{\mathrm{n}}} \tag{4.2}$$

where σ is the normal stress, u is the normal displacement of the joint, k_{n} is the stiffness of the joint, and " + " and "- " denote the mechanical parameters in front of and behind the joint, respectively.

As shown in Fig. 4.1, a single linear joint exists at $x = x_1$ for a rock mass initially in an undisturbed state. When $t = 0$, a normal stress wave $p(t)$ is applied at the left boundary of the rock mass at $x = 0$. When the stress wave propagates through the joint, the displacement discontinuity conditions Eqs. (4.1) and (4.2) are satisfied. The derivative of Eq. (4.2) with respect to time at $x = x_1$ yields

L. Fan et al., *The Method of Characteristics for Stress Wave Propagation in the Rock Mass*, https://doi.org/10.1007/978-981-97-1440-7_4

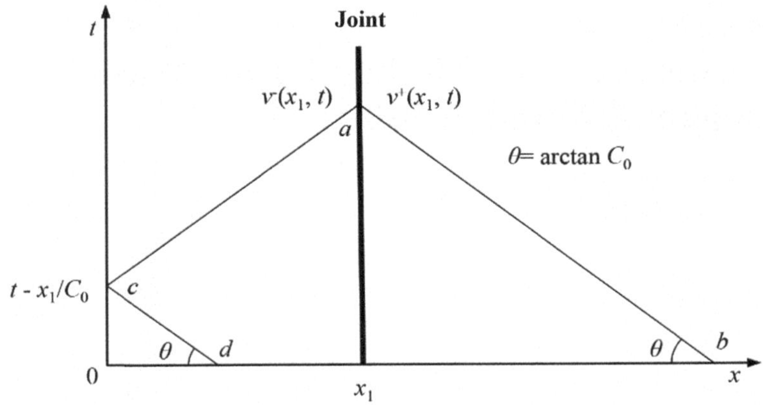

Fig. 4.1 Characteristic method

$$v^{-}(x_1, t) - v^{+}(x_1, t) = \frac{1}{k_n} \frac{\partial \sigma(x_1, t)}{\partial t} \tag{4.3}$$

where $v^{-}(x_1, t)$ and $v^{+}(x_1, t)$ are the velocities of the particle in front of and behind the joint, respectively.

The characteristic method is used to solve the fluctuation equation for stress wave propagation in a jointed rock mass. According to the fluctuation equation, two sets of characteristic lines and the corresponding characteristic compatibility relations can be solved. The left and right characteristic lines correspond to the line with slope C_0 and the line with slope $-C_0$ in the x-t plane, respectively, as shown in Fig. 4.1. The corresponding characteristic compatibility relations for the left and right characteristic lines are

$$zv(x, t) - \sigma(x, t) = \text{constant} \tag{4.4}$$

and

$$zv(x, t) + \sigma(x, t) = \text{constant} \tag{4.5}$$

The rock mass is in an undisturbed state at $t = 0$. Therefore, the mass particle velocity $v(x, t)$ and the stress $\sigma(x, t)$ at each point on the x-axis in Fig. 4.1 are equal to 0. According to Eq. (4.4), the characteristic compatibility relationship corresponding to the left characteristic line *ab* is

$$zv^{+}(x_1, t) - \sigma(x_1, t) = 0 \tag{4.6}$$

Similarly, the right characteristic line *ac* intersects the t-axis at the point $(0, t—x_1/C_0)$, and $v^{-}(t—x_1/C_0)$ is equal to $p(t—x_1/C_0)$. Therefore, according to Eq. (4.5), the characteristic compatibility relationship corresponding to the right characteristic line *ac* is

$$zv^{-}(x_1,t)+\sigma(x_1,t)=zp\left(0,t-\frac{x_1}{C_0}\right)+\sigma\left(0,t-\frac{x_1}{C_0}\right) \tag{4.7}$$

The left characteristic line *cd* corresponds to the characteristic compatibility relationship as

$$zp\left(0,t-\frac{x_1}{C_0}\right)-\sigma\left(0,t-\frac{x_1}{C_0}\right)=0 \tag{4.8}$$

Combining Eqs. (4.6) to (4.8) yields

$$v^{-}(x_1,t)+v^{+}(x_1,t)=2p\left(0,t-\frac{x_1}{C_0}\right) \tag{4.9}$$

Combining Eqs. (4.3) and (4.6) yields

$$v^{-}(x_1,t)-v^{+}(x_1,t)=\frac{z\partial v^{+}(x_1,t)}{k_{\mathrm{n}}\partial t} \tag{4.10}$$

Substituting Eq. (4.10) into Eq. (4.9) yields

$$\frac{\partial v^{+}(x_1,t)}{\partial t}=\frac{2k_{\mathrm{n}}}{z}\left[p(0,t-x_1/C_0)-v^{+}(x_1,t)\right] \tag{4.11}$$

When Δt is sufficiently small, $\partial v^{+}(x_1, t)/\partial t$ in Eq. (4.11) can be expressed as

$$\frac{\partial v^{+}(x_1,t)}{\partial t}=\frac{v^{+}(x_1,t_{j+1})-v^{+}(x_1,t_j)}{\Delta t} \tag{4.12}$$

Substituting Eq. (4.12) into Eq. (4.11) yields

$$v^{+}(x_1,t_{j+1})=\frac{2k_{\mathrm{n}}\Delta t}{z}\left[p(0,t-x_1/C_0)-v^{+}(x_1,t_j)\right]+v^{+}(x_1,t_j) \tag{4.13}$$

Given the particle velocity $p(t)$ of the incident wave and the initial condition $v^{+}(x_1, 0)$, the transmitted wave $v^{+}(x_1, t)$ can be obtained by the iterative calculation of Eq. (4.13). In the calculation process, decreasing the time step Δt can effectively improve the accuracy of the numerical solution of the transmitted wave $v^{+}(x_1, t)$. However, it reduces the computational efficiency and increases the computational cost. Therefore, it is crucial to choose a suitable time step Δt. In addition, the superimposed waveforms $v^{-}(x_1, t)$ of the incident and reflected waves can be obtained from Eq. (4.9), and the reflected waves can be obtained by subtracting the superimposed waveforms $v^{-}(x_1, t)$ from the incident waves.

The transmission coefficient is defined as

$$T_{\mathrm{v}} = \frac{\max(v_{\mathrm{t}})}{\max(v_{\mathrm{i}})} \tag{4.14}$$

The reflection coefficient is defined as

$$R_{\mathrm{v}} = \frac{\max(v_{\mathrm{r}})}{\max(v_{\mathrm{i}})} \tag{4.15}$$

where v_{t} is the mass velocity of the transmitted wave, v_{i} is the mass velocity of the incident wave, and v_{r} is the mass velocity of the reflected wave.

In the calculation, the density of intact rock ρ_0 is 2400 kg/m^3, the modulus of rock elasticity E is 48.6 GPa, the linearly deformational joint stiffness k_{n} is 3.5 GPa/m, the amplitude A_0 is 1.0 m/s, and the frequency f_0 is 50 Hz.

Figure 4.2 shows the waveform of stress wave through linearly deformational joints. It can be seen from Fig. 4.2 that when the stress wave passes through the linearly deformational joints, the amplitude of the transmitted wave attenuates significantly. The duration of the dynamic load increases, and the waveform of the transmitted wave diffuses, while there is also a hysteresis phenomenon. It can be seen from the waveform of the transmitted wave that the loading stage is less influenced by the joints, while the unloading stage is more influenced by the joints.

Figure 4.3 reveals the effect of incident wave frequency on the transmission coefficient under different joint stiffness conditions. It can be seen from Fig. 4.3 that when the incident wave frequency is small, the transmission coefficient decreases slowly from 1 as the frequency increases. As the frequency increases further, the transmission coefficient decreases sharply. When the incident wave frequency is larger, the transmission coefficient decreases slowly and tends to 0 as the frequency increases,

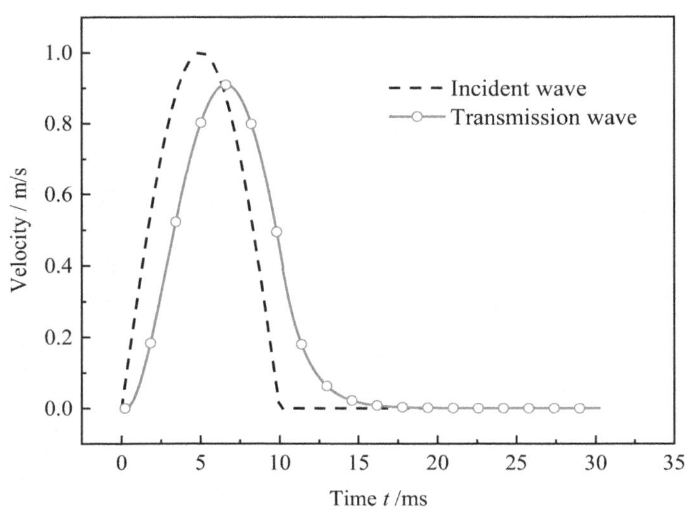

Fig. 4.2 Waveforms of stress wave propagation through linearly deformational joints

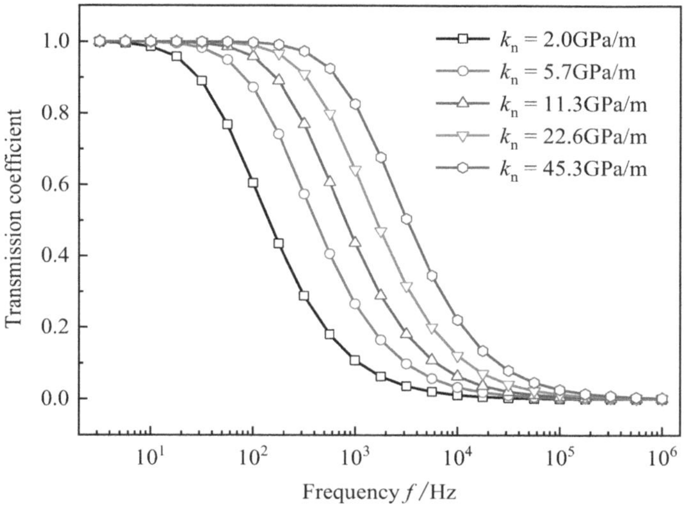

Fig. 4.3 Effect of frequency on transmission coefficient

which indicates that the joint has the effect of large-frequency filtering. In addition, for a fixed incident wave frequency, the transmission coefficient decreases as the stiffness of the joint decreases.

Figure 4.4 reveals the effect of the joint stiffness on the transmission coefficient at different incident wave frequencies. It can be observed from Fig. 4.4 that the transmission coefficient gradually increases with the increasing joint stiffness. When the incident wave frequency is small, the effect of the joint stiffness on the transmission coefficient is small and the transmission coefficient approaches 1. When the incident wave frequency is large, the effect of the joint stiffness on the transmission coefficient is large. In addition, for a fixed joint stiffness, the transmission coefficient decreases as the incident wave frequency increases.

4.2 Stress Wave Propagation Through Nonlinearly Deformational Joints

When the amplitude of the stress wave is large and the joints deform nonlinearly, the nonlinear displacement discontinuity method is usually used to study the propagation of stress waves at nonlinearly deformed joints. The BB model is used to describe the nonlinear deformation properties of the joints. Similar to the stress wave propagation through the linearly deformational joints, the joints are located at x_1, and the nonlinear displacement discontinuity condition at the joints is

$$\sigma^-(x_1, t) = \sigma^+(x_1, t) = \sigma(x_1, t) \tag{4.16}$$

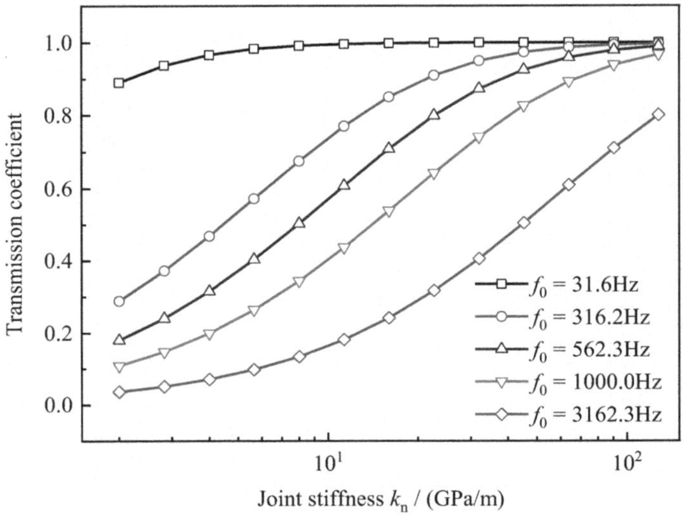

Fig. 4.4 Effect of joint stiffness on transmission coefficient

and

$$u^{-}(x_1,t) - u^{+}(x_1,t) = \frac{\sigma(x_1,t)}{k_{\mathrm{ni}} + \frac{\sigma(x_1,t)}{d_{\max}}} \tag{4.17}$$

where k_{ni} is the initial joint stiffness.

Derivation of Eq. (4.17) yields

$$v^{-}(x_1,t) - v^{+}(x_1,t) = \frac{\partial\sigma(x_1,t)}{\partial t}\frac{1}{k_{\mathrm{ni}} + \frac{\sigma(x_1,t)}{d_{\max}}} - \frac{\partial\sigma(x_1,t)}{\partial t}\frac{\sigma(x_1,t)}{d_{\max}\left[k_{\mathrm{ni}} + \frac{\sigma(x_1,t)}{d_{\max}}\right]^2} \tag{4.18}$$

The nonlinear displacement discontinuity method is combined with the characteristic line method to study the stress wave propagation through the nonlinearly deformed jointed rock mass. The characteristic method is shown in Fig. 4.1, and the characteristic lines *ab*, *ac* and *cd* correspond to the characteristic compatibility relationship of Eqs. (4.6), (4.7) and (4.8), respectively.

Combining Eqs. (4.18) and (4.6) yields

$$v^{-}(x_1,t) - v^{+}(x_1,t) = \frac{z\partial v^{+}(x_1,t)}{\partial t}\frac{1}{k_{\mathrm{ni}} + \frac{zv^{+}(x_1,t)}{d_{\max}}} - \frac{z\partial v^{+}(x_1,t)}{\partial t}\frac{zv^{+}(x_1,t)}{d_{\max}\left[k_{\mathrm{ni}} + \frac{zv^{+}(x_1,t)}{d_{\max}}\right]^2} \tag{4.19}$$

Substituting Eq. (4.19) into Eq. (4.18) yields

$$\frac{\partial v^{+}(x_1,t)}{\partial t}=\frac{2p(0,t-x_1/C_0)-2v^{+}(x_1,t)}{\frac{z}{k_{ni}+\frac{zv^{+}(x_1,t)}{d_{max}}}-\frac{z^2v^{+}(x_1,t)}{d_{max}\left[k_{ni}+\frac{zv^{+}(x_1,t)}{d_{max}}\right]^2}} \tag{4.20}$$

When Δt is sufficiently small, $\partial v^{+}(x_1, t)/\partial t$ can be expressed as

$$\frac{\partial v^{+}(x_1,t)}{\partial t}=\frac{v^{+}(x_1,t_{j+1})-v^{+}(x_1,t_j)}{\Delta t} \tag{4.21}$$

Equation (4.21) is substituted into Eq. (4.20) to obtain

$$v^{+}(x_1,t_{j+1})=\frac{\Delta t\left[2p(0,t-x_1/C_0)-2v^{+}(x_1,t_j)\right]}{\frac{z}{k_{ni}+\frac{zv^{+}(x_1,t_j)}{d_{max}}}-\frac{z^2v^{+}(x_1,t_j)}{d_{max}\left[k_{ni}+\frac{zv^{+}(x_1,t_j)}{d_{max}}\right]^2}}+v^{+}(x_1,t_j) \tag{4.22}$$

Similar to the stress wave passing through a linearly deformed jointed rock mass, the incident wave velocity $p(t)$ and the initial condition $v^{+}(x_1, 0)$ are known, and the transmitted wave $v^{+}(x_1, t)$ in the nonlinearly deformed jointed rock mass can be calculated iteratively by Eq. (4.22).

Introduction of a nonlinear coefficient γ to describe the degree of closure of the joints

$$\gamma=\frac{d}{d_{max}} \tag{4.23}$$

where d is the joint closure, and d_{max} is the maximum allowable closure of the joint.

Figure 4.5 reveals the effect of the nonlinear coefficient on the transmission coefficient for different initial joint stiffness conditions. It can be seen from Fig. 4.5 that the transmission coefficient gradually increases and converges to 1 as the nonlinear coefficient increases. For a fixed nonlinear coefficient, the transmission coefficient increases with the increase of the initial joint stiffness. In addition, the transmission coefficient of nonlinearly deformational joints increases from that of linearly deformational joints to 1 as the nonlinear coefficient increases. When the nonlinear coefficient is small enough, the transmission coefficient of nonlinearly deformational joints approaches that of linearly deformational joints. When the nonlinear coefficient is large enough, the transmission coefficient of the nonlinear deformation joint approaches 1.

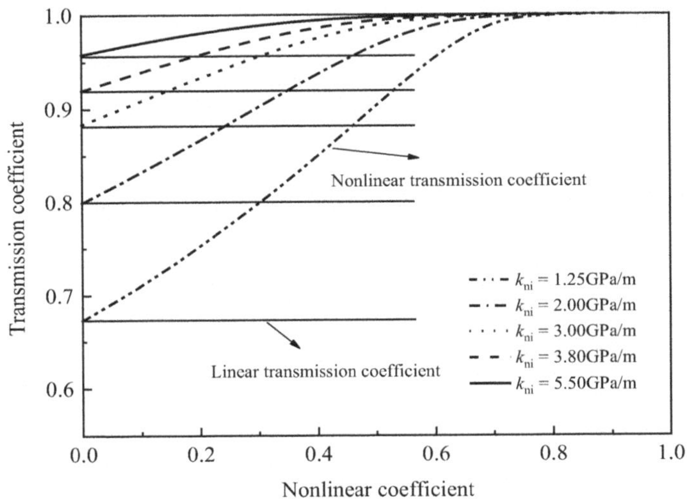

Fig. 4.5 Effect of nonlinear coefficient on transmission coefficient

4.3 Stress Wave Propagation Through Joints with Different Loading/Unloading Deformation

Rock mass commonly consists of a large number of joints, which significantly affect the mechanical properties and physical behaviors of the rock mass [1]. The seismic energy is released during an earthquake, and propagates through the Earth's crust as a seismic wave. By studying the recorded seismic wave using seismographs, the focal mechanism of an earthquake can be inferred. The Earth's reaction to the seismic wave can be interpreted for a better mitigation of the further damage caused by the earthquake. Therefore, an in-depth investigation of how the seismic energy transmits across the jointed rock mass is of great significance.

The present book investigates stress wave propagation across a joint with different nonlinear loading/unloading behavior. Two B-B models are used to describe the loading and unloading stress-closure relationship of the joint, respectively. Using the displacement discontinuity method, the influence of unloading behavior on the wave transmission is studied. The amplitude-dependence and frequency-dependence of energy transmission are investigated, and the error associated with ignoring the unloading effect is evaluated. Finally, studies of incident waves with five different waveforms and multiple pluses propagating through the rock joint are carried out.

4.3.1 Wave Propagation Equations

In this book, the joint is assumed to be planar, large in extent and small in thickness compared to the wavelength. The intact rock is linearly elastic, while the joint is nonlinearly elastic with different loading/unloading behavior. This study is concentrated on the effect of joints with different normal loading/unloading behavior on stress wave propagation. Therefore, the normal incident P-wave transmission is considered, in which the shear deformation is not involved.

(1) Rock joint model

The normal constitutive model of rock joints under cyclic loading conditions has been widely investigated [2, 3]. A total of 64 interlocked jointed rock samples subjected to a sequence of loading/unloading cycles were conducted by Bandis et al. [2]. The total deformation was recorded at regular variation of normal stress. All compression tests were conducted from an initial stress level of approximately 1 kPa. The maximum applied stresses corresponded to 1/3 to 1/2 of the uniaxial compressive strength of the various rock materials, which ensured almost complete closure of the joints [2].

A typical relationship of normal stress versus joint closure for a weathered joint in siltstone under a loading cycle is shown in Fig. 4.6, in which the rock joint has a mean joint compressive strength (JCS) of 45 MPa and a joint roughness coefficient (JRC) of 6.1 [2].

In the present book, the B-B models in the form of hyperbolic equations are adopted to fit the experimental data. The B-B models under loading/unloading situations can be uniformly expressed as

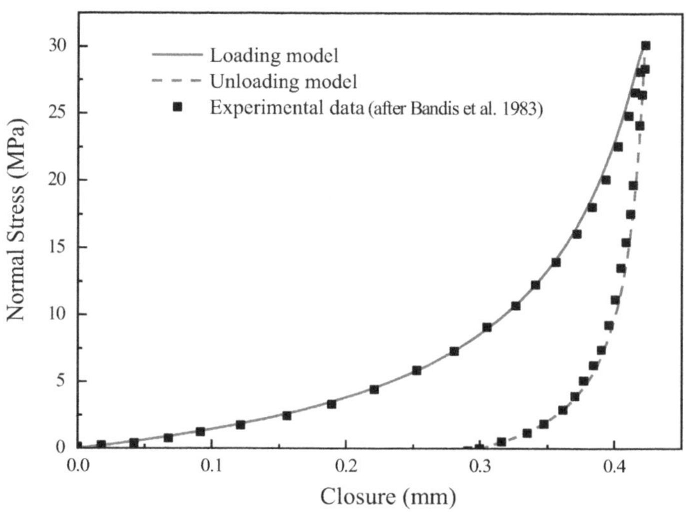

Fig. 4.6 Stress-closure relation of rock joint

$$d_n^{(l,ul)} = d_0^{(l,ul)} + \frac{\sigma_n^{(l,ul)}}{k_0^{(l,ul)} + \sigma_n^{(l,ul)}/d_{\max}^{(l,ul)}} \tag{4.24}$$

where superscript (l, ul) denotes physical quantity value under the loading and unloading conditions, respectively. i.e., $d^{(l)}$ denotes the closure under loading condition while $d^{(ul)}$ denotes the closure under unloading condition, d_n is the normal closure of the rock joint, d_{max} is the maximum allowable closure of the joint, d_0 denotes the initial closure of the joint, k_0 is the initial specific stiffness of the joint, and σ_n is the normal effective stress applied on the joint.

The joint closure (opening) and compression (tension) are assumed to be positive (negative). Two typical fitting curves for the stress-closure relations of the joint under the first cycle of loading and unloading are compared with the experimental results obtained by Bandis et al. [2] as shown in Fig. 4.6. It can be observed in Fig. 4.6 that the specific stiffness of the joint increases with normal stress. The linear rock joint model is a special case of the present B-B model when the normal stress is sufficiently small. The joint closure approaches the maximum allowable closure d_{max} provided the normal stress is sufficiently large theoretically, i.e., infinite value. Figure 4.6 also shows that under cyclic loading/unloading conditions, the B-B models used to describe the loading and unloading cycles are associated with a hysteresis. Therefore, an attenuation mechanism occurs when a stress wave propagates across such joints. The mechanical energy translates to the internal energy and raises the temperature of the rock slightly. This is quite different from the cases of wave propagation across a joint with the same loading/unloading behavior, in which the total mechanical energy is conservative.

(2) The characteristic line method

The characteristic method has been widely used to solve the wave propagation equations in a continuous or discontinuous medium. For the one-dimensional wave propagation problem, the wave propagation equation can be expressed in the form of particle wave velocity v partial differential to time t and the Cartesian coordinate along the wave propagation path x. The equations can be solved by two sets of characteristic lines with a slope of C_0 in the x-t plane as shown in Fig. 4.7, where C_0 is the wave propagation velocity in the intact rock.

The relations of particle velocity v and stress σ can be derived as $zv(x, t) + \sigma(x, t) = const$ along the right-running characteristic line (RC) and $zv(x, t) - \sigma(x, t) = const$ along the left-running characteristic line (LC), where z is the wave impendence for the intact rock.

A scheme of the characteristic method for wave propagation through a single joint is shown in Fig. 4.7. It is assumed that the joint is located at $x = x_1$, and a normal incident wave is applied at $x = 0$. Therefore, according to the displacement discontinuity theory, the stresses on the two sides of the joint are continuous while the displacements are equal to the closure (opening) of the joint, which can be respectively expressed as

$$\sigma(x_1, t) = \sigma\left(x_1 + d_n^{(l,ul)}, t\right) \tag{4.25}$$

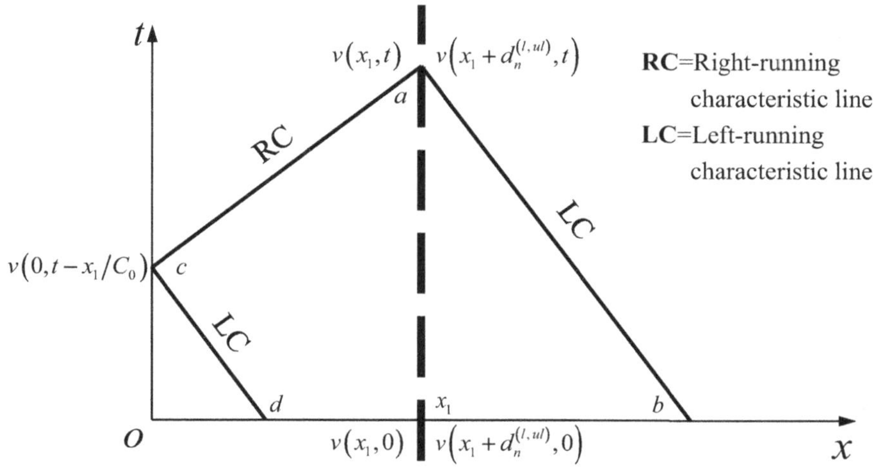

Fig. 4.7 Scheme of characteristic method

and

$$u(x_1,t) - u\left(x_1 + d_n^{(l,ul)}, t\right) = d_n^{(l,ul)} \tag{4.26}$$

where u is the displacement and $d_n^{(l,\,ul)}$ is the joint deformation as described in Eq. (4.24).

Considering Eqs. (4.24) and (4.25), Eq. (4.26) is derived as

$$\begin{aligned} v(x_1,t) - v\left(x_1 + d_n^{(l,ul)}, t\right) &= \frac{\partial d_n^{(l,ul)}}{\partial t} \\ &= \frac{\partial \sigma\left(x_1 + d_n^{(l,ul)}, t\right)}{\partial t} \frac{1}{k_0^{(l,ul)} + \frac{\sigma\left(x_1 + d_n^{(l,ul)}, t\right)}{d_{\max}^{(l,\,ul)}}} \\ &\quad - \frac{\partial \sigma\left(x_1 + d_n^{(l,\,ul)}, t\right)}{\partial t} \frac{\sigma\left(x_1 + d_n^{(l,ul)}, t\right)}{d_{\max}^{(l,ul)}\left(k_0^{(l,ul)} + \frac{\sigma\left(x_1 + d_n^{(l,ul)}, t\right)}{d_{\max}^{(l,ul)}}\right)^2} \end{aligned} \tag{4.27}$$

Zhao and Cai [4] derived the relationship between the stress and the particle velocity in front of and behind a joint as [5]

$$zv\left(x_1 + d_n^{(l,ul)}, t\right) - \sigma\left(x_1 + d_n^{(l,ul)}, t\right) = 0 \tag{4.28}$$

$$v(x_1,t) + v\left(x_1 + d_n^{(l,ul)}, t\right) = 2v(0, t - x_1/C_0) \tag{4.29}$$

According to Eq. (4.28), Eq. (4.27) becomes

$$\begin{aligned} v(x_1, t) - v\left(x_1 + d_n^{(l,ul)}, t\right) &= \frac{\partial d_n^{(l,ul)}}{\partial t} \\ &= \frac{z\partial v\left(x_1 + d_n^{(l,ul)}, t\right)}{\partial t} \frac{1}{k_0^{(l,ul)} + \frac{zv\left(x_1 + d_n^{(l,ul)}, t\right)}{d_{\max}^{(l,ul)}}} \\ &\quad - \frac{z\partial v\left(x_1 + d_n^{(l,ul)}, t\right)}{\partial t} \frac{zv\left(x_1 + d_n^{(l,ul)}, t\right)}{d_{\max}^{(l,ul)}\left(k_0^{(l,ul)} + \frac{zv\left(x_1 + d_n^{(l,ul)}, t\right)}{d_{\max}^{(l,ul)}}\right)^2} \end{aligned} \tag{4.30}$$

From Eqs. (4.29) and (4.30), the particle velocity behind the joint can be expressed as

$$\frac{\partial v\left(x_1 + d_n^{(l)}, t\right)}{\partial t} = \frac{2v(0, t - x_1/C_0) - 2v\left(x_1 + d_n^{(l)}, t\right)}{\left(\frac{z}{k_0^{(l)} + \frac{zv\left(x_1 + d_n^{(l)}, t\right)}{d_{\max}^{(l)}}} - \frac{z^2 v\left(x_1 + d_n^{(l)}, t\right)}{d_{\max}^{(l)}\left(k_0^{(l)} + \frac{zv\left(x_1 + d_n^{(l)}, t\right)}{d_{\max}^{(l)}}\right)^2}\right)}, \text{ loading} \tag{4.31}$$

and

$$\frac{\partial v\left(x_1 + d_n^{(ul)}, t\right)}{\partial t} = \frac{2v(0, t - x_1/C_0) - 2v\left(x_1 + d_n^{(ul)}, t\right)}{\left(\frac{z}{k_0^{(ul)} + \frac{zv\left(x_1 + d_n^{(ul)}, t\right)}{d_{\max}^{(ul)}}} - \frac{z^2 v\left(x_1 + d_n^{(ul)}, t\right)}{d_{\max}^{(ul)}\left(k_0^{(ul)} + \frac{zv\left(x_1 + d_n^{(ul)}, t\right)}{d_{\max}^{(ul)}}\right)^2}\right)}, \text{ unloading} \tag{4.32}$$

Equations (4.31) and (4.32) reveal that, in the practical situation, when the joint deforms differently under loading and unloading, the transmitted wave becomes different if the unloading effect is taken into consideration, hence leading to changes in energy transmission.

For computational calculation, the difference forms of Eqs. (4.31) and (4.32) are respectively expressed as

$$\begin{aligned} v\left(x_1 + d_n^{(l)}, t_{j+1}\right) &= v\left(x_1 + d_n^{(l)}, t_j\right) \\ &\quad + \frac{2v(0, t - x_1/C_0) - 2v\left(x_1 + d_n^{(l)}, t_j\right)}{\left(\frac{z}{k_0^{(l)} + \frac{zv\left(x_1 + d_n^{(l)}, t_j\right)}{d_{\max}^{(l)}}} - \frac{z^2 v\left(x_1 + d_n^{(l)}, t_j\right)}{d_{\max}^{(l)}\left(k_0^{(l)} + \frac{zv\left(x_1 + d_n^{(l)}, t_j\right)}{d_{\max}^{(l)}}\right)^2}\right)} \Delta t, \text{ loading} \end{aligned} \tag{4.33}$$

and

$$v\left(x_1 + d_n^{(ul)}, t_{j+1}\right) = v\left(x_1 + d_n^{(ul)}, t_j\right) + \frac{2v(0, t - x_1/C_0) - 2v\left(x_1 + d_n^{(ul)}, t_j\right)}{\left(\frac{z}{k_0^{(ul)} + \frac{zv\left(x_1 + d_n^{(ul)}, t_j\right)}{d_{\max}^{(ul)}}} - \frac{z^2 v\left(x_1 + d_n^{(ul)}, t_j\right)}{d_{\max}^{(ul)}\left(k_0^{(ul)} + \frac{zv\left(x_1 + d_n^{(ul)}, t_j\right)}{d_{\max}^{(ul)}}\right)^2}\right)}\Delta t, \text{ unloading} \tag{4.34}$$

The results obtained by Eqs. (4.33) and (4.34) are equivalent to the results obtained by Zhao and Cai [4] when the unloading effect is neglected.

The intact rock is assumed to be elastic in the present book, therefore $\varepsilon = v/C_0$. The transmitted wave in the form of strain can be obtained as

$$\varepsilon\left(x_1 + d_n^{(l)}, t_{j+1}\right) = \varepsilon\left(x_1 + d_n^{(l)}, t_j\right) + \frac{2\varepsilon(0, t - x_1/C_0) - 2\varepsilon\left(x_1 + d_n^{(l)}, t_j\right)}{\left(\frac{z}{k_0^{(l)} + \frac{zC_0\varepsilon\left(x_1 + d_n^{(l)}, t_j\right)}{d_{\max}^{(l)}}} - \frac{z^2 C_0 \varepsilon\left(x_1 + d_n^{(l)}, t_j\right)}{d_{\max}^{(l)}\left(k_0^{(l)} + \frac{zC_0\varepsilon\left(x_1 + d_n^{(l)}, t_j\right)}{d_{\max}^{(l)}}\right)^2}\right)}\Delta t, \text{ loading} \tag{4.35}$$

and

$$\varepsilon\left(x_1 + d_n^{(ul)}, t_{j+1}\right) = \varepsilon\left(x_1 + d_n^{(ul)}, t_j\right) + \frac{2\varepsilon(0, t - x_1/C_0) - 2\varepsilon\left(x_1 + d_n^{(ul)}, t_j\right)}{\left(\frac{z}{k_0^{(ul)} + \frac{zC_0\varepsilon\left(x_1 + d_n^{(ul)}, t_j\right)}{d_{\max}^{(ul)}}} - \frac{z^2 C_0 \varepsilon\left(x_1 + d_n^{(ul)}, t_j\right)}{d_{\max}^{(ul)}\left(k_0^{(ul)} + \frac{zC_0\varepsilon\left(x_1 + d_n^{(ul)}, t_j\right)}{d_{\max}^{(ul)}}\right)^2}\right)}\Delta t, \text{ unloading} \tag{4.36}$$

where $\varepsilon(x, t)$ is the wave expressed in the form of strain.

To check the wave attenuation and the conservation of mechanical energy of a stress wave propagating across the rock mass, the wave transmission coefficient (T_v) and the energy transmission coefficient (T_e) are respectively defined as

$$T_v = \frac{\max(v_{tran})}{\max(v_{inc})} \tag{4.37}$$

and

$$T_e = \frac{energy_{inc}}{energy_{tran}} = \frac{\sum_{j=t_{tran}^0}^{j=t_{tran}^1} z(v_{tran})^2 \Delta t}{\sum_{j=t_{inc}^0}^{j=t_{inc}^1} z(v_{inc})^2 \Delta t} \tag{4.38}$$

where subscripts "*inc*" and "*tran*" refer to incident wave and transmitted wave, respectively, $t^0{}_{inc}$ and $t^0{}_{tran}$ are the initial time of incident waves and transmitted waves, and $t^1{}_{inc}$ and $t^1{}_{tran}$ are the final time of incident waves and transmitted waves.

4.3.2 Comparison of Results with and Without Unloading Effect

This section presents an example of investigating the unloading effect on stress transmission across a rock joint in siltstone. Two B-B models in the form of Eq. (4.24) are used to fit the experimental data for the loading and unloading behavior of a rock joint, respectively. Figure 4.6 shows the experimental data and fitting results of stress versus closure. The values of parameters in the fitting B-B models are shown in Table 4.1.

The incident wave is assumed to be half-sinusoidal as

$$v(0,t) = \begin{cases} A_0 \sin(2\pi f_0 \times t) \\ 0 \end{cases}, \text{ when } \begin{matrix} 0 \le t \le 1/(2f) \\ others \end{matrix} \tag{4.39}$$

where A_0 is the amplitude of the incident wave and f_0 is the frequency.

The incident wave is assumed to be applied on the surface of the rock joint, therefore $x_1 = 0$ in the calculation (as shown in Fig. 4.7). The joint is assumed to follow the same B-B model if the unloading effect is neglected.

Figure 4.8 compares the transmitted waves considering the unloading effect and without considering the unloading effect when the amplitude of the incident wave is $A_0 = 0.5$ m/s and the frequency is $f_0 = 400$ Hz. As shown in Fig. 4.8, for the same incident wave, the transmitted waves have similar waveforms with or without considering the unloading effect. In addition, the same attenuation coefficient can be obtained with or without considering the unloading effect for a single pulse incident wave propagating across a rock joint.

It also can be observed from Fig. 4.8 that the transmitted wave considering the unloading effect has much narrower waveforms. The mechanical energy is proportional to the filled area under the second power of waveforms from Eq. (4.38). Therefore, higher mechanical energy dissipation occurs when the unloading effect is considered. This phenomenon can be validated from Fig. 4.6 that if the joint has different loading/unloading behavior, a hysteresis occurs during a loading/unloading cycle, which causes mechanical energy dissipation.

Table 4.1 Parameters of B-B model

	d_0 (mm)	k_0 (GPa)	d_{max} (mm)
Loading	0	11.44	0.505
Unloading	0.30	25.83	0.137

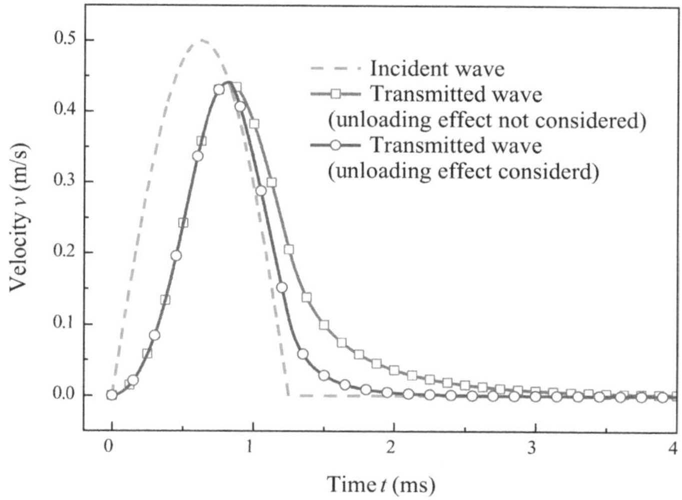

Fig. 4.8 Incident wave and corresponding transmitted wave with/without considering unloading effect

4.3.3 Stress Wave Propagation Properties

(1) Amplitude-dependence of energy transmission

The amplitude-dependence of energy transmission is investigated in this section. The incident waves are in the form of Eq. (4.39). The variation of energy transmission T_e with respect to the amplitude of incident wave for different frequencies of $f_0 =$ 300, 500, 900 and 1500 Hz are compared. The energy transmission coefficient T_e is defined in Eq. (4.38). Figure 4.9 shows the energy transmission for the stress wave propagating across the rock joint with or without considering the unloading effect.

Figure 4.9 reveals that, for the stress wave propagating across a rock joint with nonlinearly deformational behavior, the energy transmission increases as the incident wave amplitude increases. For the rock joint with nonlinear behavior, the specific stiffness increases with the amplitude of the incident stress wave (as shown in Fig. 4.6), which leads to a smaller difference between the stiffnesses of the joint and the intact rock, resulting in a larger energy transmission. Thus, the energy transmission coefficient approaches 1 for the incident waves with an infinitely large amplitude theoretically. It also can be observed from Fig. 4.9 that the energy transmission coefficients considering the unloading effect are smaller than those without considering the unloading effect, which is validated by Fig. 4.8.

The unloading effect on the energy transmission is evaluated by comparing the results with or without considering it. The difference is defined as

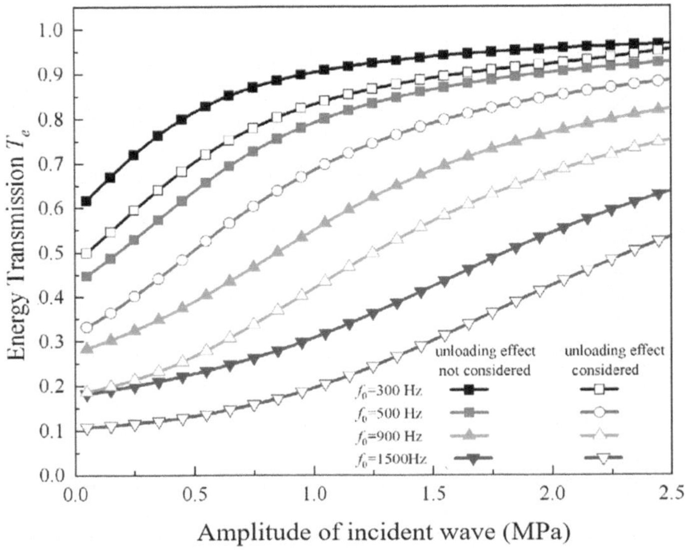

Fig. 4.9 The effect of amplitude of incident wave on energy transmission

$$\Delta T_e = \frac{\left| energy_{tran}^{different\ loading/unloading} - energy_{tran}^{same\ loading/unloading} \right|}{energy_{inc}} \tag{4.40}$$

where $energy_{tran}{}^{different\ loading/unloading}$ and $\text{energy}_{tran}{}^{same\ loading/unloading}$ denote the transmitted energy with and without considering the unloading effect, respectively. It can be observed from Fig. 4.10 that the difference ΔT_e increases with the increase of the amplitude of the incident wave at the initial stage and then decreases with a further increase of amplitude after its peak value. Figure 4.10 also reveals that when the amplitude of the incident wave is sufficiently small, i.e., $A_0 < 0.25$ m/s, the incident wave with a larger frequency has a smaller unloading effect on the energy transmission than the incident wave with a smaller frequency, while for the cases when the amplitude of incident waves is sufficiently large, i.e., $A_0 > 1.75$ m/s, the incident wave with a larger frequency has a larger unloading effect on the energy transmission than the incident wave with a smaller frequency.

When the amplitude of the incident wave is sufficiently large, the error approaches zero. In that case, the mechanical behavior of a joint is equal to the mechanical behavior of a rock. Therefore, all of the incident wave energy can be transmitted (as shown in Fig. 4.9). The unloading effect hence becomes zero and can be ignored.

(2) Frequency-dependence of energy transmission

The frequency-dependence of energy transmission for the incident wave with the waveform of Eq. (4.39) is investigated in this section. The incident waves are assumed to have a varying frequency and fixed amplitudes of $A_0 = 0.1$, 1.0 and 1.35 m/s.

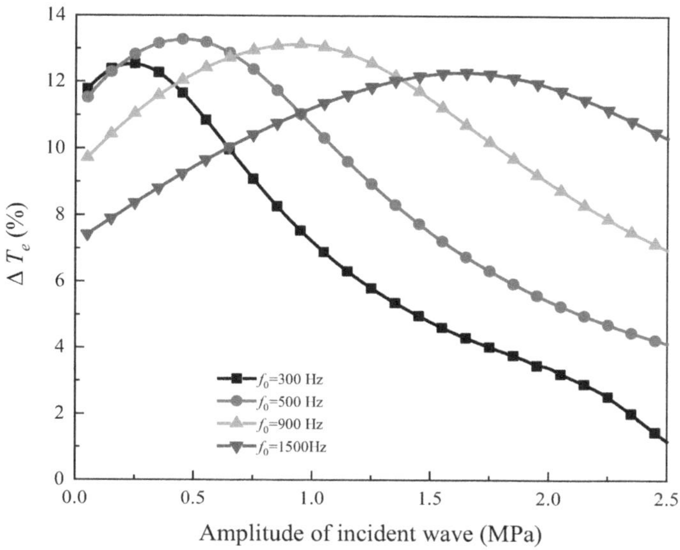

Fig. 4.10 Amplitude-dependence of unloading effect

The relations of energy transmission versus frequency of an incident wave are shown in Fig. 4.11. The results show that energy transmission decreases as frequency increases. Therefore, the small-frequency waves can propagate a longer distance than the large-frequency blast waves. For the quasi-static situation, when the frequency approaches zero, the energy transmission approaches 1.

It can also be observed from Fig. 4.11 that larger energy transmission coefficients can be obtained if the unloading effect is ignored. The frequency-dependence of the unloading effect on the energy transmission is investigated in Fig. 4.12.

Figure 4.12 reveals that the unloading effect on the energy transmission increases as the frequency increases to a peak value, which then decreases as the frequency further increases. For the incident waves with a small frequency, the unloading is more significant for the incident wave with a smaller amplitude. However, for the incident waves with a large frequency, the unloading effect is more significant for the incident wave with a larger amplitude.

Figures 4.11 and 4.12 also show that, for the incident wave with a sufficiently small frequency, most of the energy can be transmitted across the joint (Fig. 4.11). In that case, the unloading effect can be neglected for an approximate calculation (Fig. 4.12). On the other hand, when the incident wave has a sufficiently large frequency, most of the incident energy is scattered by the joint (Fig. 4.11). The total energy transmission coefficient approaches zero. Therefore, the unloading effect can also be neglected. Between these two extremes, the unloading effect on the energy transmission should be taken into consideration.

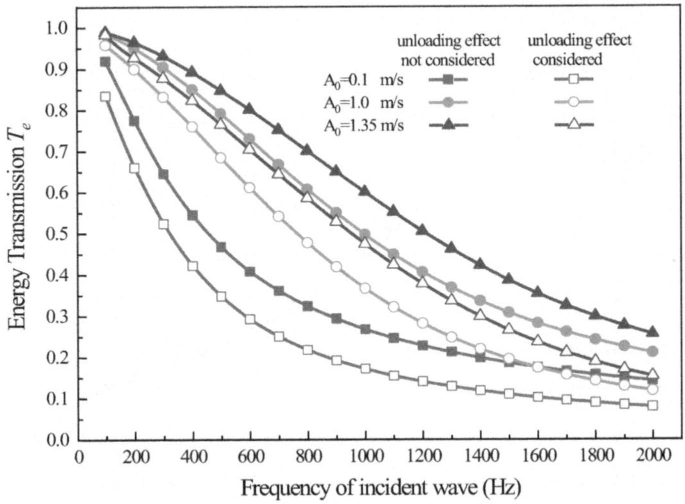

Fig. 4.11 The effect of frequency of incident wave on energy transmission

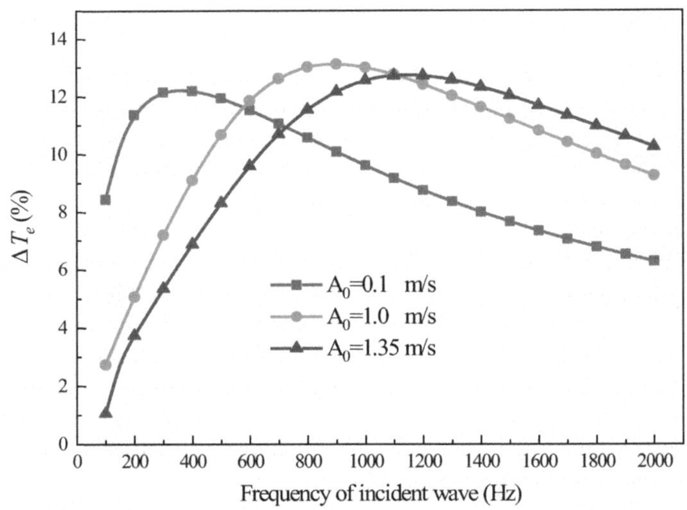

Fig. 4.12 Frequency-dependence of unloading effect

(3) Unloading effect on the waves with different waveforms

For an arbitrary incident waveform, the transmitted waveforms can be directly calculated by using Eqs. (4.33) and (4.34). A blast wave is normally simplified as a half-cycle sinusoidal, triangular or rectangular pulse by defining the peak value, rise time and duration for the convenience of analysis.

To investigate the unloading effect on the stress wave with different waveforms, the incident waves with five basic waveforms, namely half-sinusoidal, rectangular, triangular, right-triangular and left-triangular, are assumed. The waves are assumed to have the same amplitude $A_0 = 0.5$ m/s and the same duration $T_0 = 1.25$ ms. Figures 4.8 and 4.13 show the results when the incident waves propagate across the rock joint with or without the unloading effect. The parameters of the joint are the same as shown in Table 4.1. It can be observed from Figs. 4.8 and 4.13 that transmitted waves for the half-sinusoidal, triangular and right-triangular incident waves are sleekly continuous (Figs. 4.8, 4.13b and d), while the transmitted waves have an inflection point for the incident waves with rectangular and left-triangular waveforms (Fig. 4.13a and c). The effect of different loading/unloading behavior on the energy transmission appears sharply after the unloading point of transmitted waves. Table 4.2 calculates the energy transmissions for the present five types of incident wave propagation across a rock joint with or without considering the unloading effect. The comparison shows that the incident wave with a rectangular waveform is less affected by different loading/unloading behavior.

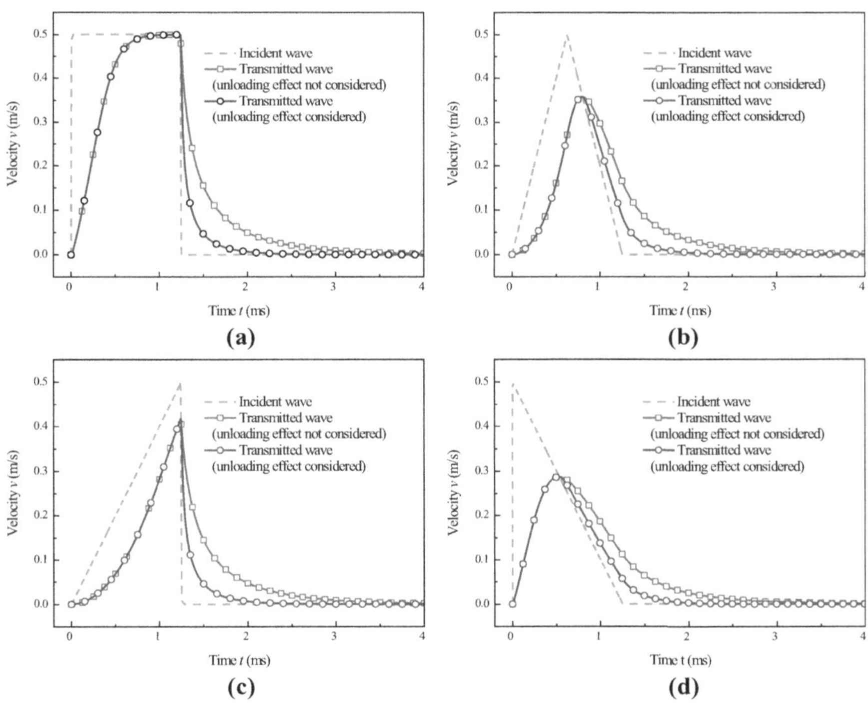

Fig. 4.13 Unloading effect on the stress wave with different waveforms. **a** Rectangular. **b** Triangular. **c** Left triangular. **d** Right triangular

Table 4.2 Unloading effect on the wave with different waveforms

Waveform	Half-sinusoidal	Rectangular	Triangular	Right triangular	Left-triangular
T_e (%)(unloading effect not considered)	72.39	75.99	66.49	56.34	64.50
T_e (%)(unloading effect considered)	59.68	70.32	52.27	45.40	50.85
Error (%)	12.71	5.67	14.22	10.94	13.65

(4) Unloading effect on the multiple pulse transmission

In most practical applications, stress waves with multiple pulses are always generated. A modified split Hopkinson press bar (SHPB) test was performed by Li and Ma [6] to study the wave propagation across a filled rock joint. A sand layer is sandwiched between the incident and transmitted granite pressure bars to simulate the deformation of the joint. The filled sand joint is 3.0 mm in thickness. The lengths of the incident granite bar and the transmitted granite bar are 97.0 and 100.5 cm, respectively. Both bars are 5.0 cm in diameter. The density of the granite is 2650.0 kg/m^3 and the P-wave velocity is 4758.0 m/s. Using a wave separation method [6] based on the measured wave recordings, the incident P-wave with pulses generated by a pendulum hammer at a swing-angle of 40° is shown in Fig. 4.14. By respectively using B-B models with the parameters $k_0 = 5.6$ GPa, $d_0 = 0$ mm and $d_{max} = 0.48$ mm for the loading path, $k_0 = 28.8$ GPa, $d_0 = 0.236$ mm and $d_{max} = 0.12$ mm for the first unloading and reloading path and $k_0 = 140.0$ GPa, $d_0 = 0.331$ mm and $d_{max} = 0.03$ mm for the second unloading path, the transmitted waves with and without considering the unloading effect are obtained. Figure 4.14 compares the theoretical results and experimental results. The unloading behavior is found to affect the waveform since the commencement of the first unloading. For the first transmitted pulse, the unloading effect influences the wave dissipation only. However, for the later transmitted pulses, the unloading effect not only influences the wave dissipation, but also the wave amplitude attenuation.

Without losing generality, an incident wave in the form of two half-sinusoidal pulses with different time intervals is introduced as shown in Fig. 4.15 to evaluate the unloading effect. The amplitude of both pulses is $\varepsilon_0 = 2.0 \times 10^{-4}$ s^{-1} and the frequency is $f_0 = 2667$ Hz, so the duration time can be calculated as $t_0 = 0.1875$ ms. The time interval for the first incident wave is $\Delta t = 0.1875$ ms and that for the second incident wave is $\Delta t = 1.6875$ ms. The same joint models as shown in Table 4.2 are used to evaluate the unloading effect on the multiple pulse transmission for the first pulse. The joint model for the second pulse is fitted with the parameters of $k_0 = 35.85$ GPa and $d_{max} = 0.13$ mm for the loading path and $k_0 = 91.86$ GPa and $d_{max} = 0.064$ mm for the unloading path, respectively.

Figure 4.15a reveals that when the time interval is sufficiently small, the later transmitted pulses can be superposed by the former transmitted pulses. The unloading

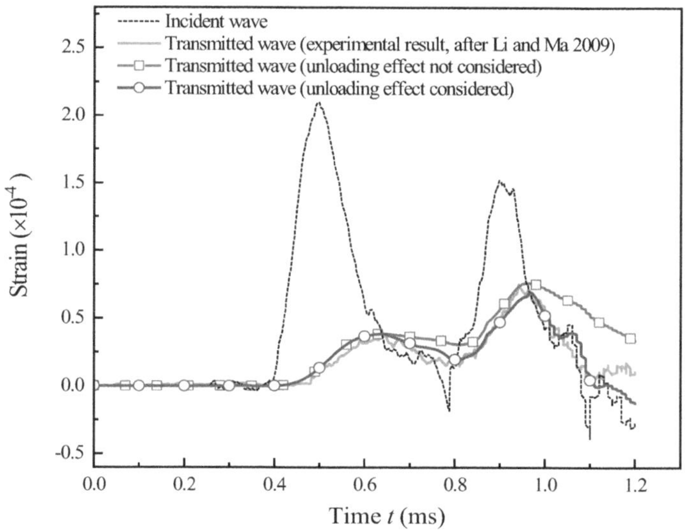

Fig. 4.14 Unloading effect on multiple pulse transmission

influences both the energy transmission (related to waveform) and wave attenuation (related to wave amplitude). Figure 4.15b reveals that when the time interval is sufficiently large, the unloading effect influences the transmitted waveform only provided the later transmitted pulses are not superposed by the tail of the former transmitted pluses. It is noticed from Fig. 4.15 that dissipation happens when the stress wave propagates across the filled joint. Therefore, the time interval required to avoid the wave superposition should be larger than that of the incident wave duration time $1/(2f_0)$.

Figure 4.15 also shows that the second pulse has a significantly larger transmission coefficient than that of the first pulse. This phenomenon reveals that the joint can be regarded as a compressible material. The joint is compressed due to the first pulse, which increases the stiffness of the joint and decreases the stiffness difference between the joint and intact rock. The transmission rate of the latter pulses thus subsequently increases when it propagates across the joint.

4.4 Stress Wave Propagation Through Joints Under In Situ Stress

Within the Earth's crust, the rock mass is initially subjected to high stress. Moreover, joints usually exist in the crustal rock mass, which often control its mechanical behavior and significantly influence seismic wave propagation [7–9]. Therefore, it is important in geophysics, earthquake engineering, non-destructive evaluation and

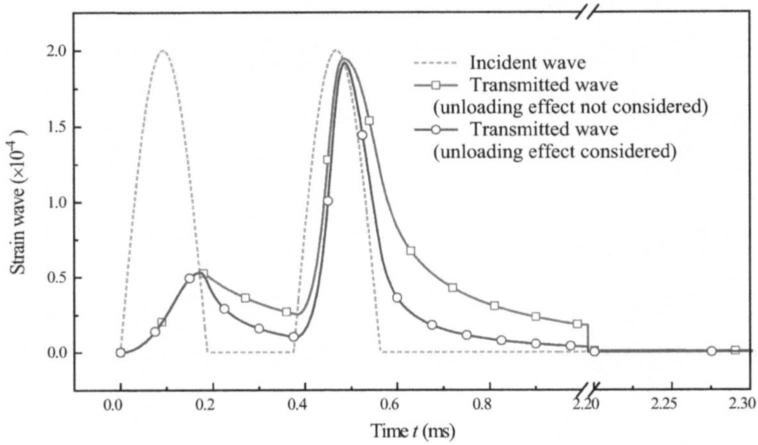

(a) Time interval $\Delta t = 0.1875$ ms

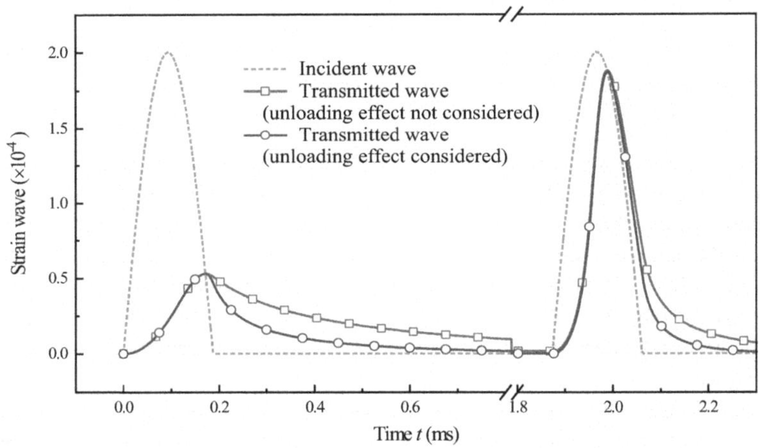

(b) Time interval $\Delta t = 1.6875$ ms

Fig. 4.15 Unloading effect on multiple pluse transmission with different time intervals

rock mechanics to study the effect of in situ stress on seismic wave propagation through the rock mass.

This section presents a study on seismic wave propagation through an in situ stressed rock mass by DDM, in which the joint is treated as a special nonlinearly deformational layer. The amplitude- and frequency-dependence of seismic wave propagation through in situ stressed rock mass are analyzed. Subsequently, the effect of in situ stress on seismic attenuation, energy transmission and effective seismic velocity is discussed.

4.4.1 A Displacement Discontinuity Method for Considering In Situ Stress

The displacement discontinuity method (DDM) is a successful method for investigating stress wave propagation through a rock mass. It has been used in many studies, such as Zhao et al. [10]. The present book systemically investigates the effect of in situ stress on wave attenuation, energy transmission and effect velocity, and discusses the amplitude- and frequency-dependence of the influence of in situ stress.

By contrast with conventional studies on elastic wave propagation through a medium with welded interfaces, the joint is assumed to be a special layer according to preview investigations, which is planar, large in extent and small in thickness compared to the wavelength [2, 11]. The intact rock is linearly elastic, while the joint is nonlinearly elastic. The rock mass is initially under high static stress [12].

Figure 4.16 shows a typical relationship of normal stress versus joint closure (opening), which can be expressed by the B-B model in the form of a hyperbolic equation:

$$d = \frac{\sigma}{k_0 + \sigma / d_{\max}} \tag{4.41}$$

where k_0 is the initial specific stiffness of the joint and σ is the normal effective stress applied to the joint, d and d_{max} respectively denote normal closure and maximum allowable closure of the joint [2, 3].

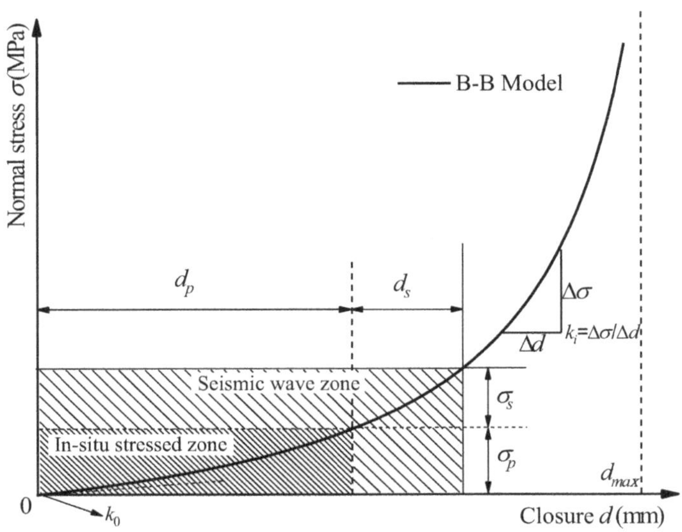

Fig. 4.16 Illustration of normal behavior of joint considering vertical in situ stress

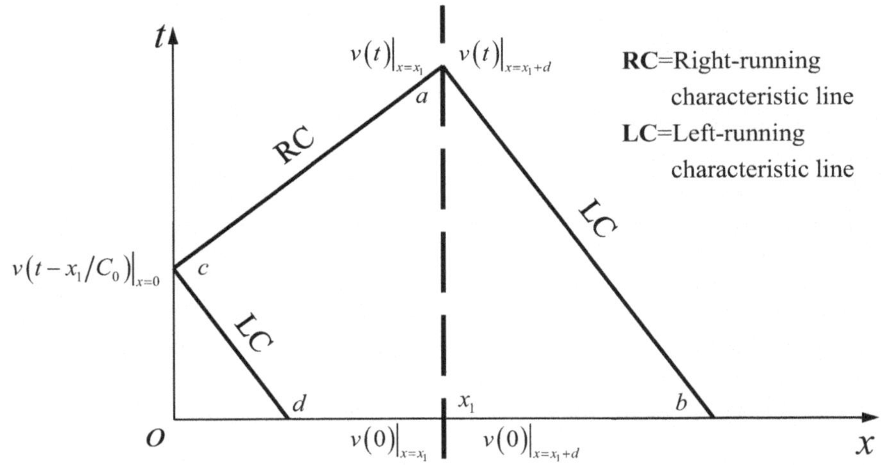

Fig. 4.17 Scheme of characteristic method for stress wave propagation through a joint

It is observed from Fig. 4.16 that the deformation of joint d can be divided into two parts when a seismic wave propagates through an in situ stressed rock mass. d_p denotes the static deformation of the joint caused by an in situ stress σ_p and d_s denotes the dynamic deformation of the joint caused by a seismic stress σ_s.

A method has been introduced for investigating seismic wave propagation through an initially in situ stressed rock mass as shown in Fig. 4.17. For the one-dimensional wave propagation problem, the wave propagation equation can be expressed in the form of a particle wave velocity v partial differential to the time t and the Cartesian coordinate along the wave propagation path x. The equations can be solved by two sets of characteristic lines with the slope of C_0 in the x-t plane as shown in Fig. 4.17, where C_0 is the seismic velocity in intact rock.

It is assumed that the joint is located at $x = x_1$, and a normal incident wave is applied at $x = 0$. Therefore, according to the displacement discontinuity theory, the stresses on the two sides of the joint are continuous while the difference in displacements equals the closure (opening) of the joint, which can be respectively expressed as

$$\sigma(t)|_{x=x_1} = \sigma(t)|_{x=x_1+d} \tag{4.42}$$

and

$$u(t)|_{x=x_1} - u(t)|_{x=x_1+d} = d \tag{4.43}$$

where u is the displacement and d is the joint deformation as described in Eq. (4.41).

For an in situ stressed rock mass under the seismic wave, the stress also can be divided into two parts, the initial in situ stress σ_p caused by gravity and the dynamic seismic stress σ_s caused by the seismic wave. Therefore, Eq. (4.42) can be expressed as

$$\sigma = \left[\sigma_p(t) + \sigma_s(t)\right]\Big|_{x=x_1} = \left[\sigma_p(t) + \sigma_s(t)\right]\Big|_{x=x_1+d} \tag{4.44}$$

where σ_p and σ_s denote static in situ stress and dynamic seismic stress, respectively.

The in situ stress caused by gravity generally remains constant during seismic wave propagation. Therefore, substituting Eqs. (4.41) and (4.44) into the derivation of Eq. (4.43) yields

$$\begin{aligned} v(t)|_{x=x_1} - v(t)|_{x=x_1+d} &= \frac{\partial d}{\partial t} \\ &= \frac{\partial \sigma_s(t)}{\partial t} \Big/ \left\{ k_0 + \frac{\left[\sigma_p(t) + \sigma_s(t)\right]}{d_{\max}} \right\}\Bigg|_{x=x_1+d} \\ &- \frac{\partial \sigma_s(t)}{\partial t}\left[\sigma_p(t) + \sigma_s(t)\right]/d_{\max}\left(k_0 + \frac{\left[\sigma_p(t) + \sigma_s(t)\right]}{d_{\max}}\right)^2\Bigg|_{x=x_1+d} \end{aligned} \tag{4.45}$$

The relationship of seismic stress σ_s and particle seismic velocity v in front of and behind a joint can be obtained as

$$zv(t)|_{x=x_1+d} - \sigma_s(t)|_{x=x_1+d} = 0 \tag{4.46}$$

and

$$v(t)|_{x=x_1} + v(t)|_{x=x_1+d} = 2v\left(t - x_1 / C_0\right)\big|_{x=0} \tag{4.47}$$

where $z = \rho C_0$ is the seismic wave impendence of intact rock and ρ is the density of intact rock.

According to Eqs. (4.45, 4.46) becomes

$$\begin{aligned} v(t)|_{x=x_1} - v(t)|_{x=x_1+d} &= \frac{\partial d}{\partial t} \\ &= \frac{z\partial v(t)}{\partial t} \Big/ \left[k_0 + \frac{\left(\sigma_p(t) + zv(t)\right)}{d_{\max}} \right]\Bigg|_{x=x_1+d} \\ &- \frac{z\partial v(t)}{\partial t}\left(\sigma_p(t) + zv(t)\right)/d_{\max}\left[k_0 + \frac{\left(\sigma_p(t) + zv(t)\right)}{d_{\max}}\right]^2\Bigg|_{x=x_1+d} \end{aligned} \tag{4.48}$$

From Eqs. (4.47) and (4.48), the particle velocity behind the joint can be expressed as

$$\left.\frac{z\partial v(t)}{\partial t}\right|_{x=x_1+d} = \frac{2v(t-x_1/C_0)|_{x=0} - 2v(t)|_{x=x_1+d}}{\left.\frac{1}{k_0+\frac{\sigma_p(t)+zv(t)}{d_{\max}}}\right|_{x=x_1+d} - \left.\frac{\sigma_p(t)+zv(t)}{d_{\max}\left(k_0+\frac{\sigma_p(t)+zv(t)}{d_{\max}}\right)^2}\right|_{x=x_1+d}} \tag{4.49}$$

Equation (4.49) reveals the consideration of in situ stress during seismic wave propagation. For computational calculation, Eq. (4.49) is expressed in the difference form as

$$v(t_{j+1})\big|_{x=x_1+d} = v(t_j)\big|_{x=x_1+d} + \frac{2v(t-x_1/C_0)|_{x=0} - 2v(t_j)\big|_{x=x_1+d}}{\left(\left.\frac{z}{k_0+\frac{\sigma_p(t_j)+zv(t_j)}{d_{\max}}}\right|_{x=x_1+d} - \left.\frac{z(\sigma_p(t_j)+zv(t_j))}{d_{\max}\left(k_0+\frac{\sigma_p(t_j)+zv(t_j)}{d_{\max}}\right)^2}\right|_{x=x_1+d}\right)}\Delta t \tag{4.50}$$

In order to investigate the seismic attenuation and energy transmission of seismic wave propagation through an in situ stressed rock mass consists of joints with nonlinear deformation. The seismic wave transmission coefficient T_v and the energy transmission coefficient T_e are respectively defined as

$$T_v = \frac{\max(v_{tran})}{\max(v_{inc})} \tag{4.51}$$

and

$$T_e = \frac{energy_{inc}}{energy_{tran}} = \frac{\sum_{j=t^0_{tran}}^{j=t^1_{tran}} z(v_{tran})^2\Delta t}{\sum_{j=t^0_{inc}}^{j=t^1_{inc}} z(v_{inc})^2\Delta t} \tag{4.52}$$

where the subscripts "*inc*" and "*tran*" refer to the incident wave and transmitted wave respectively, $t^0{}_{inc}$ and $t^0{}_{tran}$ are the initial times of incident waves and transmitted waves, and $t^1{}_{inc}$ and $t^1{}_{tran}$ are the final times of incident waves and transmitted waves.

4.4.2 Amplitude-Dependence of Seismic Wave Propagation

This section presents the amplitude- and frequency-dependence of the wave transmission coefficient, energy transmission coefficient and effective seismic velocity. The wave transmission coefficient and the energy transmission coefficient are respectively defined by Eqs. (4.51) and (4.52). The effective velocity C_{eff} for a seismic wave in a rock mass is a function of the representative elementary volume (REV) length to the time difference between the two peak velocities of incident and transmitted waves as

$$C_{eff} = \frac{S}{t_{tran} - t_{inc}} \tag{4.53}$$

where t_{inc} and t_{tran} are the time spots for peak velocities of incident wave v_{inc} and transmitted wave v_{tran}, respectively. S is the length of representative elementary volume (REV) [13].

In the present presentation, the density of intact rock ρ is 2830 kg/m^3, the seismic wave velocity in intact rock C_0 is 5830 m/s, Young's modulus of intact rock is 90.0 GPa, and the length of RVE is selected as $S = 1.0$ m.

The nonlinear behavior of the joint is assumed to follow the study of Li et al. [13] and Zhao et al. [14], in which the initial specific stiffness of the joint is $k_0 = 3.5$ GPa/s, and the initial closure and the maximum allowance closure of joint are $d_0 =$ 0 mm and $d_{max} = 1.0$ mm, respectively.

The incident wave is assumed to be half-sinusoidal as

$$v(0, t) = \begin{cases} A_0 \sin(2\pi f_0 \times t) \\ 0 \end{cases}, \text{ when } \begin{matrix} 0 \le t \le 1/(2f_0) \\ \text{others} \end{matrix} \tag{4.54}$$

where A_0 and f_0 denote the amplitude and frequency respectively of the incident wave.

The incident wave is assumed to be applied on the surface of the rock joint, therefore $x_1 = 0$ mm in the calculation (as shown in Fig. 4.17).

Amplitude-dependence is an important property of the propagation of seismic waves through a rock mass with nonlinearly deformational joints. By fixing the frequency of the incident wave at $f_0 = 100$ Hz, Figs. 4.18, 4.19 and 4.20 show the amplitude-dependence of the wave transmission coefficient, the energy transmission coefficient and the effective seismic velocity of seismic wave propagation through the rock mass under the in situ stresses of $\sigma_p = 0$, 2.0, 4.0, 8.0 and 15.0 MPa, respectively.

For earthquake engineering, the damage criteria of rock structures are generally regulated according to the threshold values of wave amplitude, such as peak particle velocity (PPV). Therefore, the prediction of the wave transmission coefficient through the jointed rock mass is important in safety assessment. Figure 4.18 reveals that, for seismic wave propagation through an in situ stressed rock mass with a nonlinearly deformational joint, the wave transmission coefficient increases with an increase in the amplitude of the incident wave. Since, for a rock joint with nonlinear behavior, the specific stiffness increases with the amplitude of the incident seismic wave, leading to a smaller difference between the stiffnesses of the joint and intact rock, it results in a large wave transmission. These increasing tendencies agree with the preview studies [13, 15]. Figure 4.18 also reveals that the wave transmission coefficient increases as the in situ stress increases. The wave transmission coefficient approaches 1 when the in situ stress is sufficiently large, e.g., $P_0 \ge 8.0$ MPa in the present study. This explains the phenomenon that the specific stiffness of the joint

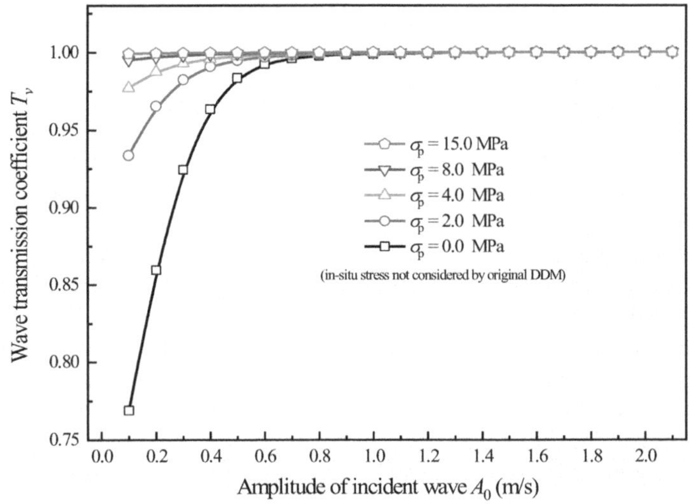

Fig. 4.18 Amplitude-dependence of seismic wave transmission

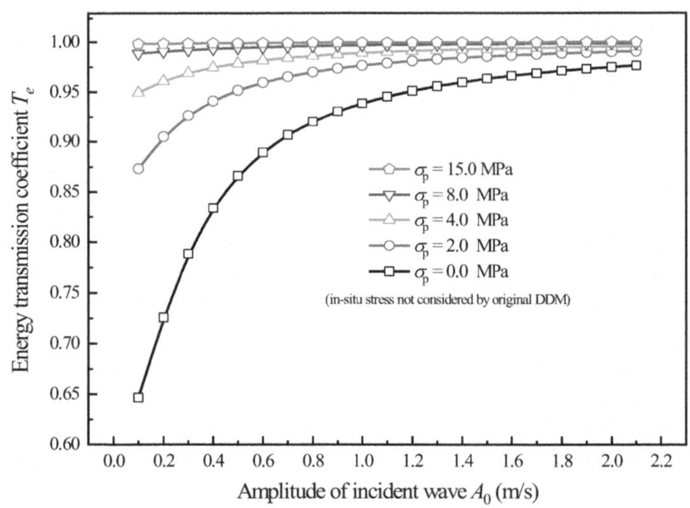

Fig. 4.19 Amplitude-dependence of energy transmission

also increases as the in situ stress increases, and theoretically approaches the stiffness of intact rock when the in situ stress approaches infinity, resulting in a complete wave transmission without reflection (i.e., the wave transmission coefficient equals 1).

Energy transmission is important for seismic analysis. Equation (4.52) shows that the energy of a seismic wave is proportional to the filled area under the curve $v(t)^2$.

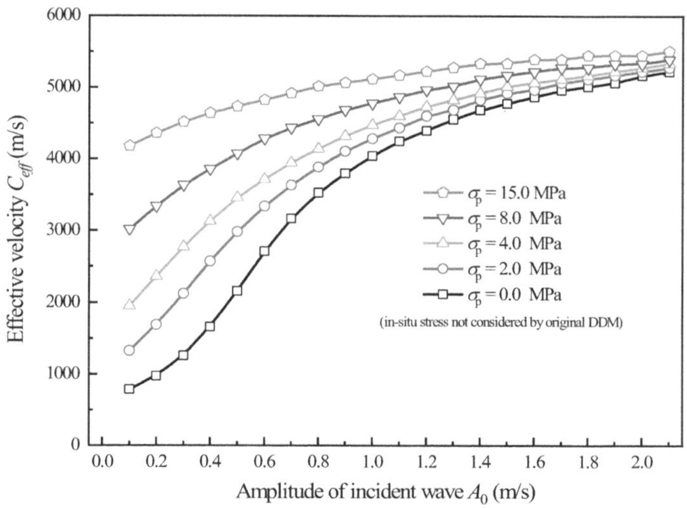

Fig. 4.20 Amplitude-dependence of effective velocity

The energy transmission coefficient is therefore closely associated with the transmission waveform. Higher seismic energy dissipation occurs when the transmission waveform is narrower even with the same wave transmission coefficient. Similar to the amplitude-dependence behavior shown in Figs. 4.18, 4.19 shows that the energy transmission coefficient increases as the amplitude of the incident wave increases. The value of the energy transmission coefficient is larger than the square of the wave transmission coefficient, showing that energy dissipation occurs during seismic wave propagation through a joint.

Figure 4.20 shows the relationship between the effective seismic velocity and the amplitude of the incident wave for rock mass under different in situ stresses. It is observed that the effective seismic velocity increases as the amplitude of the incident wave increases. The effective velocity also increases as the in situ stress increases, which theoretically approaches the seismic wave velocity $C_0 = 5830$ m/s when the rock mass under an infinitely large in situ stress or the amplitude of an incident wave is sufficiently large.

Figures 4.18, 4.19 and 4.20 show that seismic wave attenuation, the energy transmission coefficient and the effective seismic velocity are highly amplitude-dependent for cases with small in situ stresses. However, the amplitude-dependence decreases as the in situ stress increases. It can be neglected when the in situ stress is sufficiently large. In addition, the amplitude-dependence of seismic waves obtained by conventional DDM is a special case of that obtained when the in situ stress is neglected.

4.4.3 *Frequency-Dependence of Seismic Wave Propagation*

The frequency-dependence is another important dynamic property of seismic wave propagation through the rock mass. By fixing the amplitude of the incident wave at $A_0 = 0.1$ m/s Hz, Figs. 4.21, 4.22 and 4.23 show the frequency-dependence of the wave transmission coefficient, the energy transmission coefficient and the effective seismic velocity of seismic wave propagation through a rock mass under the in situ stresses of $\sigma_p = 0$, 2.0, 4.0, 8.0 and 15.0 MPa, respectively.

The relation of the wave transmission coefficient and the energy transmission coefficient to the frequency of an incident wave are shown in Figs. 4.21 and 4.22, respectively. The results show that both the wave transmission coefficient and the energy transmission coefficient decrease as the frequency increases. Therefore, a seismic wave with a smaller frequency can propagate a longer distance than a seismic wave with a large frequency. For the quasi-static situation, when the frequency approaches zero, both the wave transmission coefficient and the energy transmission coefficient approach 1. It is also observed that the energy transmission coefficient in Fig. 4.22 is smaller than the square of the corresponding wave transmission coefficient in Fig. 4.21, which reveals seismic dissipation during wave propagation.

Figures 4.21 and 4.22 also show that both the wave transmission coefficient and the energy transmission coefficient increase as in situ stress increases. Similar to the amplitude-dependence behavior as shown in Figs. 4.18 and 4.19, the frequency-dependence obtained by the conventional DDM is a special case when the in situ stress is neglected. However, it is different from the amplitude-dependence shown

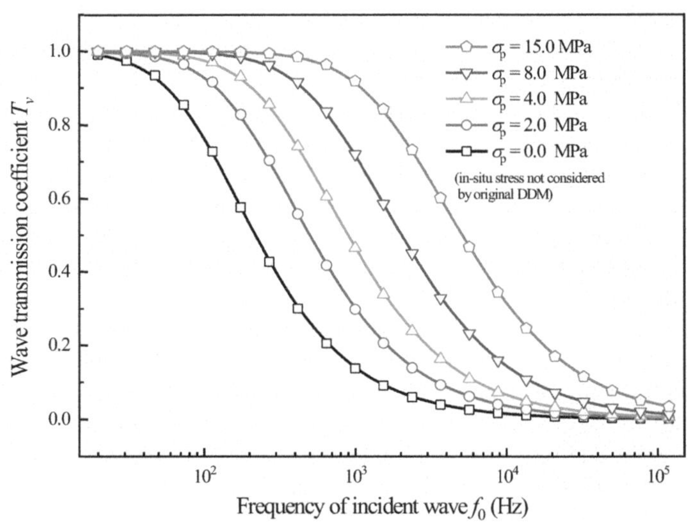

Fig. 4.21 Frequency-dependence of seismic wave transmission

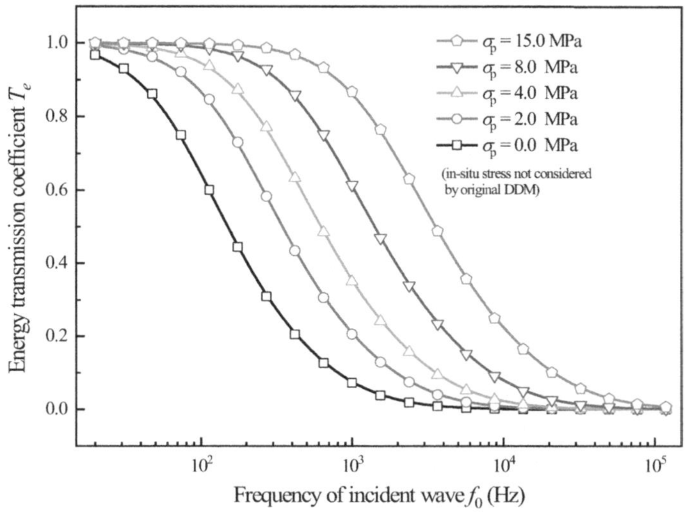

Fig. 4.22 Frequency-dependence of energy transmission

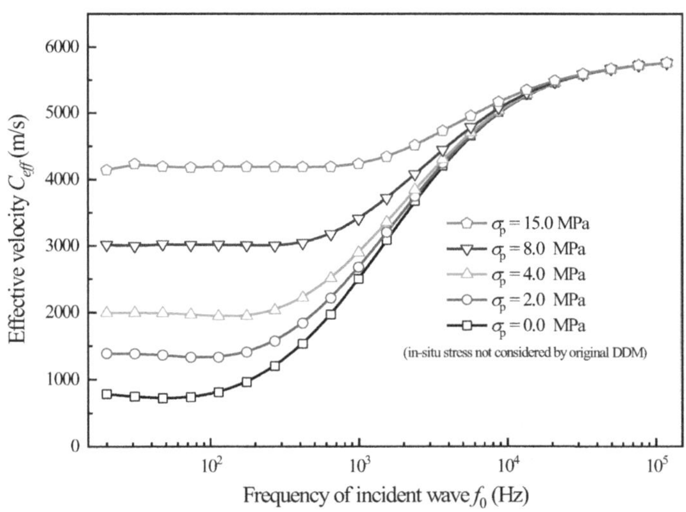

Fig. 4.23 Frequency-dependence of effective velocity

in Figs. 4.18 and 4.19 in that frequency-dependence should not be ignored even for the seismic wave propagation through the rock mass under infinite in situ stress.

Figure 4.23 shows the frequency-dependence of the effective seismic velocity. It can be observed that the effective seismic velocity remains constant at the initial stage when the frequency is small, but it increases sharply with increasing frequency at the transition stage and approaches the effective seismic velocity in intact rock C_0 when the frequency is sufficiently large. The effective seismic velocity increases as the in situ stress increases; it also theoretically approaches the effective seismic velocity in intact rock C_0 when the in situ stress approaches infinity. In that case, the stiffness of the joint approaches the stiffness of the intact rock, and the rock mass behaves as an intact rock and the frequency-dependence of effective seismic velocity can be ignored.

4.4.4 Effect of In Situ Stress on Wave Attenuation

Figure 4.24 shows the variation of the wave transmission coefficient T_v with respect to the in situ stress P_0 for the incident wave with a frequency of $f_0 = 100$ Hz and different amplitudes of $A_0 = 0.1, 0.4, 0.7$, 1.1 and 2.1 m/s, respectively. The incident wave is in the form of Eq. (4.54). It can be observed from Fig. 4.24 that the wave transmission coefficient increases with increasing in situ stress at the initial stage, and subsequently remains nearly constant as the in situ stress further increases. The wave transmission coefficient approaches 1 for seismic wave propagation through rock mass under an infinite in situ stress as observed from Fig. 4.18. For a seismic wave with a large amplitude, the in situ stress has less effect on the wave transmission coefficient, resulting in a large wave transmission coefficient. The effect of in situ stress can be ignored for the incident wave with a sufficiently large amplitude, e.g., $A_0 \geq 2.1$ m/s in the present study.

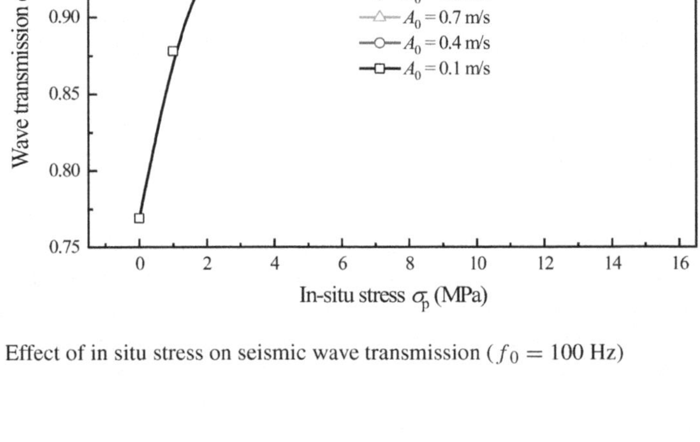

Fig. 4.24 Effect of in situ stress on seismic wave transmission ($f_0 = 100$ Hz)

Figure 4.25 shows the effect of in situ stress on the wave transmission coefficient for incident seismic waves with a fixed amplitude of $A_0 = 0.1$ m/s and varying frequencies of $f_0 = 30.9$, 997.4, 2377.8, 8752.3 and 20,865.1 Hz, respectively. As observed in Fig. 4.21, the results show that the wave transmission coefficient increases as the in situ stress increases. The wave transmission coefficient theoretically approaches 1 for an infinite in situ stress. It is also observed from Fig. 4.25 that the in situ stress has a more significant effect on the seismic wave with a medial frequency (e.g., $f_0 = 997.4$ Hz in Fig. 4.25). For a seismic wave with a small frequency (e.g., $f_0 \leq 30.9$ Hz), the wave transmission coefficient remains constant with the increasing in situ stress. In that case, most parts of the seismic wave can be transmitted and the rock mass behaves as an intact rock, so the effect of in situ stress can be neglected. For the seismic wave with a large frequency (e.g., $f_0 \geq 20{,}865.1$ Hz), the wave transmission coefficient approaches 0. In that case, the seismic wave is filtered out by the joint, so the effect of in situ stress can also be neglected.

(1) Effect of in situ stress on energy transmission

The seismic waves in the form of Eq. (4.54) with a fixed frequency of $f_0 = 100$ Hz and amplitudes of $A_0 = 0.1$, 0.4, 0.7, 1.1 and 2.1 m/s respectively are applied to the rock mass. Based on the transmitted waveforms, the energy transmission is analyzed. It is observed from Fig. 4.26 that the energy transmission coefficient increases as the in situ stress increases, and approaches 1 for a sufficiently large in situ stress. For an in situ stress larger than 10.0 MPa, the effect of in situ stress remains constant, revealing that the stiffness of the joint approaches the stiffness of intact rock. For the rock mass under the same in situ stress, the energy transmission coefficient increases

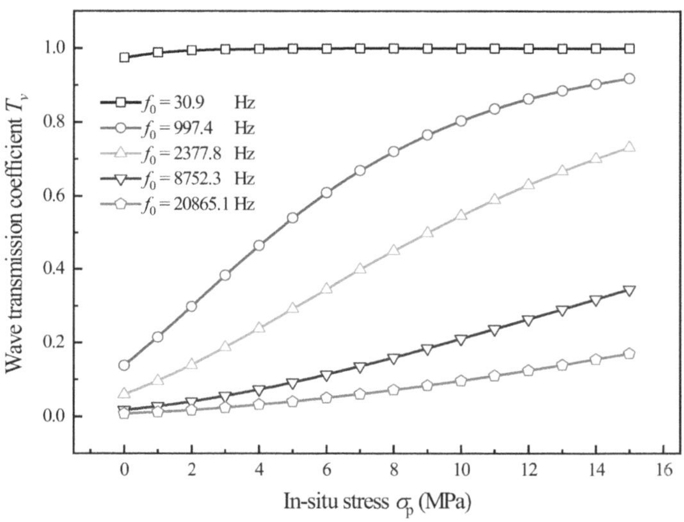

Fig. 4.25 Effect of in situ stress on seismic wave transmission ($A_0 = 0.1$ m/s)

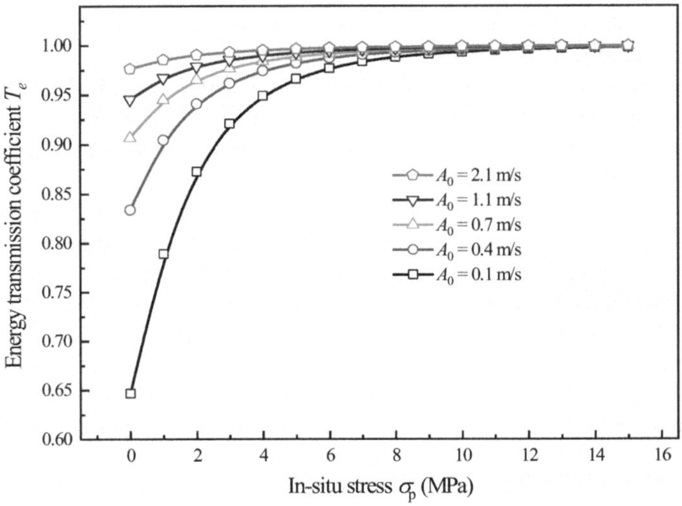

Fig. 4.26 Effect of in situ stress on energy transmission ($f_0 = 100$ Hz)

as the amplitude of the seismic wave increases; however, the increasing tendency is smaller than that of the wave transmission coefficient as shown in Fig. 4.25.

Figure 4.27 shows the effect of in situ stress on the energy transmission coefficient for seismic waves in the form of Eq. (4.54) with a fixed amplitude of $A_0 = 0.1$ m/s and frequencies of $f_0 = 30.9$, 997.4, 2377.8, 8752.3 and 20,865.1 Hz, respectively. It is observed from Fig. 4.27 that the energy transmission coefficient increases as the in situ stress increases. The effect of in situ stress on the energy transmission coefficient can be ignored when the frequency is sufficiently small or sufficiently large. Figure 4.27 also shows that, for the seismic wave propagation through the rock mass under the same in situ stress, the energy transmission coefficient increases as the frequency decreases and explains the phenomenon that a seismic wave with a small frequency can propagate a longer distance than a seismic wave with a large frequency.

(2) Effect of in situ stress on effective seismic velocity

Figure 4.28 shows the effect of in situ stress on the effective velocity of seismic waves in the form of Eq. (4.54) with a fixed frequency of $f_0 = 100$ Hz and amplitudes of $A_0 = 0.1, 0.4, 0.7, 1.1$ and 2.1 m/s, respectively. The definition of effective velocity is as given by Eq. (4.53). It is observed from Fig. 4.28 that the effective velocity is significantly affected by in situ stress. The effective velocity of a seismic wave increases as the in situ stress increases. Under the same in situ stress, the effective velocity increases as the amplitude of the seismic wave increases and the effective seismic velocity approaches a constant value, which is the effective velocity in the intact rock $C_{eff} = 5830$ m/s when the amplitude is sufficiently large.

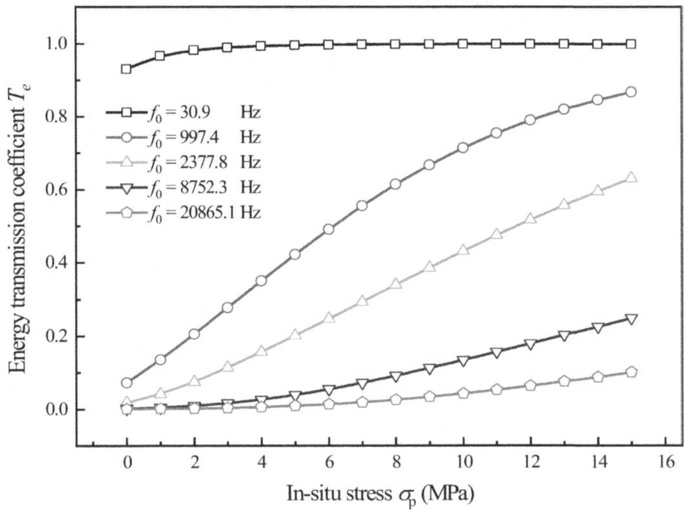

Fig. 4.27 Effect of in situ stress on energy transmission ($A_0 = 0.1$ m/s)

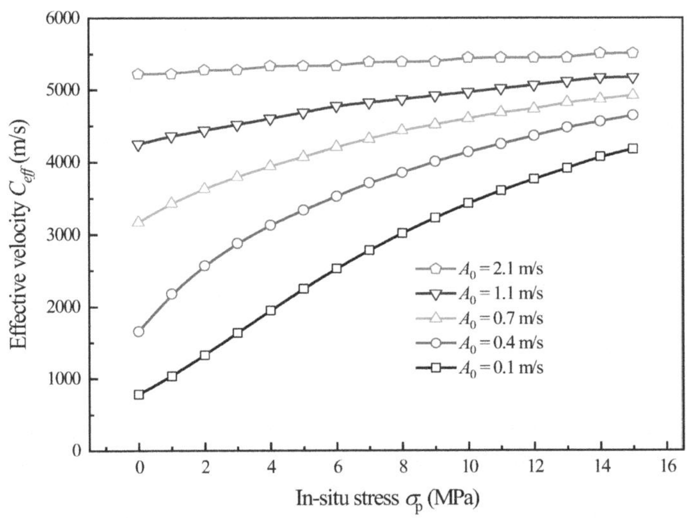

Fig. 4.28 Effect of in situ stress on effective velocity ($f_0 = 100$ Hz)

Figure 4.29 shows the effect of in situ stress on the effective velocity of seismic waves in the form of Eq. (4.54) with a fixed amplitude of $A_0 = 0.1$ m/s and frequencies of $f_0 = 30.9$, 997.4, 2377.8, 8752.3 and 20,865.1 Hz, respectively. It can be observed from Fig. 4.29 that the effective velocity also increases as the in situ stress increases for seismic waves with the same frequency. For seismic waves under the

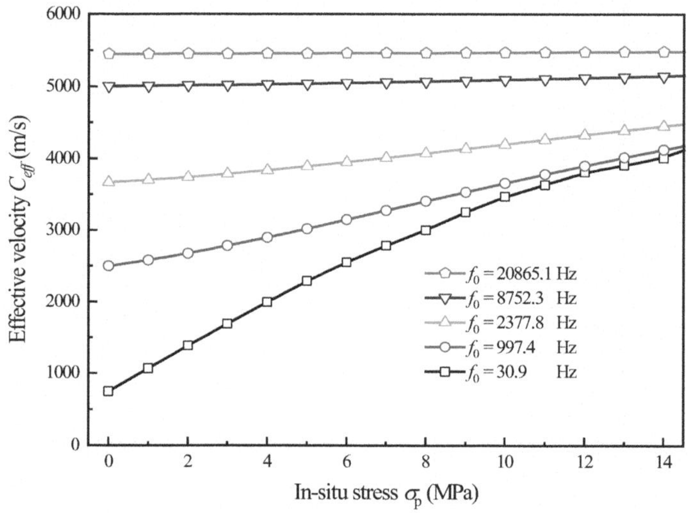

Fig. 4.29 Effect of in situ stress on effective velocity ($A_0 = 0.1$ m/s)

same in situ stress, the effective seismic velocity increases as the frequency increases, and approaches the effective velocity in intact rock as the frequency approaches infinity. Figure 4.29 shows that, for the situation when the frequency is sufficiently large (e.g., $f_0 \geq 8752.3$ Hz), the effect of in situ stress can be ignored as even the effective seismic velocity is much smaller than the seismic velocity in intact rock C_0.

4.5 Conclusions

(1) Wave propagation across a rock joint with different loading/unloading behavior is performed. Two-characteristic method is proposed to solve the wave propagation equations across the joint, and the unloading effect on the stress wave propagation is evaluated.

For a single loading/unloading pulse, the energy transmission increases as the amplitude of the incident wave increases while it decreases as the frequency of the incident wave increases. The unloading effect influences energy transmission significantly. However, when the amplitude of the incident wave approaches zero or the frequency of the incident wave approaches infinity, the unloading effect can be ignored.

For a single loading/unloading pulse with five different basic waveforms, the unloading effect has the least influence on the rectangular waveform, while it has the largest influence on the incident pulse with the triangular waveform.

For multiple loading/unloading pulses, the unloading influences the energy transmission only when the time interval is large enough to avoid the superposition between the transmitted pulses. Otherwise, the unloading influences both the energy transmission and wave attenuation when the later transmitted pulses are contaminated by the former transmitted pulses.

(2) A DDM method considering in situ stress is introduced to study seismic wave propagation through in situ stressed rock mass. The amplitude- and frequency-dependence of seismic waves under in situ stresses are analyzed. The effect of in situ stress on the wave transmission coefficient, energy transmission coefficient and effective seismic velocity is discussed.

It is concluded that both the wave transmission coefficient and the energy transmission coefficient increase as the amplitude of a seismic wave increases, and decrease as the frequency of the incident wave increases. The effect of in situ stress on the amplitude- and frequency-dependence of seismic wave cannot be neglected.

Three important parameters, wave transmission coefficient, energy transmission coefficient and effective seismic velocity, are distinctly affected by in situ stress. The wave transmission coefficient, energy transmission coefficient and effective seismic velocity increase fast in the initial stage of in situ stress increase. When the in situ stress is sufficiently large, the wave transmission coefficient and energy transmission coefficient approach 1 and the effective velocity approaches the seismic velocity in the intact rock.

It should be mentioned that the present study gives only preliminary results on seismic wave propagation through in situ stressed rock mass. Further research are still necessary to consider more complicated cases, for example, obliquely incident wave propagation through in situ stressed rock mass, and seismic wave propagation through in situ stressed rock mass with sets of joints.

References

1. Jaeger JC, Cook NGW, Zimmerman R. Fundamentals of rock mechanics. 4th ed. Malden: Blackwell Publishing; 2007.
2. Bandis SC, Lumsden AC, Barton NR. Fundamentals of rock joint deformation. Int J Rock Mech Min. 1983;20(6):249–68.
3. Barton N, Bandis S, Bakhtar K. Strength, deformation and conductivity coupling of rock joints. Int J Rock Mech Min Sci & Geomech Abstr. 1985;22(3):121–40.
4. Zhao J, Cai JG. Transmission of elastic p-waves across single fractures with a nonlinear normal deformational behavior. Rock Mech Rock Eng. 2001;34(1):3–22.
5. Ma GW, Li JC, Zhao J. Three-phase medium model for filled rock joint and interaction with stress waves. Int J Numer Anal Met. 2011;35(1):97–110.
6. Li JC, Ma GW. Experimental study of stress wave propagation across a filled rock joint. Int J Rock Mech Min. 2009;46(3):471–8.
7. Pyrak-Nolte LJ, Myer LR, Cook NGW. Anisotropy in seismic velocities and amplitudes from multiple parallel fractures. J Geophys Res. 1990;95(B7):11345–58.

8. Pyrak-Nolte LJ, Myer LR, Cook NGW. Transmission of seismic waves across single natural fractures. J Geophys Res. 1990;95(B6):8617–38.
9. Li JC. Wave propagation across non-linear rock joints based on time-domain recursive method. Geophys J Int. 2013;193(2):970–85.
10. Zhao J, Zhao XB, Cai JG. A further study of p-wave attenuation across parallel fractures with linear deformational behaviour. Int J Rock Mech Min. 2006;43(5):776–88.
11. Cook NGW. Natural joints in rock—mechanical, hydraulic and seismic behavior and properties under normal stress. Int J Rock Mech Min Sci & Geomech Abstr. 1992;29(3):198–223.
12. Li J, Ma G, Huang X. Analysis of wave propagation through a filled rock joint. Rock Mech Rock Eng. 2009;43(6):789–98.
13. Li JC, Ma GW, Zhao J. An equivalent viscoelastic model for rock mass with parallel joints. J Geophys Res-Solid Ea. 2010;115(B3):B03305.
14. Zhao XB, Zhao J, Cai JG. P-wave transmission across fractures with nonlinear deformational behaviour. Int J Numer Anal Met. 2006;30(11):1097–112.
15. Zhao XB, Zhao J, Hefny AM, et al. Normal transmission of s-wave across parallel fractures with coulomb slip behavior. J Eng Mech. 2006;132(6):641–50.

Chapter 5
Two-Characteristic Method for the Wave Propagation Through Complex Stratus Rock Mass

5.1 Stress Wave Propagation Through Complex Rock Mass with Linearly Deformational Joints

Wave propagation in the rock mass plays an important role in mining safety assessment, oil recovery, underground construction and the study of earthquakes. Joints that exist in natural rock mass significantly influence wave transmission. Therefore, it is important to study the effect of joints on wave propagation in the rock mass [1].

The present section investigates the stress wave propagation through the rock mass with different mechanical properties in front of and behind the joint. The conventional characteristic method is modified to analyze wave transmission through the complex rock mass. Two typical cases, of "soft-to-hard" and "hard-to-soft" rock mass, are illustrated. The "soft-to-hard" rock mass case means that the stress waves propagate from rocks with small wave impedances to rocks with large wave resistances, while the "hard-to-soft" rock mass case means that stress waves propagate from rocks with large wave impedances to rocks with small wave impedances. The particle velocity transmission ratio, the stress transmission ratio and the effective velocity are investigated by changing the frequency of the incident wave, the joint stiffness and the wave impedance ratio.

5.1.1 A Modified Characteristic Method

In complex rock mass, joints are widely distributed among different rock layers, as shown in Fig. 5.1a. Figure 5.1b shows a representative elementary volume (REV) that has different rock types in front of and behind the joint.

The wave equation for a one-dimensional elastic wave is

L. Fan et al., *The Method of Characteristics for Stress Wave Propagation in the Rock Mass*, https://doi.org/10.1007/978-981-97-1440-7_5

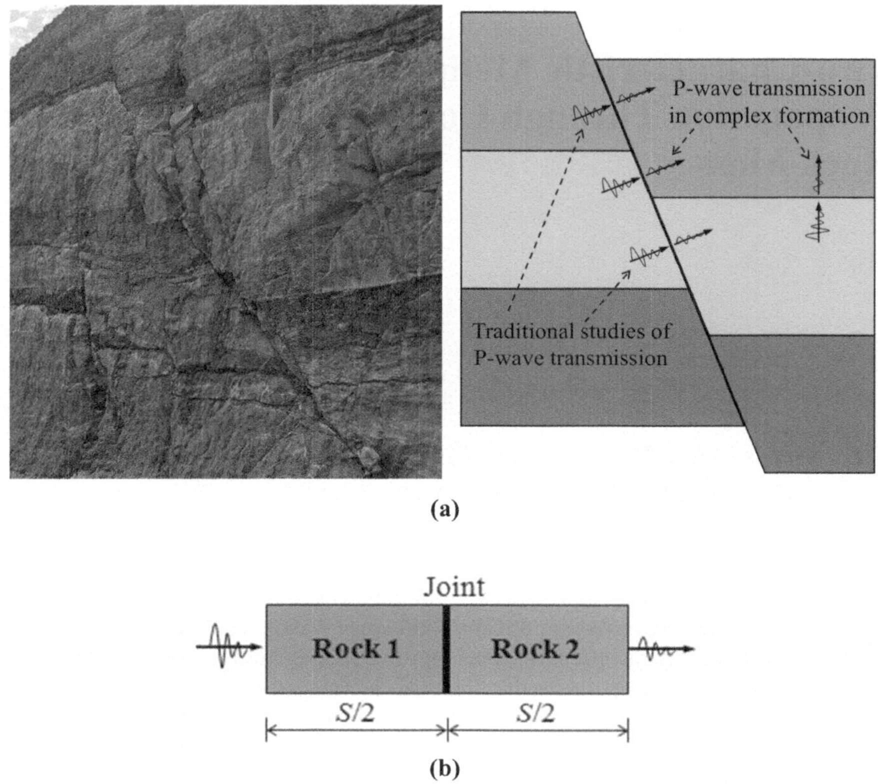

Fig. 5.1 Illustration of the complex rock mass. **a** Joints in a complex formation. **b** Representative elementary volume

$$\frac{\partial^2 u}{\partial t^2} - C_0^2 \frac{\partial^2 u}{\partial x^2} = 0 \tag{5.1}$$

where C_0 is the velocity of wave propagation, u is the displacement, and t is the time.

Equation (5.1) is a hyperbolic partial differential equation with two distinct real characteristic lines at any point in the (x, t) plane. The characteristic line is the basis of the characteristic method and is one of the main approaches used to solve hyperbolic partial differential equations. The characteristic method is widely used in the propagation of one-dimensional waves [2, 3].

To find the characteristic lines in the (x, t) plane, differential expressions of u are expressed as

$$dv = \frac{\partial v}{\partial x} dx + \frac{\partial v}{\partial t} dt = \frac{\partial^2 u}{\partial x \partial t} dx + \frac{\partial^2 u}{\partial t^2} dt \tag{5.2}$$

$$d\varepsilon = \frac{\partial \varepsilon}{\partial x}dx + \frac{\partial \varepsilon}{\partial t}dt = \frac{\partial^2 u}{\partial x^2}dx + \frac{\partial^2 u}{\partial x \partial t}dt \tag{5.3}$$

Equations (5.2) and (5.3) are linearly combined to obtain

$$dv + \lambda d\varepsilon = \frac{\partial^2 u}{\partial t^2}dt + (\lambda dt + dx)\frac{\partial^2 u}{\partial x \partial t} + \lambda \frac{\partial^2 u}{\partial x^2}dx = 0 \tag{5.4}$$

where λ is an undetermined coefficient.

By comparing Eq. (5.4) with Eq. (5.1), the following expression is yielded:

$$\frac{1}{dt} = \frac{0}{\lambda dt + dx} = -\frac{C^2}{\lambda dx} \tag{5.5}$$

From Eq. (5.5),

$$C_0 = \pm \frac{dx}{dt} \tag{5.6}$$

Therefore, the slopes of the two sets of characteristic lines are given by C_0.

In the conventional characteristic method, the wave propagation equations can be solved by two sets of characteristic lines with slopes of C_0 and $-C_0$ in the $x-t$ plane, as shown in Fig. 5.2a, where C_0 is the wave propagation velocity in the intact rock. To study the propagation of waves in the complex rock mass, a modified characteristic model is proposed, which includes four sets of characteristic lines with slopes of C_{01}, $-C_{01}$, C_{02} and $-C_{02}$, as shown in Fig. 5.2b, where C_{01} and C_{02} are the wave propagation velocities of the intact rocks in front of and behind the joint, respectively.

The relations for the particle velocity v and stress σ can be derived as $zv(x, t) + \sigma(x, t) =$ constant along the right-running characteristic line (RC) and $zv(x, t) - \sigma(x, t) =$ constant along the left-running characteristic line (LC), where z is the wave impedance of the intact rock. It should be noted that normal stress is defined to be positive for compressive stresses and negative for tensile stress.

As shown in Fig. 5.1b, a normally incident wave is applied at the left boundary of the representative elementary volume. During the wave transmission across the joint, the discontinuous displacement model is applied at $x = x_1$, as shown in Fig. 5.3, which yields

$$\sigma^-(x_1, t) = \sigma^+(x_1, t) = \sigma(x_1, t) \tag{5.7}$$

$$u^-(x_1, t) - u^+(x_1, t) = \frac{\sigma(x_1, t)}{k_n} \tag{5.8}$$

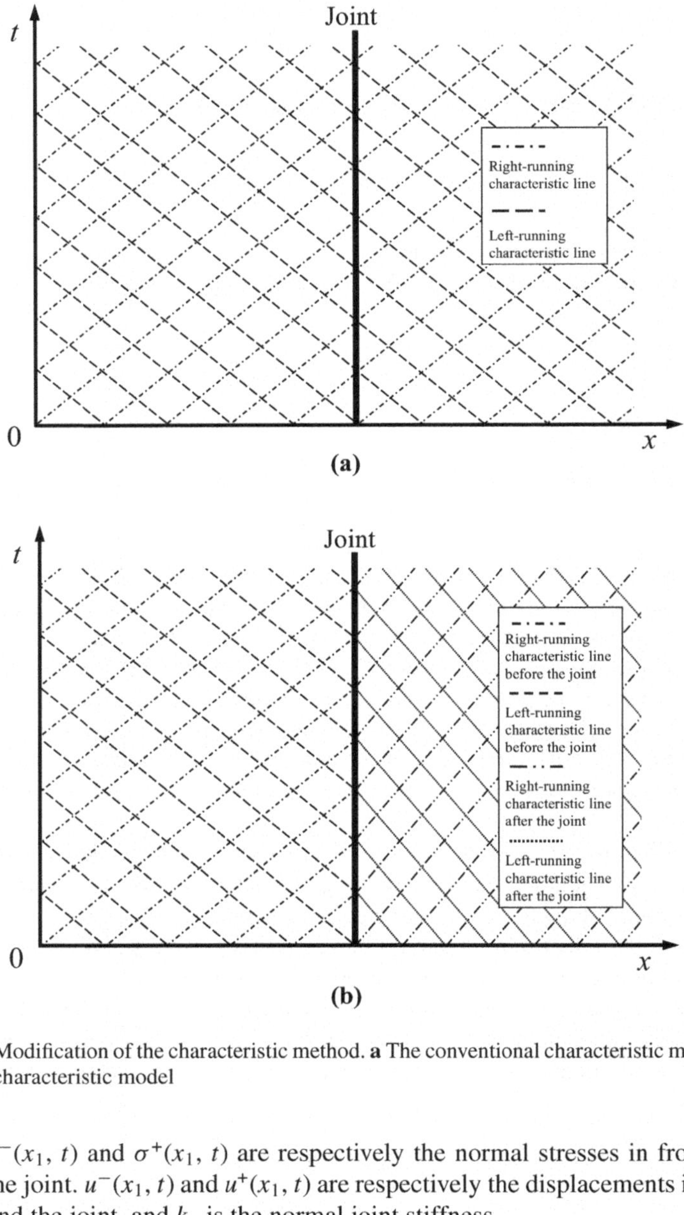

Fig. 5.2 Modification of the characteristic method. **a** The conventional characteristic model. **b** The modified characteristic model

where $\sigma^{-}(x_1, t)$ and $\sigma^{+}(x_1, t)$ are respectively the normal stresses in front of and behind the joint. $u^{-}(x_1, t)$ and $u^{+}(x_1, t)$ are respectively the displacements in front of and behind the joint, and k_n is the normal joint stiffness.

The differential expression of Eq. (5.8) is

$$v^{-}(x_1, t) - v^{+}(x_1, t) = \frac{1}{k_n}\frac{\partial \sigma(x_1, t)}{\partial t} \tag{5.9}$$

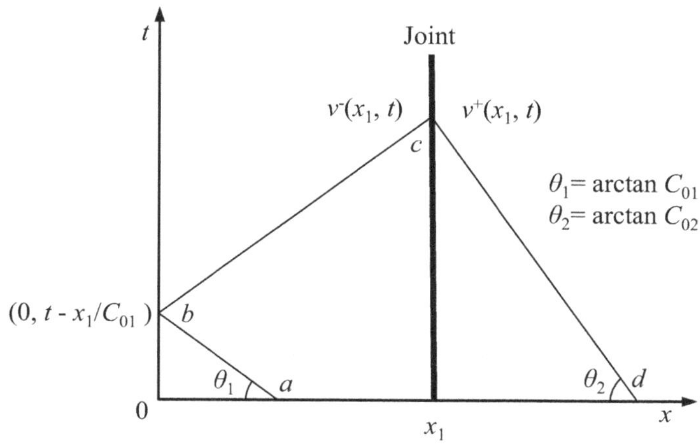

Fig. 5.3 Scheme of the characteristic method

where $v^-(x_1, t)$ and $v^+(x_1, t)$ are the particle velocities in front of and behind the joint, respectively.

Figure 5.3 illustrates the left-running characteristic *cd* behind the joint. When $t = 0$, the rock mass is undisturbed. Therefore, $v(x, 0) = 0$ and $\sigma(x, 0) = 0$ at point *d*. Then, along the line *cd*,

$$z_2 v^+(x_1, t) - \sigma(x_1, t) = 0 \tag{5.10}$$

where z_2 is the wave impedance of the rock mass behind the joint.

The right-running characteristic *cb* is in front of the joint and intersects the *t*-axis at the point $(0, t - x_1/C_{01})$. Therefore, along the right-running characteristic *cb* can be obtained as

$$z_1 v^-(x_1, t) + \sigma(x_1, t) = z_1 v(0, t - x_1/C_{01}) + \sigma(0, t - x_1/C_{01}) \tag{5.11}$$

where z_1 is the wave impedance of the rock mass in front of the joint.

Along the left-running characteristic *ba*, we can obtain

$$z_1 v(0, t - x_1/C_{01}) - \sigma(0, t - x_1/C_{01}) = 0 \tag{5.12}$$

Substituting Eq. (5.12) into Eq. (5.11) gives

$$z_1 v^-(x_1, t) + \sigma(x_1, t) = 2 z_1 v(0, t - x_1/C_{01}) \tag{5.13}$$

Summing Eqs. (5.10) and (5.13) yields

$$z_1 v^-(x_1, t) + z_2 v^+(x_1, t) = 2 z_1 v(0, t - x_1/C_{01}) \tag{5.14}$$

Substituting Eqs. (5.9) and (5.10) into Eq. (5.14) gives

$$v^-(x_1,t) - v^+(x_1,t) = \frac{z_2 \partial v^+(x_1,t)}{k_n \partial t} \tag{5.15}$$

Substituting Eq. (5.15) into Eq. (5.14) yields

$$\frac{\partial v^+(x_1,t)}{\partial t} = \frac{2k_n}{z_2} v(0, t - x_1/C_{01}) - k_n(1/z_2 + 1/z_1)v^+(x_1,t) \tag{5.16}$$

If Δt is sufficiently small, the derivative in Eq. (5.16) can be replaced by a differential expression of

$$\frac{\partial v^+(x_1,t)}{\partial t} = \frac{v^+(x_1,t_{j+1}) - v^+(x_1,t_j)}{\Delta t} \tag{5.17}$$

Therefore, Eq. (5.16) is rewritten as

$$v^+(x_1,t_{j+1}) = \frac{2k_n \Delta t}{z_2} v(0, t - x_1/C_{01}) + [1 - k_n \Delta t(1/z_1 + 1/z_2)]v^+(x_1,t_j) \tag{5.18}$$

Given an input velocity of $v(0, t - x_1/C_{01})$ and the initial particle velocity $v^+(x_1, 0)$, the particle velocity at any time t in front of the joint, $v^+(x_1, t)$, can be determined through an iterative process.

To study the transmitted wave across the joint between different rock mass, the particle velocity transmission ratio (T_v) and the stress transmission ratio (T_σ) are respectively defined as

$$T_v = \frac{\max(v_T)}{\max(v_I)} \tag{5.19}$$

and

$$T_\sigma = \frac{\max(\sigma_T)}{\max(\sigma_I)} \tag{5.20}$$

The effective velocity C_{eff} for a stress wave in the rock mass is a function of the representative elementary volume length to the time difference between the two peak velocities of the incident and transmitted waves as

$$C_{\text{eff}} = \frac{S}{t_T - t_I} \tag{5.21}$$

where t_I and t_T are the times for peak velocities of the incident wave v_I and transmitted wave v_T, respectively, and S is the length of the representative elementary volume.

5.1.2 Wave Propagation Through Complex Rock Mass

The incident wave is assumed to be half-sinusoidal as

$$v_I(0,t) = \begin{cases} A_0 \sin(2\pi f_0 \times t) \\ 0 \end{cases}, \quad \text{when} \quad \begin{matrix} 0 \le t \le 1/(2f_0) \\ \text{others} \end{matrix} \tag{5.22}$$

where A_0 is the amplitude of the incident wave and f_0 is the frequency of the incident wave.

The incident wave is applied on the left boundary of the representative elementary volume, x_1 is set to be $S/2$ and the amplitude A_0 is set to 1.0 m/s in the present study, as was done in previous research [4]. The length of the representative elementary volume is selected as $S = 1.0$ m. The mechanical parameters of these intact rocks are shown in Table 5.1.

(1) "soft-to-hard" rock mass

To study the wave transmission across the joint from the "soft" rock to the "hard" rock, Rock *III* from Table 5.1 is selected for the rock in front of the joint, while Rock *I* and Rock *II* are selected for the rock behind the joint. Therefore, the wave impedance ratios are $n = z_1/z_2 = 0.306$ and 0.468 with corresponding Young's modulus ratios $m = E_1/E_2 = 0.125$ and 0.25, respectively. The frequency of the incident wave is $f_0 = 316.2$ Hz, and the joint stiffness is $k_n = 8.0$ GPa/m, as in previous research [5].

Figure 5.4a and b shows the particle velocity and stress variations when the wave propagates through the "soft-to-hard" rock mass with $n = 0.306$ ($m = 0.125$) and $n = 0.468$ ($m = 0.25$), respectively. It is observed from Fig. 5.4a that when the wave impedance ratio $n = 0.306$, the particle velocity transmission ratio $T_v = 0.414$. However, when the wave impedance ratio $n = 0.468$, the particle velocity transmission ratio T_v is 0.523. Similar to the results of wave propagation through welded joints in the previous research [6], the particle velocity decreases when the wave propagates through "soft-to-hard" rock mass. In addition, the particle velocity attenuation increases as the wave impedance ratio decreases.

Figure 5.4b shows the stress transmission across the representative elementary volume of the complex rock mass. The incident stress waves are calculated based on Fig. 5.4a. Figure 5.4b shows that when the wave impedance ratio is $n = 0.306$ ($m = 0.125$), the stress transmission ratio T_σ is 1.350, while when the wave impedance ratio is $n = 0.468$ ($m = 0.25$), the stress transmission ratio T_σ is 1.117. It can be

Table 5.1 Parameters of the intact rocks

	Rock *I*	Rock *II*	Rock *III*	Rock *IV*	Rock *V*
E (GPa)	5	10	40	60	80
ρ (kg/m^3)	1800	2100	2400	2600	3000

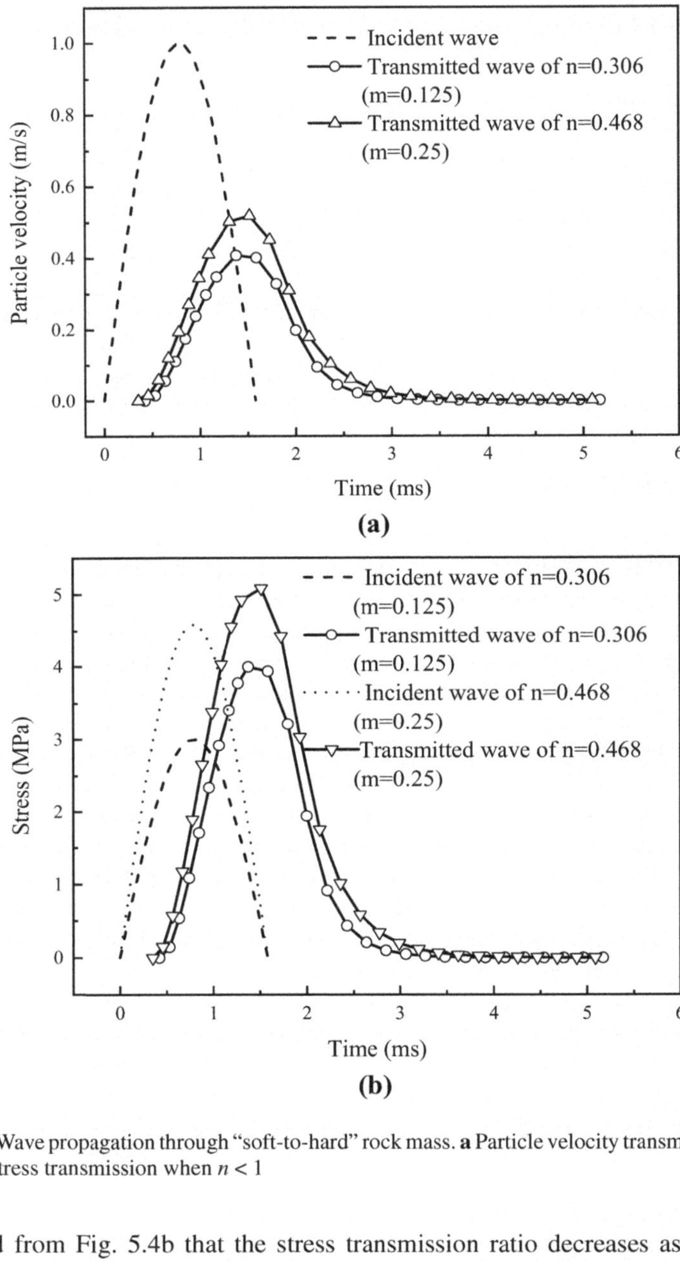

Fig. 5.4 Wave propagation through "soft-to-hard" rock mass. **a** Particle velocity transmission when $n < 1$. **b** Stress transmission when $n < 1$

observed from Fig. 5.4b that the stress transmission ratio decreases as the wave impedance ratio increases, which is similar to the wave propagation in ordinary rock mass. However, it is interesting to observe from Fig. 5.4b that the transmitted stress waves have larger amplitudes than the incident waves when crossing the joint from the "soft" rock to the "hard" rock.

(2) "Hard-to-soft" rock mass

Similar to the study in the previous sections, to study the wave transmission across the joint from the "hard" rock to the "soft" rock, Rock *III* from Table 5.1 was selected for the rock in front of the joint, while Rock *IV* and Rock *V* are selected for the rock behind the joint. Therefore, the wave impedance ratios are $n = z_1/z_2 = 1.275$ and 1.581 with corresponding Young's modulus ratios $m = E_1/E_2 = 1.5$ and 2.0, respectively. The frequency of the incident wave is $f_0 = 562.3$ Hz, and the joint stiffness is $k_n = 64.0$ GPa/m.

Figure 5.5a shows the incident velocity wave and transmitted wave through the right boundary of the representative elementary volume when $n = 1.275$ and $n = 1.581$. Figure 5.5b shows the corresponding incident stress waves and transmitted stress waves. In Fig. 5.5a, when the wave impedance ratio is $n = 1.275$, the particle velocity transmission ratio T_v is 1.076, and when the wave impedance ratio is $n = 1.581$, the particle velocity transmission ratio T_v is 1.167. The particle velocity of the transmitted waves is larger than that of the incident waves when crossing the joint from the "hard" rock to the "soft" rock. In addition, the particle velocity of the transmitted wave increases as the wave impedance ratio increases.

In Fig. 5.5b, when the wave impedance ratio is $n = 1.275$, the stress transmission ratio T_σ is 0.844, while when the wave impedance ratio is $n = 1.581$, the stress transmission ratio T_σ is 0.738. The transmitted stress waves attenuate when the wave crosses the joint from the "hard" rock to the "soft" rock. It also can be observed from Fig. 5.5b that the stress attenuation increases as the wave impedance ratio increases.

5.1.3 Wave Transmission and Effective Velocity

(1) Effect of frequency on wave transmission and effective velocity

The case of "soft-to-hard" rock mass is utilized to study the effect of frequency on the wave transmission across linearly jointed complex rock mass. The incident wave is in the form of Eq. (5.22). Figure 5.6a–c shows the variation in the particle velocity transmission ratio T_v, the stress transmission ratio T_σ and effective velocity C_{eff} with respect to the frequency f_0 with wave impedance ratios of $n = 0.306$ and joint stiffnesses of $k_n = 2.0$, 5.7, 11.3, 22.6 and 45.3 GPa/m, respectively.

It is seen from Fig. 5.6a, b that both the particle velocity transmission ratio and the stress transmission ratio decrease as the frequency increases. When the frequency is sufficiently small, such as $f_0 < 10.0$ Hz, both the particle velocity transmission ratio and stress transmission ratio remain stable as the frequency increases. However, they decrease sharply as the frequency further increases in the transitional frequency range from 10 to 10^5 Hz. The particle velocity transmission ratio and stress transmission ratio approach zero for the cases when the frequency is sufficiently large, which indicates that the waves are completely reflected and slightly transmitted in those cases. Figure 5.6a, b also shows that the transmission ratios increase as the joint

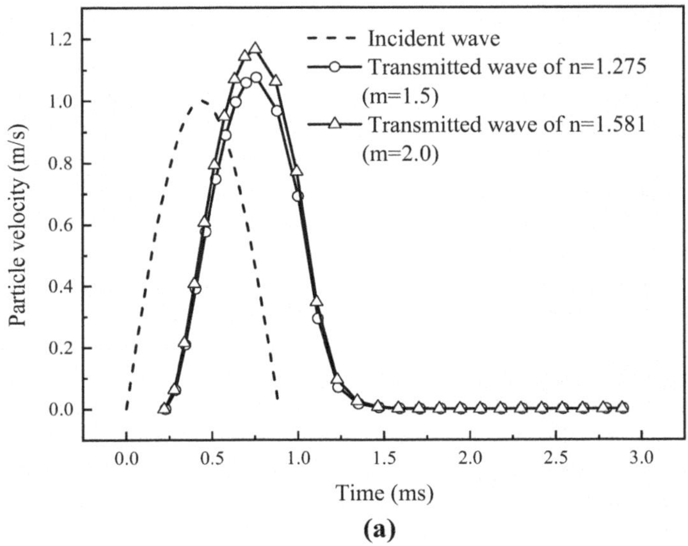

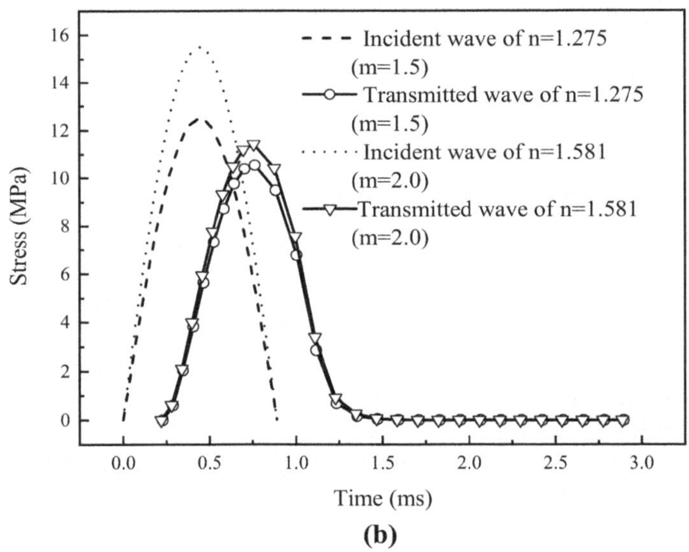

Fig. 5.5 Wave propagation through "hard-to-soft" rock mass. **a** Particle velocity transmission when $n > 1$. **b** Stress transmission when $n > 1$

stiffness increases for waves with the same frequency, which coincides with previous research on wave propagation through jointed rock mass [7].

It should be noted that, in Fig. 5.6b, the stress transmission ratio may be larger than 1.0 when the frequency of the incident wave is sufficiently small, which is caused by the difference in the wave impedances on either side of the joint. For

Fig. 5.6 Effect of frequency on wave transmission. **a** Particle velocity transmission ratio. **b** Stress transmission ratio. **c** Effective velocity

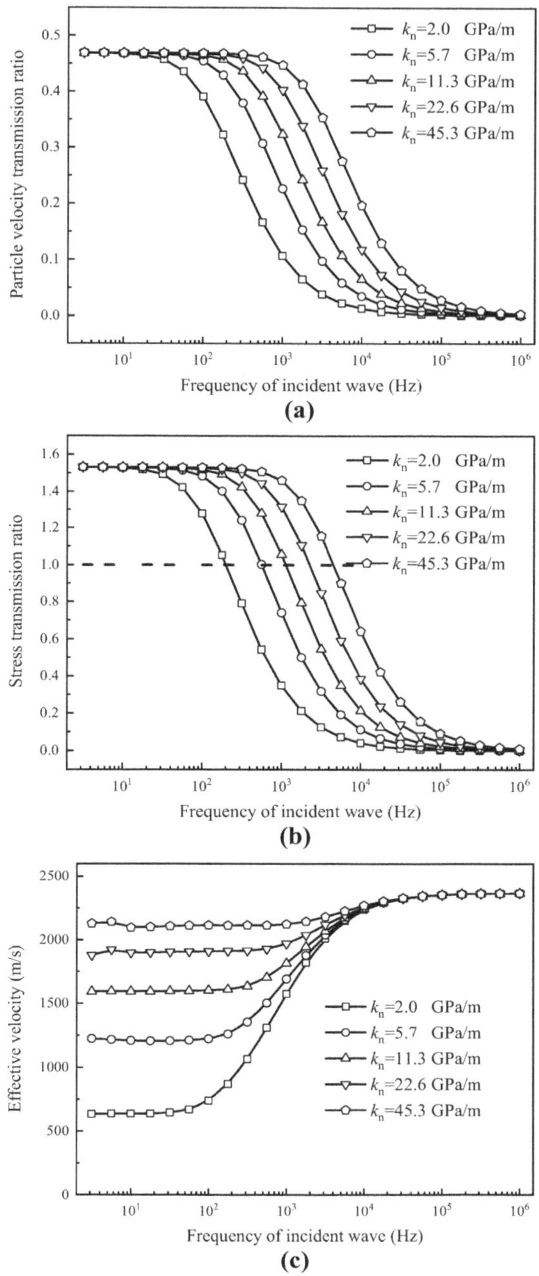

the case of wave propagation through "soft-to-hard" rock mass, the particle velocity transmission ratio is always smaller than 1.0, while the stress transmission ratio may be larger than 1.0. On the other hand, in the case of wave propagation through "hard-to-soft" rock mass, the stress transmission ratio is always smaller than 1.0, and the particle velocity transmission ratio may be larger than 1.0.

Figure 5.6c shows the effect of frequency on the effective velocity. The relationship between the effective velocity and frequency of the incident wave propagation through complex rock mass is similar to the effective velocity and frequency of wave propagation through conventional jointed rock mass. It is seen that the effective velocity remains constant when the frequency is small. Subsequently, the velocity increases sharply as the frequency increases, and finally approaches a constant as the frequency further increases. Moreover, it is also observed from Fig. 5.6c that the effective velocity increases as the joint stiffness increases for the incident waves with the same frequency.

(2) Effect of joint stiffness on wave transmission and effective velocity

Similar to the previous section, the case of "soft-to-hard" rock mass ($n = 0.306$) is utilized to study the effect of joint stiffness on the wave transmission across linearly jointed complex rock mass. Figure 5.7a–c shows the effect of the joint stiffness on the particle velocity transmission ratio, stress transmission ratio and effective velocity when a wave in the form of Eq. (5.22) propagates through complex rock mass with frequencies of $f_0 = 31.6$, 316.2, 562.3, 1000.0 and 3162.3 Hz.

Similar to the wave propagation through conventional jointed rock mass, it is seen from Fig. 5.7a, b that both the particle velocity transmission ratio and stress transmission ratio increase as the joint stiffness increases when a wave propagates through a complex rock mass. Moreover, the wave with a smaller frequency is slightly influenced by the joint stiffness. Therefore, the curve for $f_0 = 31.6$ Hz is more linear than the curve for $f_0 = 3162.3$ Hz. In addition, the wave with a larger frequency results in a smaller particle velocity transmission ratio and smaller stress transmission ratio once the joint stiffness is fixed.

Different from the wave propagation through conventional jointed rock mass, it is seen from Fig. 5.7a, b that the particle velocity transmission ratio approaches a constant smaller than 1.0 and the stress transmission ratio approaches a constant larger than 1.0 as joint stiffness increases for the case of "soft-to-hard" rock mass. In contrast, both the particle velocity transmission ratio and stress transmission ratio approach 1.0 as the joint stiffness increases when the wave propagates through the conventional jointed rock mass. On the other hand, the particle velocity transmission ratio approaches a constant larger than 1.0 and the stress transmission ratio approaches a constant smaller than 1.0 as the joint stiffness increases for the case of "hard-to-soft" rock mass.

Figure 5.7c shows the effect of joint stiffness on the effective velocity. It is seen that the effective velocity increases as the joint stiffness increases, then approaches a constant equal to that of the wave propagation in the welded jointed rock mass. It is also seen from Fig. 5.7c that the wave with a larger frequency is slightly influenced

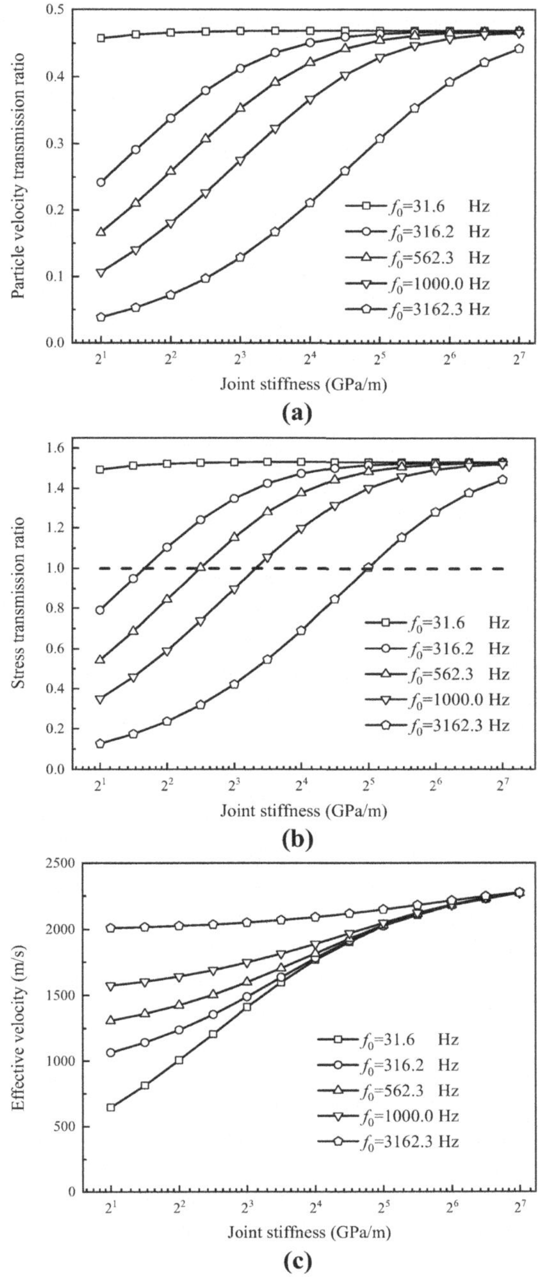

Fig. 5.7 Effect of joint stiffness on wave transmission. **a** Particle velocity transmission ratio. **b** Stress transmission ratio. **c** Effective velocity

by the joint stiffness and results in a larger effective velocity once the joint stiffness is fixed.

(3) Effect of wave impedance ratio on wave transmission and effective velocity

Figure 5.8a, b shows the effect of the wave impedance ratio on the particle velocity transmission ratio and stress transmission ratio, respectively. The joint stiffness is 8.0 GPa/m, and the waves are in the form of Eq. (5.22) with frequencies of $f_0 =$ 31.6, 316.2, 562.3, 1000.0 and 3162.3 Hz.

Figure 5.8a shows that the particle velocity transmission ratio increases as the wave impedance ratio increases, while Fig. 5.8b shows that the stress transmission ratio decreases as the wave impedance increases. It is seen from Fig. 5.8a, b that the curve of the wave impedance ratio $n = 1.0$ denotes the wave propagation in the conventional jointed rock mass, as seen in previous research. Therefore, the left regions in Fig. 5.8a, b denote the cases of wave transmission in the "soft-to-hard" rock mass, while the right regions denote the cases of wave transmission in the "hard-to-soft" rock mass. It is observed from Fig. 5.8a, b that the particle velocity transmission ratios in the "soft-to-hard" rock mass are always smaller than those in the "hard-to-soft" rock mass. However, the stress transmission ratios in the "soft-to-hard" rock mass are always larger than those in the "hard-to-soft" rock mass. Moreover, it can be observed from Fig. 5.8a that the particle velocity transmission ratio may be larger than 1.0 when the frequency is sufficiently small and the wave impedance ratio is large enough. On the other hand, it is observed from Fig. 5.8b that the stress transmission ratio may be larger than 1.0 when both the frequency and wave impedance are sufficiently small. It is also observed from Fig. 5.8a, b that the cases of wave propagation through conventional jointed rock mass are special cases of the present study when the wave impedance ratio equals 1.0.

For the case when a wave propagates across the joint from the "soft" rock side to the "hard" rock side, if we assume that the energy can be completely transmitted, the deformation in the "soft" rock will be larger than the deformation in the "hard" rock. This results in the stresses for the "soft" rock being smaller than those in the "hard" rock. However, in practical engineering, the energy cannot be completely transmitted due to the existence of reflection induced by the joint, which results in an attenuation of the transmitted wave.

Therefore, in traditional studies, the wave attenuates when it propagates through rock mass with the same properties in front of and behind the joint. However, when a wave propagates through a complex rock mass, once the decreasing degree caused by the joint is smaller than the increasing degree caused by different wave impedances for the wave propagation through "soft-to-hard" rock mass, the stress behind the joint in the "hard" rock may be larger than the stress in front of the joint in the "soft" rock, which indicates that the stress transmission coefficient is larger than 1.0.

A similar explanation can be utilized for the fact that the particle velocity transmission coefficient is larger than 1.0 when the wave propagates through the "hard-to-soft" rock mass.

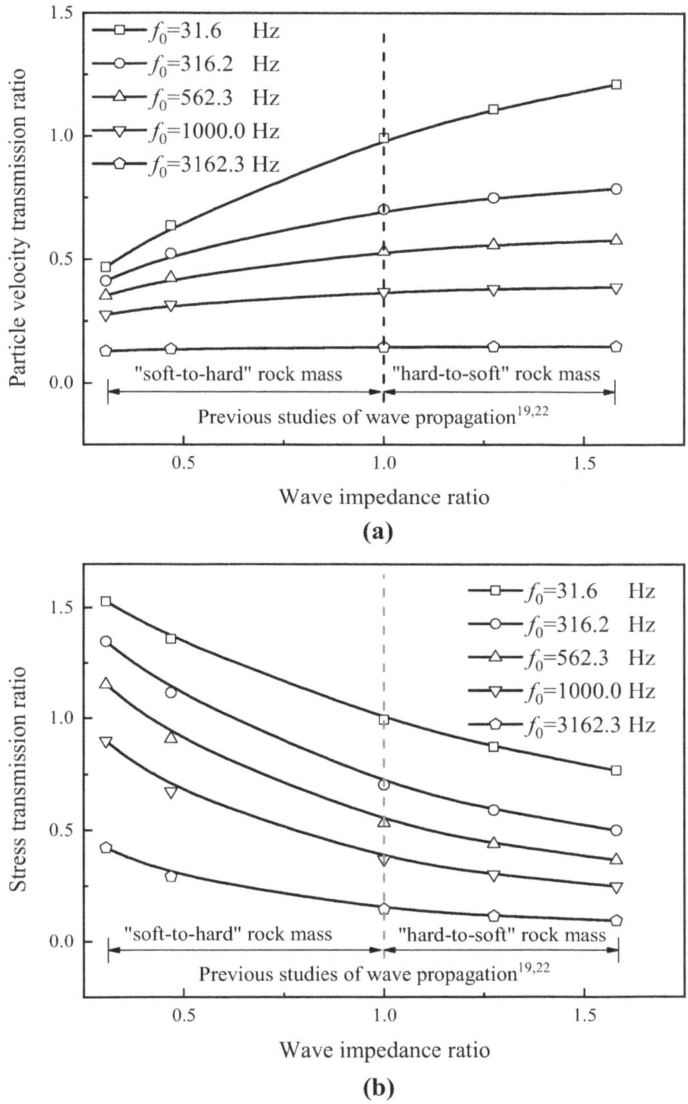

Fig. 5.8 Effect of wave impedance on wave transmission. **a** Particle velocity transmission ratio. **b** Stress transmission

5.2 Stress Wave Propagation Through Complex Rock Mass with Nonlinearly Deformational Joints

Filled joints commonly exist in the rock mass with different material properties on two sides of the joint, and they respond nonlinearly when the amplitude of a stress wave is sufficiently large (e.g., the region close to the wave source) [8–10]. It is essential to study stress wave propagation through a nonlinearly deforming jointed complex rock mass (different wave impedances in front of and behind the joint) to fully understand rock mechanics, geology engineering, seismic engineering, etc. [11–13].

The present section investigates the stress wave propagating through a complex rock mass with an individual nonlinearly deforming joint. Rock with a larger wave impedance is noted as "hard" rock, while rock with a smaller wave impedance is noted as "soft" rock. The conventional method of characteristics is modified. The particle velocity transmission coefficient and energy transmission coefficient are obtained. The effect of the frequency of the incident wave, the amplitude of the incident wave, the initial joint stiffness and the wave impedance ratio on the wave transmission are discussed.

5.2.1 *Theoretical Model*

(1) Nonlinear displacement discontinuity model

In a jointed complex rock mass, the complete deformation process of a filled joint is nonlinear, as shown in Fig. 5.9. The mechanical behaviors of the rocks in front of and behind the joint are different. The complete deformational behavior of the filled joint can be described by the dynamic B-B model as [14]

$$d = \frac{\sigma}{k_{ni} + \frac{\sigma}{d_{\max}}} \tag{5.23}$$

where σ is the normal stress, d is the joint closure, k_{ni} is the initial joint stiffness, and $d_{\max}$ is the maximum allowable closure of the joint.

The displacement discontinuity method (DDM) considers the joint as a nonwelded interface. The stress σ across the joint is continuous, while the displacement u across the joint is discontinuous. Therefore, it can be analytically expressed as

$$\begin{cases} \sigma_1 = \sigma_2 = \sigma \\ u_1 - u_2 = d = \frac{\sigma}{k_{ni} + \frac{\sigma}{d_{\max}}} \end{cases} \tag{5.24}$$

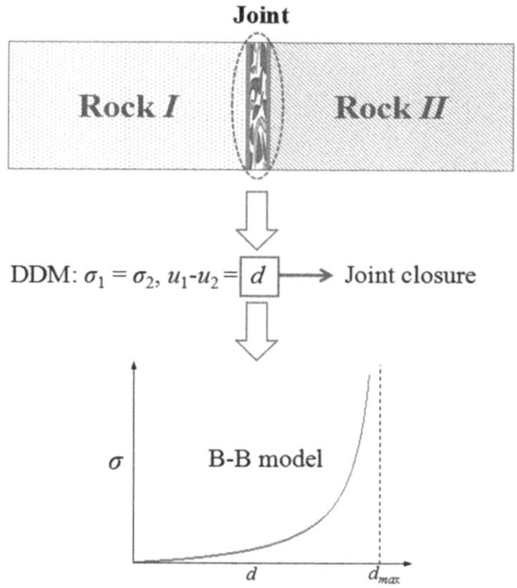

Fig. 5.9 The nonlinear displacement discontinuity model

where σ_1 and σ_2 respectively are the normal stresses in front of and behind the joint. u_1 and u_2 respectively are the displacements in front of and behind the joint, as described in Eq. (5.24) and shown in Fig. 5.9.

(2) Modified characteristic method in a complex rock mass

Since the wave impedances of the rock on two sides of the joint are different in this complex rock mass, the traditional characteristic method of analyzing a jointed rock mass (wave impedances on two sides of the joint are the same) is not applicable. We improve the method as shown in Fig. 5.10a, b. Figure 5.10a is applicable to a "soft-to-hard" rock mass and Fig. 5.10b is applicable to a "hard-to-soft" rock mass. The slopes of the right-running and left-running characteristic lines in front of the joint respectively are C_{01} and $-C_{01}$, where C_{01} is the wave velocity of the rock in front of the joint. Similarly, the slopes of the right-running and left-running characteristic lines behind the joint are C_{02} and $-C_{02}$, respectively, where C_{02} is the wave velocity of the rock behind the joint. Figure 5.10 shows that $C_{01} < C_{02}$ for the "soft-to-hard" rock mass (Fig. 5.10a), while $C_{01} > C_{02}$ for the "hard-to-soft" rock mass (Fig. 5.10b).

It is assumed that a complex rock mass with a single nonlinearly deforming joint is undisturbed at $t = 0$ when a stress wave $p(0, t)$ is applied at $x = 0$. The initial condition is

$$p(0, t) = p_0(\tau), \quad t \geq 0 \tag{5.25}$$

and the boundary condition is

$$v(x, 0) = \sigma(x, 0) = 0, \quad 0 < x < l \tag{5.26}$$

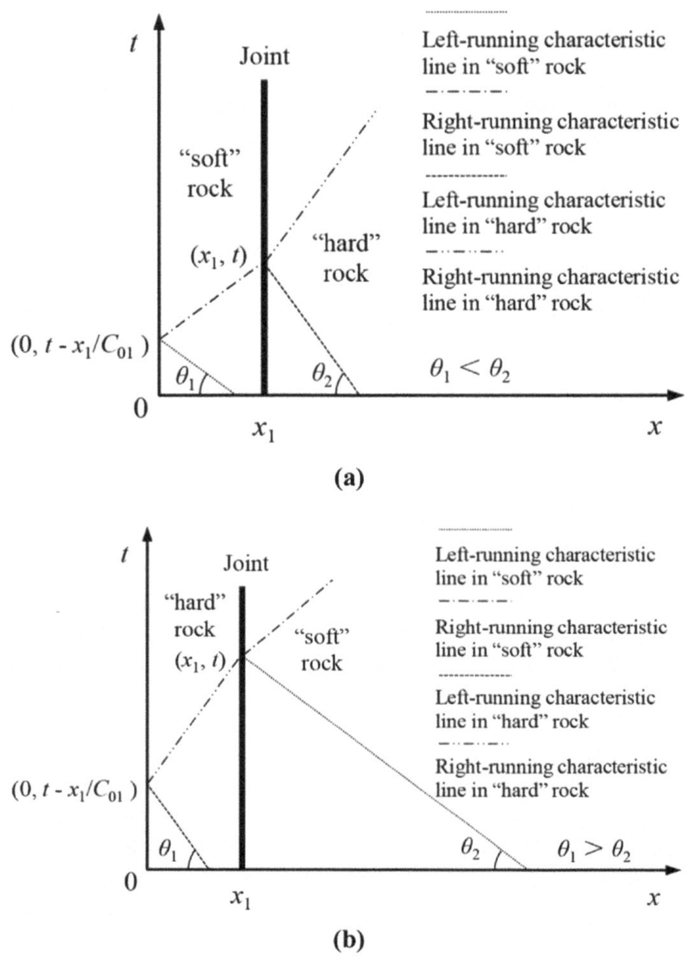

Fig. 5.10 Right-running and left-running characteristic lines in complex rock mass. **a** "soft-to-hard" rock mass. **b** "hard-to-soft" rock mass

where l is the length of the complex rock mass, as shown in Fig. 5.9. $v(x, 0)$ and $\sigma(x, 0)$ are the particle velocity and normal stress along the x-axis, respectively.

The characteristic method assumes that $zv \pm \sigma =$ constant along the characteristic lines, where z is the wave impedance of the rock. Therefore, in Fig. 5.10a, b, along the left-running characteristic line in front of the joint can be obtained as

$$z_1 p_0(0, t - x_1/C_{01}) - \sigma(0, t - x_1/C_{01}) = 0 \tag{5.27}$$

where z_1 is the wave impedance of the rock in front of the joint.

Along the right-running characteristic line in front of the joint yields

$$z_1 v^-(x_1, t) + \sigma(x_1, t) = z_1 p_0(0, t - x_1/C_{01}) + \sigma(0, t - x_1/C_{01}) \tag{5.28}$$

where $v^-(x_1, t)$ is the particle velocity in front of the joint, which is the superposition of the reflected particle velocity and the incident particle velocity.

Along the left-running characteristic line behind the joint yields

$$z_2 v^+(x_1, t) - \sigma(x_1, t) = 0 \tag{5.29}$$

where z_2 is the wave impedance of the rock behind the joint and $v^+(x_1, t)$ is the particle velocity behind the joint as the transmitted particle velocity.

Substituting Eqs. (5.27) and (5.29) into Eq. (5.28) gives

$$z_1 v^-(x_1, t) + z_2 v^+(x_1, t) = 2 z_1 p_0(0, t - x_1/C_{01}) \tag{5.30}$$

where $v^-(x_1, t)$ and $v^+(x_1, t)$ in Eq. (5.30) express the relationship between the particle velocities in front of and behind the joint.

Based on the DDM and B-B model, the displacement at the joint is determined. Substituting Eq. (5.29) into the derivation of displacement yields

$$\begin{aligned} v^-(x_1, t) - v^+(x_1, t) &= \frac{\partial\sigma(x_1, t)}{\partial t}\frac{1}{k_{ni} + \frac{\sigma(x_1,t)}{d_{\max}}} - \frac{\partial\sigma(x_1, t)}{\partial t}\frac{\sigma(x_1, t)}{d_{\max}\left(k_{ni} + \frac{\sigma(x_1,t)}{d_{\max}}\right)^2} \\ &= \frac{z_2 \partial v^+(x_1, t)}{\partial t}\frac{1}{k_{ni} + \frac{z_2 v^+(x_1,t)}{d_{\max}}} \\ &\quad - \frac{z_2^2 \partial v^+(x_1, t)}{\partial t}\frac{v^+(x_1, t)}{d_{\max}\left(k_{ni} + \frac{z_2 v^+(x_1,t)}{d_{\max}}\right)^2} \end{aligned} \tag{5.31}$$

According to Eqs. (5.30) and (5.31), the particle velocity behind the joint can be expressed as

$$\frac{\partial v^+(x_1, t)}{\partial t} = \frac{2 z_1 p_0(0, t - x_1/C_{01}) - (z_1 + z_2) v^+(x_1, t)}{\frac{z_1 z_2}{k_{ni} + \frac{z_2 v^+(x_1,t)}{d_{\max}}} - \frac{z_1 z_2^2 v^+(x_1,t)}{d_{\max}\left(k_{ni} + \frac{z_2 v^+(x_1,t)}{d_{\max}}\right)^2}} \tag{5.32}$$

If Δt is small enough, it is determined that

$$\frac{\partial v^+(x_1, t)}{\partial t} = \frac{v^+(x_1, t_{j+1}) - v^+(x_1, t_j)}{\Delta t} \tag{5.33}$$

Substituting Eq. (5.33) into Eq. (5.32) yields

$$v^{+}\left(x_{1},t_{j+1}\right)=\frac{\Delta t\left(2z_{1}p_{0}\left(0,t-x_{1}/C_{01}\right)-\left(z_{1}+z_{2}\right)v^{+}\left(x_{1},t_{j}\right)\right)}{\frac{z_{1}z_{2}}{k_{ni}+\frac{z_{2}v^{+}\left(x_{1},t_{j}\right)}{d_{\max}}}-\frac{z_{1}z_{2}^{2}v^{+}\left(x_{1},t_{j}\right)}{d_{\max}\left(k_{ni}+\frac{z_{2}v^{+}\left(x_{1},t_{j}\right)}{d_{\max}}\right)^{2}}}+v^{+}\left(x_{1},t_{j}\right) \quad (5.34)$$

Equation (5.34) is the expression of the transmitted particle velocity of a nonlinearly deforming joint in the complex rock mass.

Therefore, the transmission coefficient can be defined as

$$T_{v}=\frac{\max\left(v^{+}\right)}{\max(p)} \quad (5.35)$$

The energy of the stress wave is

$$E_{p}=\int_{0}^{t_{n}}zv^{2}(t)dt \quad (5.36)$$

where t_n is the computation time.

The energy of the stress wave can be approximately calculated in the differential form as

$$E_{p}=\sum_{i=1}^{n}zv^{2}(t_{i})\Delta t \quad (5.37)$$

Thus, the energies of the incident wave and transmitted wave respectively are

$$E_{inc}=\sum_{i=1}^{n}z_{1}p^{2}(t_{i})\Delta t \quad (5.38)$$

and

$$E_{tra}=\sum_{i=1}^{n}z_{2}\left(v^{+}(t_{i})\right)^{2}\Delta t \quad (5.39)$$

Similarly, the energy transmission coefficient is defined as

$$T_{e}=\frac{E_{tra}}{E_{inc}} \quad (5.40)$$

5.2.2 *Stress Wave Transmission in a Nonlinearly Jointed Complex Rock Mass*

The incident wave is assumed to be half-sinusoidal and the intensity is expressed in terms of the velocity amplitude as

$$p(t) = V_0 \sin(2\pi f_0 \cdot t), \quad 0 \leq t \leq \frac{1}{2f_0} \tag{5.41}$$

where V_0 is the amplitude and f_0 is the frequency of the incident wave.

In the present section, n is defined as the ratio of the wave impedance in front of the joint to the wave impedance behind the joint, $n = z_1/z_2$. Five groups of rock mass are illustrated with wave impedance ratios of $n = 0.633$, 0.785, 1.000, 2.139 and 3.266. The parameters of these rock mass are shown in Table 5.2 in detail. For the cases of $n < 1.0$, the stress wave propagates through the "soft-to-hard" rock mass. For the cases of $n > 1.0$, the stress wave propagates through the "hard-to-soft" rock mass. $n = 1.0$ is a special case when the material properties of the rock on two sides of the joint are equal. The amplitude of the incident wave is $V_0 = 0.1$ m/s, the frequency is $f_0 = 50.0$ Hz, and the initial joint stiffness is $k_{ni} = 3.5$ GPa/m. The maximum allowable closure of the joint is $d_{\max} = 0.5$ mm.

Figure 5.11 shows the incident wave and the corresponding transmitted waves of the five groups of rock mass calculated by Eq. (5.34). The transmitted waves of different rock mass are significantly different. For the cases of $n < 1.0$, the amplitudes of the transmitted wave are smaller than that of the case of $n = 1.0$. For the cases of $n > 1.0$, the amplitudes of the transmitted wave are larger than that of the case of $n = 1.0$. Moreover, it is interesting to observe that the amplitudes of the transmitted wave in the cases of $n > 1.0$ are larger than that of the incident wave. In general, the stress of the reflected wave has a sign opposite to that of the incident wave due to the effect of reflection unloading, when a stress wave propagates across an interface from "hard" rock to "soft" rock. According to the continuous conditions and conservation of momentum at the wavefronts, the particle velocity of the reflected wave has the same sign as the incident wave; then, the particle velocity of the transmitted wave v_T ($v_T = v_I + v_R$, where v_I is the particle velocity of the incident wave and v_R is the particle velocity of the reflected wave) is higher than that of the incident wave, which will result in a particle velocity transmission coefficient larger than 1.0. However, the joint always slows the particle velocity of the stress wave. Therefore, when a wave propagates through the "hard-to-soft" complex rock mass, if the degree of

Table 5.2 Parameters of the rock mass

	Rock mass *I*	Rock mass *II*	Rock mass *III*	Rock mass *IV*	Rock mass *V*
z_1 (kg/(m^2·s)	9.80×10^6	9.80×10^6	9.80×10^6	9.80×10^6	9.80×10^6
z_2 (kg/(m^2·s)	15.49×10^6	12.49×10^6	9.80×10^6	4.58×10^6	3.00×10^6
n	0.633	0.785	1.000	2.139	3.266

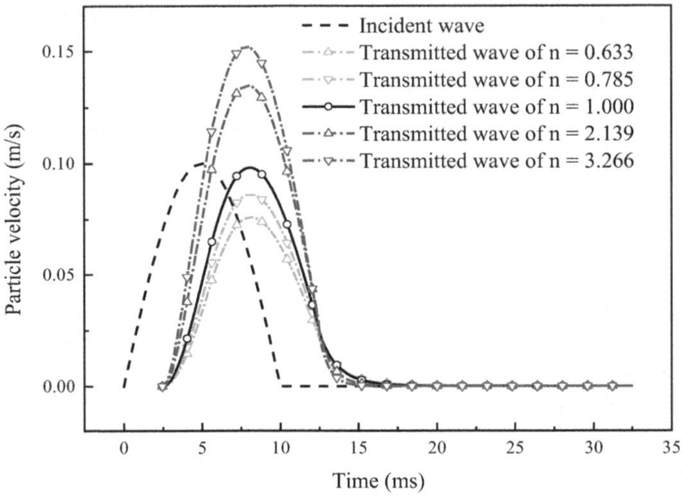

Fig. 5.11 Incident wave and corresponding transmitted waves in the rock mass

amplification is larger than that of attenuation caused by the joint, the particle velocity of the "soft" rock behind the joint will be higher than the particle velocity of the "hard" rock in front of the joint, which will result in a particle velocity transmission coefficient larger than 1.0.

5.2.3 Results and Discussion

To study the transmission behaviors of stress waves in a complex rock mass with nonlinearly deforming joints, three categories of parameters, namely stress wave parameters (frequency and amplitude of the incident wave), joint parameters (initial joint stiffness) and material parameters (wave impedance ratio), are taken into consideration.

(1) Effect of the nonlinear coefficient of a joint

The nonlinear coefficient γ is introduced to describe the degree of joint closure under the effect of the stress wave as [2]

$$\gamma = \frac{d}{d_{\max}} \tag{5.42}$$

The joint is slightly compressed when the nonlinear coefficient γ is relatively small, and the joint is tightly compressed when the nonlinear coefficient γ is relatively large.

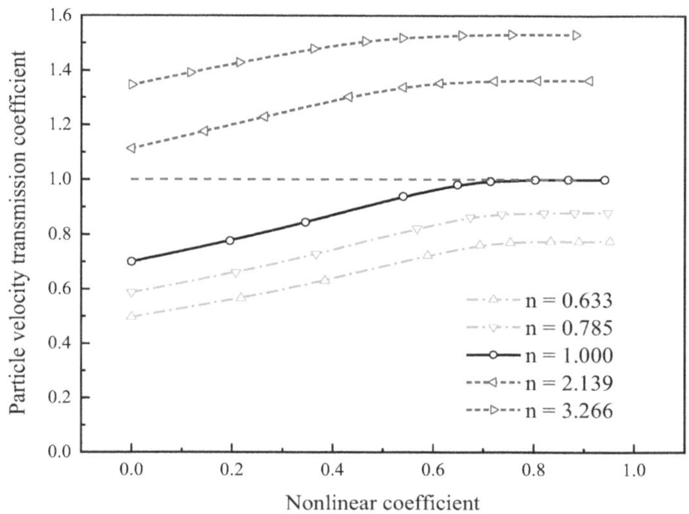

Fig. 5.12 Effect of nonlinear coefficient on particle velocity transmission coefficient

Figure 5.12 shows the particle velocity transmission coefficient for five groups of rock mass as a function of the nonlinear coefficient with the incident frequency $f_0 = 50.0$ Hz, and the initial joint stiffness $k_{ni} = 1.25$ GPa/m. The particle velocity transmission coefficient is strongly dependent on the nonlinear coefficient. The particle velocity transmission coefficient approaches the linear particle velocity transmission coefficient when the nonlinear coefficient γ approaches zero, and increases with increasing nonlinear coefficient. When the nonlinear coefficient γ is relatively small, the particle velocity transmission coefficient increases as the nonlinear coefficient γ increases. Subsequently, the particle velocity transmission coefficient remains constant as the nonlinear coefficient γ further increases.

As the nonlinear coefficient γ increases, the joint stiffness K_n ($K_n = k_{ni}/(1-\gamma)^2$) increases. Similarly, the particle velocity transmission coefficient increases. When the nonlinear coefficient γ approaches 1.0 (the joint closure approaches the maximum allowable closure of the joint), the joint is tightly compressed, and the joint stiffness K_n approaches infinity. The particle velocity transmission coefficient approaches the maximum, which is consistent with the stress wave propagating through the welded interface. It can be observed that the nonlinearity of the joint has an important influence on the propagation and transmission of stress waves in a complex rock mass, and nonlinearly deforming joints are more conducive to the propagation of waves than linearly deforming joints are.

(2) Effect of the frequency of an incident wave

Figure 5.13a, b shows the particle velocity transmission coefficient T_v and energy transmission coefficient T_e for the five groups of rock mass as a function of frequency f_0 with amplitude $V_0 = 0.3$ m/s and initial joint stiffness $k_{ni} = 3.5$ GPa/m.

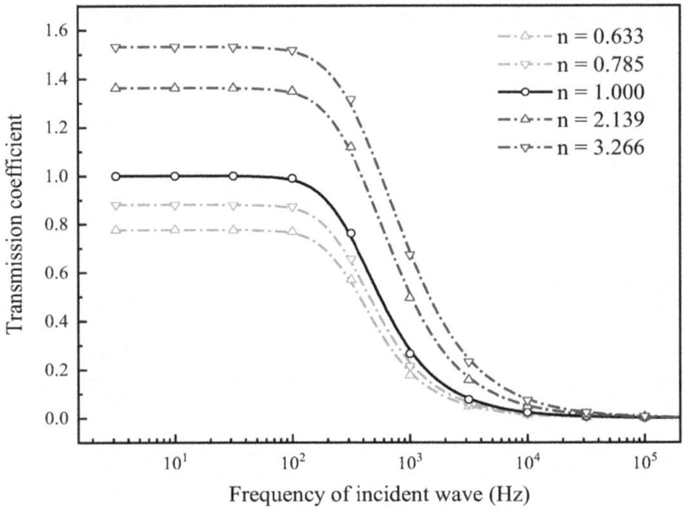

(a)

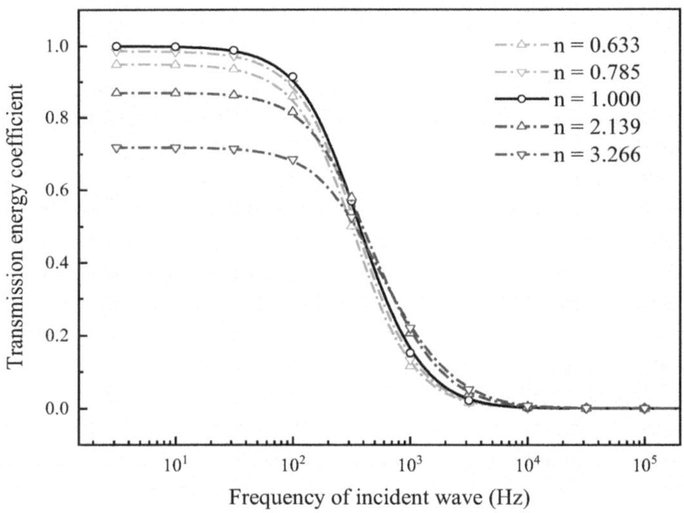

(b)

Fig. 5.13 Effect of frequency on wave transmission. **a** Particle velocity transmission coefficient. **b** Energy transmission coefficient

It is observed from Fig. 5.13a that, for relatively small incident wave frequencies ($f_0 = 0$ to 100 Hz), the particle velocity transmission coefficient remains constant, approaching those of the stress waves propagating through the welded interface. For incident waves with frequency $f_0 > 100$ Hz, the particle velocity transmission coefficient decreases sharply with increasing frequency. For incident waves with considerably higher frequencies ($f_0 > 10^5$ Hz), the particle velocity transmission coefficients approach zero, which reveals that the stress waves are reflected completely. This finding further validates that the higher frequencies are preferentially filtered out. In addition, the particle velocity transmission coefficient increases with increasing wave impedance ratio for a fixed frequency.

Figure 5.13b shows the effect of the frequency of incident wave f_0 on the energy transmission coefficient T_e. Figure 5.13b shows that the energy transmission coefficient remains approximately constant when the frequency $f_0 < 100$ Hz, as shown in Fig. 5.13a. However, the results in Fig. 5.13b are different from those in Fig. 5.13a because both the energy transmission coefficients in the cases of $n < 1.0$ and $n > 1.0$ are smaller than 1.0 and the energy transmission coefficients in the cases of $n > 1.0$ are the smallest in Fig. 5.13b. Subsequently, the energy transmission coefficient decreases sharply as the frequency increases when the frequency of the incident wave is 100 Hz $< f_0 < 10^4$ Hz. In this frequency region, the energy transmission coefficients for the cases of $n < 1.0$ are always smaller than that of $n = 1.0$, while the energy transmission coefficients for the cases of $n > 1.0$ may be larger than those of $n = 1.0$ and $n < 1.0$. Finally, in the present study, the energy transmission coefficient converges gradually and approaches zero when the frequency is $f_0 > 10^4$ Hz. Notably, for stress wave propagation through a complex rock mass (e.g., the four categories of rock mass considered in the present study), the amplitude of the transmitted wave can be larger than that of the incident wave (Fig. 5.13a), but the transmitted energy always attenuates due to the different material properties on the two sides of the joint, even when the frequency is sufficiently small. This phenomenon has not been predicted by traditional studies, which are special cases of the present study (when $n = 1.0$). In general, waves with higher frequencies are filtered out by the joint, and waves with small frequencies (e.g., seismic waves) travel farther and carry more energy.

(3) Effect of the amplitude of an incident wave

Figure 5.14a, b shows the particle velocity transmission coefficient T_v and energy transmission coefficient T_e as functions of the amplitude of incident wave V_0 with a fixed frequency $f_0 = 150$ Hz and the initial joint stiffness $k_{ni} = 3.5$ GPa/m.

For linearly deforming joints, the joint stiffness is constant; therefore, the transmission behaviors are insensitive to the amplitude of the incident wave. However, for nonlinearly deforming joints, a higher load amplitude increases the normal stiffness; thus, the particle velocity transmission coefficient and energy transmission coefficient are amplitude-dependent.

In Fig. 5.14a, the particle velocity transmission coefficient increases rapidly with increasing amplitude of the incident wave when the amplitude is sufficiently small, e.g., $V_0 < 0.4$ m/s in the present study. When the amplitude increases (e.g., $V_0 > 0.4$ m/s in the present study), the transmission coefficients of the particle velocity approach constants. It can be determined that the amplitude of the incident wave ranges from 0 to 0.4 m/s and has a significant effect on the transmission coefficient

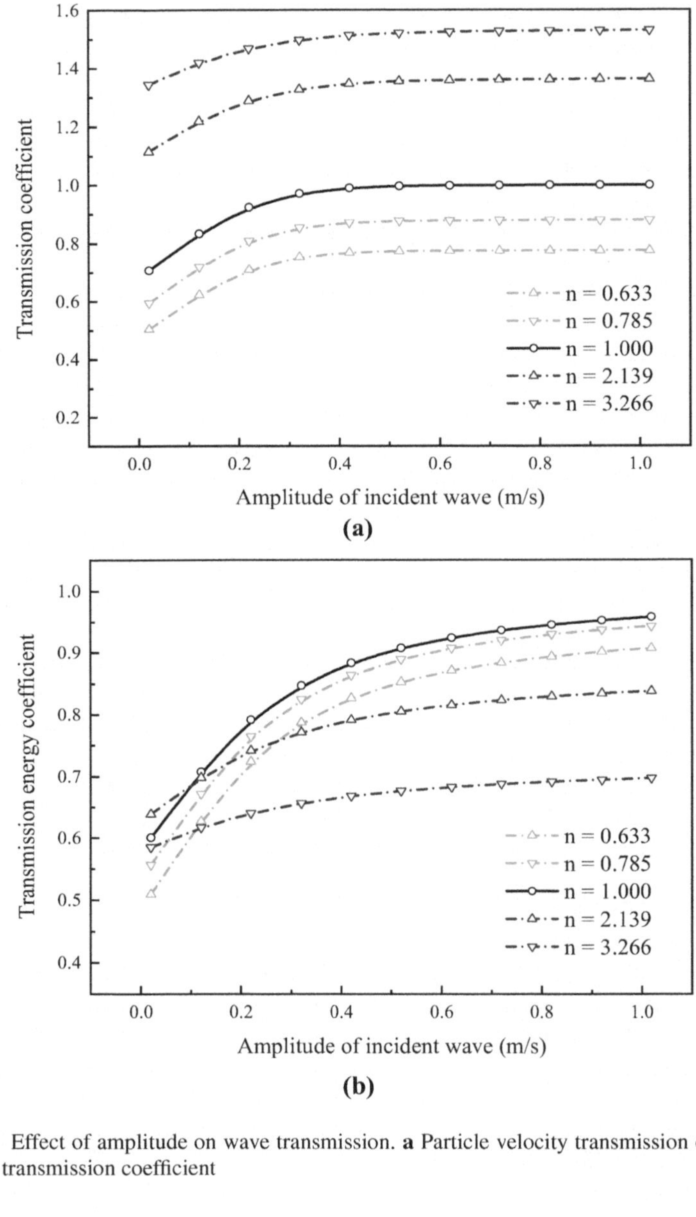

Fig. 5.14 Effect of amplitude on wave transmission. **a** Particle velocity transmission coefficient. **b** Energy transmission coefficient

of the particle velocity. For an incident wave with a fixed amplitude, the particle velocity transmission coefficient increases as the wave impedance ratio increases.

In Fig. 5.14b, the energy transmission coefficient increases with increasing amplitude of the incident wave. In the first stage (e.g., the amplitude $V_0 < 0.4$ m/s), the energy transmission coefficient increases rapidly as the amplitude increases.

Subsequently, the energy transmission coefficient increases slowly and approaches a constant as the amplitude of the incident wave further increases, which is different from the effect of the amplitude of the incident wave on the particle velocity transmission coefficient (as shown in Fig. 5.14a). The energy transmission coefficient decreases as the wave impedance ratio decreases for the cases of $n \leq 1.0$, while the energy transmission coefficient decreases as the wave impedance ratio increases for the cases of $n > 1.0$. It is also observed from Fig. 5.14(b) that the increasing rates of the energy transmission coefficient for the cases of $n > 1.0$ are smaller than those for the cases of $n \leq 1.0$. When the amplitude of the incident wave is sufficiently small, the energy transmission coefficients for the cases of $n > 1.0$ may be larger than those for the cases of $n \leq 1.0$. When the amplitude of the incident wave is sufficiently large, the energy transmission coefficients for the cases of $n > 1.0$ are smaller than those for the cases of $n \leq 1.0$.

(4) Effect of the initial joint stiffness

The initial joint stiffness is an important parameter of nonlinearly deforming joints. The illustrated incident waves have a fixed amplitude of incident wave $V_0 = 0.1$ m/s and a fixed frequency $f_0 = 150$ Hz.

Figure 5.15a shows the influence of the initial joint stiffness k_{ni} on the particle velocity transmission coefficient T_v. The particle velocity transmission coefficient increases as the initial joint stiffness increases. However, the increasing rate decreases as the initial joint stiffness increases. Therefore, the particle velocity transmission coefficient remains almost constant when the initial joint stiffness is sufficiently large. Similar to the effect of the amplitude of the incident wave on the particle velocity transmission coefficient, Fig. 5.15a also shows that the particle velocity transmission coefficients for the cases of $n > 1.0$ are larger than those for the cases of $n \leq 1.0$ for a fixed initial joint stiffness. Moreover, when the initial joint stiffness is sufficiently large, the results agree well with those of stress waves propagating through the complex materials (a welded jointed rock mass). The traditional studies of waves propagating through rock mass can also be regarded as special cases of the present study as the cases of $n = 1.0$.

Figure 5.15b shows the effect of the initial joint stiffness k_{ni} on the energy transmission coefficient T_e. It can be observed that the energy transmission coefficient firstly increases sharply as the initial joint stiffness increases and then approaches a constant as the initial joint stiffness further increases. The energy transmission coefficients for the cases of $n < 1.0$ are always smaller than those of $n = 1.0$, while the energy transmission coefficients for the cases of $n > 1.0$ may be larger than those of $n \leq 1.0$ when the initial joint stiffness is small, while the energy transmission coefficients for the cases of n > 1.0 are smaller than those of $n \leq 1.0$ when the initial joint stiffness is sufficiently large. Both the energy transmission coefficients for the cases of $n < 1.0$ and $n > 1.0$ are smaller than those for the case of $n = 1.0$ when the initial joint stiffness is sufficiently large, which reveals that stress waves in a complex rock mass ($n > 1.0$ or $n < 1.0$) cannot be completely transmitted as those for the cases of n

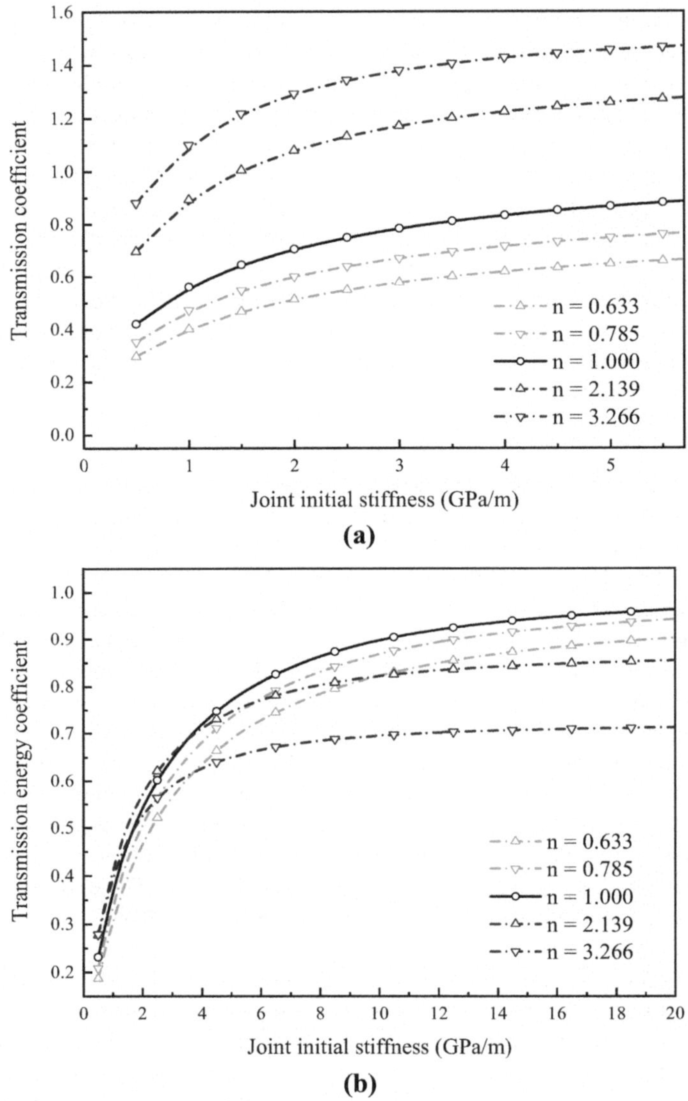

Fig. 5.15 Effect of initial joint stiffness on wave transmission. **a** Particle velocity transmission coefficient. **b** Energy transmission coefficient

= 1.0 in the traditional studies. When the contacts between the rock layers are weak, the joint may transmit more energy when the wave propagates through "hard-to-soft" rock mass (e.g., $n = 2.139$ and $n = 3.266$) than through "soft-to-hard" rock mass. As the initial joint stiffness increases, the joint transmits more energy when the wave propagates through "soft-to-hard" rock mass.

(5) Effect of the wave impedance ratio

The wave impedance ratio n is an important parameter used to analyze the propagation of stress waves in the complex rock mass. In a complex rock mass, the wave impedance ratio n may be larger or smaller than 1.0 depending on the wave impedances on two sides of the joint. In this section, a wave with a fixed amplitude of $V_0 = 0.1$ m/s propagating through the complex rock mass is illustrated. The initial joint stiffness is $k_{ni} = 3.5$ GPa/m.

In Fig. 5.16a, the particle velocity transmission coefficient T_v increases with increasing wave impedance ratio n under different frequency conditions. It is revealed from Fig. 5.16a that the wave reflects at the free boundary when the wave impedance ratio approaches infinity. In these cases, the transmission coefficient of the particle velocity approaches 2.0. The variation in the particle velocity transmission coefficient with respect to the wave impedance ratio agrees well with the results of previous studies. It is also observed from Fig. 5.16a that the particle velocity transmission coefficient is always smaller than 1.0 for the "soft-to-hard" rock mass ($n < 1.0$), while it can be larger than 1.0 for the "hard-to-soft" rock mass ($n > 1.0$) as the frequency of the incident wave decreases and the wave impedance ratio increases. Moreover, it can also be observed from Fig. 5.16a that the particle velocity transmission coefficient increases as the frequency decreases for a fixed wave impedance ratio. The results are in accordance with those obtained in Fig. 5.13a.

Figure 5.16b shows the effect of wave impedance ratio n on the energy transmission coefficient T_e for the wave propagating through a complex rock mass with different frequencies. It is clearly seen that, for the cases of incident waves with small frequencies (e.g., $f_0 < 150$ Hz), the energy transmission coefficient increases to a maximum as the wave impedance ratio increases to 1.0 and then decreases as the wave impedance ratio further increases. This result is consistent with that shown in Fig. 5.13b, the energy transmission coefficients are largest when $n = 1.000$, and both the energy transmission coefficients for $n < 1.0$ and $n > 1.0$ are smaller than that for $n = 1.000$ when the frequencies are relatively small. With increasing frequency, the wave impedance ratio corresponding to the peak energy transmission coefficient increases, which results in the energy transmission coefficients of $n > 1.0$ (e.g., $n = 2.139$ and $n = 3.266$) being equal to or larger than that of $n \leq 1.0$. In the present study, when the frequency is $f_0 > 400$ Hz, the wave impedance ratio corresponding to the peak energy transmission coefficient is larger than the energy transmission coefficient of $n = 3.266$. Therefore, the energy transmission coefficient of the five groups of rock mass increases as the wave impedance increases as shown in Fig. 5.16b. It can be predicted that the energy transmission coefficient approaches zero when the wave impedance ratio is sufficiently large (free interface). It is also shown in Fig. 5.16b that the energy transmission coefficient is significantly influenced by the wave impedance ratio for the incident wave with a small frequency than that with a large frequency. Moreover, the energy transmission coefficient increases as the frequency decreases for a fixed wave impedance ratio.

When the wave impedances on two sides of the joint are the same, Zhao and Cai [2] derived the wave propagation equation for a single joint with nonlinearly

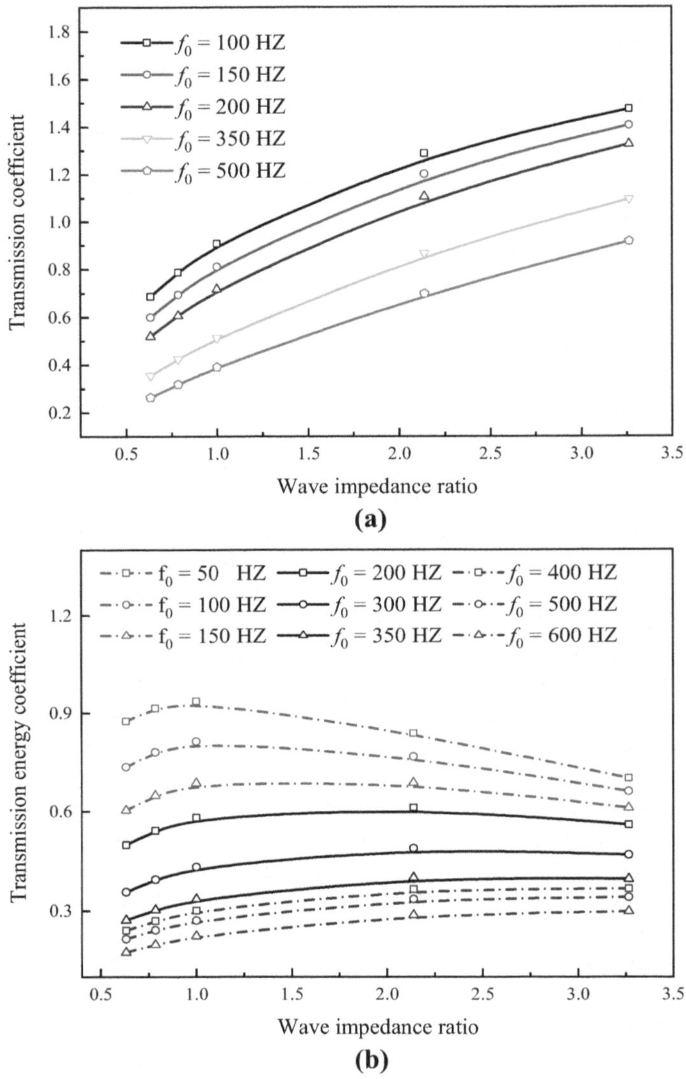

Fig. 5.16 Effect of wave impedance ratio on wave transmission. **a** Particle velocity transmission coefficient. **b** Energy transmission coefficient

deformational behavior based on the characteristic method and the displacement discontinuity model. Their results are shown as the scattered points in Fig. 5.17. We assumed that $z1 = z2$ in the modified characteristic method, and the calculation results are shown as the solid line in Fig. 5.17. By comparison, it is found that the results from the modified characteristic method agree well with that of the traditional

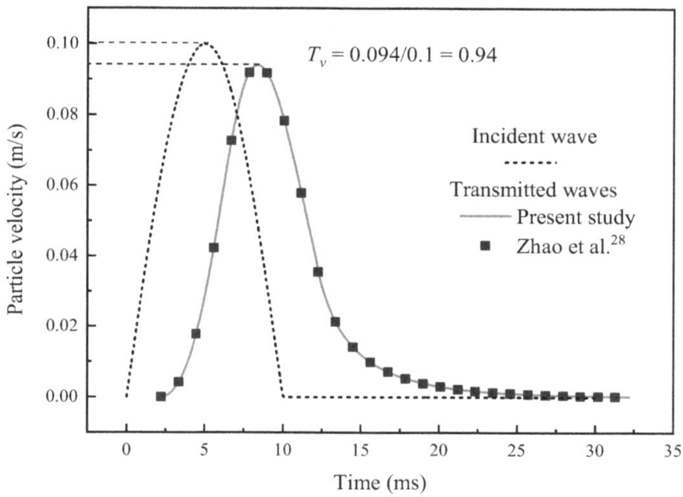

Fig. 5.17 Verification of modified characteristic method

characteristic method. This traditional characteristic method can be regarded as a special case of the modified characteristic method when $z1 = z2$.

5.3 Reflection of Stress Wave Through Complex Jointed Rock Mass

In natural geological structures, joints exist widely with different wave impedances on two sides, which greatly affect the reflection of stress waves [15–17]. Rock is a brittle material with a tensile strength far smaller than its compressive strength, which makes the tensile failure induced by wave reflection can't be ignored in the stability analysis of rock engineering [18, 19]. Therefore, it is of great significance to study the tensile wave reflection in the rock mass with different rock properties on two sides of the joint.

A theoretical study on amplitude attenuation of reflected tensile wave in the rock mass with different wave impedances on two sides of the joint is introduced. The equation of the reflected wave considering the variation of wave impedance ratio is derived based on the combination of the displacement discontinuity model and the method of characteristics. The parametric studies with respect to the incident frequency, joint stiffness and wave impedance ratio on reflection coefficient are carried out.

5.3.1 Theoretical Study of Reflected Wave in Jointed Rock Mass

Figure 5.19 is a representative elementary volume (REV) of rock mass with different wave impedances on two sides of the joint from Fig. 5.18. Rock *I* and Rock *II* are the intact rocks in front of and behind the joint, respectively. When the stress wave propagates across the rock joint, wave reflection is generated.

The jointed rock mass is a discontinuous medium. In order to solve the stress wave propagation in the discontinuous medium, the following displacement discontinuity model is proposed to study wave transmission and reflection [7, 20–22]:

$$\sigma^- = \sigma^+ = \sigma \tag{5.43}$$

and

$$u^- - u^+ = \frac{\sigma}{k_n} \tag{5.44}$$

where σ is the normal stress, and σ^- and σ^+ respectively are the normal stresses in front of and behind the joint. u^- and u^+ are respectively the displacements in front of and behind the joint and k_n is the normal joint stiffness.

The differential expression of Eq. (5.44) is

Fig. 5.18 Rock mass with different material properties on two sides of the joint

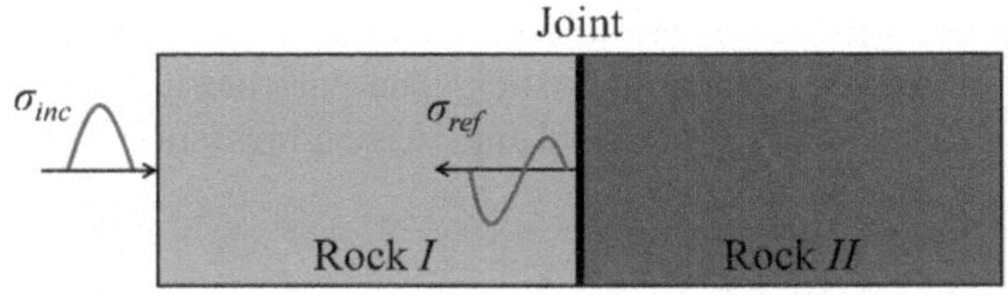

Fig. 5.19 Representative elementary volume (REV) of rock mass

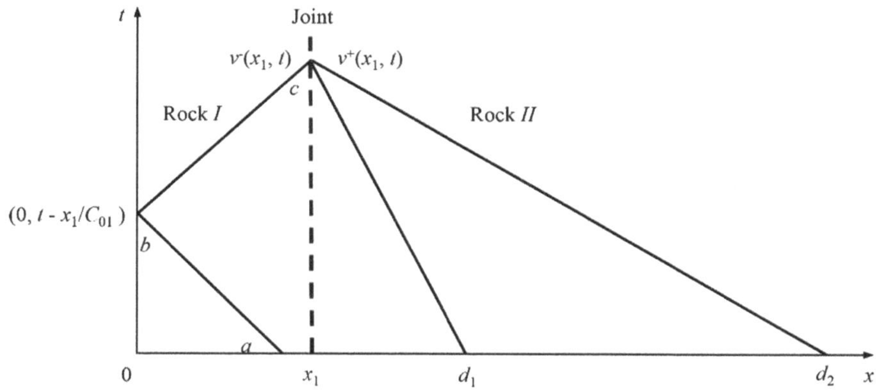

Fig. 5.20 The method of characteristics

$$v^- - v^+ = \frac{1}{k_n}\frac{\partial \sigma}{\partial t} \tag{5.45}$$

where v^- and v^+ are the particle velocities in front of and behind the joint, respectively.

The method of characteristics is widely used to solve the wave propagation equation in a jointed rock mass with the same wave impedances on two sides of the joint [2, 4] and is further used in a jointed rock mass with different wave impedances on two sides of the joint [23], as shown in Fig. 5.20.

In Fig. 5.20, a joint is located at $x = x_1$ in the rock mass. Rock *I* and Rock *II* in front of and behind the joint are linearly elastic, homogeneous and isotropic. When a wave propagates along the x-axis and reaches the joint, both the reflection and transmission take place. Lines *ab*, *bc*, cd_1 and cd_2 are the characteristic lines of wave in the $x - t$ plane. The absolute values of the slopes of these characteristic lines equal the wave velocities in the rocks. It is seen from Fig. 5.20 that the absolute values of the slope of cd_1 are larger than that of *ab*, while the absolute value of the slope of cd_2 is smaller than that of *ab*. Therefore, the characteristic lines "a-b-c-d_1" is for the case of the stress wave propagating from the rock with a large wave velocity to a small wave velocity while the characteristic lines "a-b-c-d_2" is for the case of the stress wave propagating from the rock with a small wave velocity to a large wave velocity, respectively.

According to the method of characteristics, the relations for the particle velocity v and stress σ can be derived as $zv(x, t) + \sigma(x, t) =$ constant along the right-running characteristic line and $zv(x, t) - \sigma(x, t) =$ constant along the left-running characteristic line, where z is the wave impedance of the intact rock. Therefore, along characteristic lines "a-b-c-d_1" or "a-b-c-d_2",

$$\left.\begin{array}{l} z_1 v_{inc}(t - x_1/C_{01}) - \sigma_{inc}(t - x_1/C_{01}) = 0 \\ z_1 v^-(x_1, t) + \sigma(x_1, t) = z_1 v_{inc}(t - x_1/C_{01}) + \sigma_{inc}(t - x_1/C_{01}) \\ z_2 v^+(x_1, t) - \sigma(x_1, t) = 0 \end{array}\right\} \tag{5.46}$$

where C_{01} is the wave velocity of the rock in front of the joint, and $v_{inc}(t - x_1/C_{01})$ and $\sigma_{inc}(t - x_1/C_{01})$ respectively are particle velocity and stress of the incident wave at $x = 0$. $v^{+}(x_1, t)$ and $v^{-}(x_1, t)$ respectively are particle velocities in front of and behind the joint. $\sigma\ (x_1, t)$ is normal stress at the joint, and z_1 and z_2 are the wave impedances of the rocks in front of and behind the joint, respectively.

With Eq. (5.46), the relationship between the particle velocity in front of and behind the joint is obtained as

$$z_1 v^{-}(x_1, t) + z_2 v^{+}(x_1, t) = 2z_1 v_{inc}(t - x_1/C_{01}) \tag{5.47}$$

Substituting Eq. (5.46) into Eq. (5.45) gives

$$v^{-}(x_1, t) - v^{+}(x_1, t) = \frac{z_2 \partial v^{+}(x_1, t)}{k_n \partial t} \tag{5.48}$$

Substituting Eq. (5.48) into Eq. (5.47) yields

$$\frac{\partial v^{+}(x_1, t)}{\partial t} = \frac{2k_n}{z_2} v_{inc}(t - x_1/C_{01}) - k_n(1/z_2 + 1/z_1)v^{+}(x_1, t) \tag{5.49}$$

If Δt is sufficiently small, the derivative in Eq. (5.49) can be replaced by a differential expression of

$$\frac{\partial v^{+}(x_1, t_j)}{\partial t} = \frac{v^{+}(x_1, t_{j+1}) - v^{+}(x_1, t_j)}{\Delta t} \tag{5.50}$$

Therefore, the particle velocity behind the joint as the transmitted wave is

$$\begin{aligned} v_{\text{tran}} &= v^{+}(x_1, t_{j+1}) \\ &= \frac{2k_n \Delta t}{z_2} v_{inc}(t_j - x_1/C_{01}) + [1 - k_n \Delta t(1/z_1 + 1/z_2)]v^{+}(x_1, t_j) \end{aligned} \tag{5.51}$$

Equation (5.51) is a recurrence equation, so the transmitted wave v_{tran} can be determined in an iterative calculating process if the incident wave, the initial particle velocity condition, the joint parameter and the material parameter are given.

Equation (5.47) yields

$$v^{-}(x_1, t) = 2v_{inc}(t - x_1/C_{01}) - \frac{z_2 v^{+}(x_1, t)}{z_1} \tag{5.52}$$

where $v^{-}(x_1, t)$ is the particle velocity superposed by the incident and reflected waves. Then, the particle velocity of the reflected wave is

$$v_{ref} = v^{-}(x_1, t) - v_{inc}(t - x_1/C_{01}) = v_{inc}(t - x_1/C_{01}) - \frac{v^{+}(x_1, t)}{n} \tag{5.53}$$

where n is the wave impedance ratio which is defined as

$$n = \frac{z_1}{z_2} \tag{5.54}$$

For the rock mass with different wave impedances on two sides of the joint, the wave impedance ratio is $n \neq 1.0$. The wave impedance ratio $n < 1.0$ denotes the stress wave propagating from the rock with a small wave impedance to the rock with a large wave impedance, while the wave impedance ratio $n > 1.0$ denotes the stress wave propagating from the rock with large wave impedance to the rock with small wave impedance. Similarly, the wave impedance ratio of rock mass with the same wave impedance on two sides of the joint is $n = 1.0$.

According to the conservation of momentum on the wavefronts

$$\sigma = \pm \rho C v = \pm z v \tag{5.55}$$

where ρ is the density of the rock, and C is wave velocity. Positive sign is for left-traveling wave, and negative sign is for right-traveling wave.

Therefore, the stress of the incident wave is

$$\sigma_{inc} = -\rho_{01} C_{01} v_{inc} = -z_1 v_{inc} \tag{5.56}$$

where ρ_{01} is the density of the rock in front of the joint.

The stress of the reflected wave is

$$\sigma_{ref} = \rho_{01} C_{01} v_{ref} = z_1 v_{ref} \tag{5.57}$$

It should be noted that σ is defined to be positive for compression and negative for tension which is consistent with the common definition used in rock mechanics.

Thus, the reflection coefficient R can be defined to characterize the amplitude magnitude of tensile wave as

$$R = \frac{\max\left|\sigma_{\text{ref},t}\right|}{\max\left|\sigma_{inc}\right|} \tag{5.58}$$

where $\sigma_{\text{ref},t}$ is the tensile stress of the reflected wave.

5.3.2 Results

For an arbitrary incident wave, the reflected wave can be calculated from Eqs. (5.53) and (5.57). The incident wave is assumed to be a half-cycle sinusoidal wave with an amplitude of $\sigma_0 = 1.00$ MPa. The rock density and wave velocity in front of the joint are $\rho_{01} = 2400$ kg/m^3 and $C_{01} = 4500$ m/s, respectively.

(1) Incident and reflected waveforms at different frequencies

Frequency is one of the most important parameters of the incident wave. Figure 5.21a, b shows the incident waves and the corresponding reflected waves in the rock mass with joint stiffness of $k_n = 3.5$ GPa/m and incident frequencies of $f_0 = 50.0$, 100.0 and 150.0 Hz, respectively. The wave impedance ratios of rock mass are $n = 0.4$ and $n = 1.6$ in Fig. 5.21a, b, respectively. It can be observed that when the wave propagates across the rock joint, wave reflection takes place, and both the tensile wave and the compression wave are generated in the reflected wave. Figure 5.21a, b also shows that the amplitude of tensile wave increases as the incident frequency increases from 50.0 to 150.0 Hz. The tensile wave is frequency-dependent in the rock mass with different wave impedances on two sides of the joint. Therefore, more attention should be paid to the rock mass with large incident frequency, which may be subject to large tensile stress.

(2) Incident and reflected waveforms at different joint stiffnesses

Figure 5.22a, b shows the incident waves and the corresponding reflected waves in the rock mass with a frequency of $f_0 = 50.0$ Hz and joint stiffness of $k_n = 1.5$, 3.5 and 5.5 GPa/m, respectively. The wave impedance ratios of rock mass are $n = 0.4$ and $n = 1.6$ in Fig. 5.22a, b, respectively. It is seen from Fig. 5.22a, b that the reflected tensile waves at the joint with different wave impedance ratios (e.g., $n = 0.4$ or $n = 1.6$) are both influenced by the joint stiffness significantly. Small joint stiffness ($k_n = 1.5$ GPa/m) results in a reflected tensile wave with a large amplitude, while large joint stiffness ($k_n = 5.5$ GPa/m) results in reflected tensile wave with a small amplitude. Therefore, the rock mass may be subjected to large tensile stress when the joint stiffness is small.

(3) Incident and reflected waveforms at different wave impedance ratios

Figure 5.23 shows the incident wave and the corresponding reflected waves in the rock mass with a joint stiffness of $k_n = 3.5$ GPa/m, and frequency of incident wave $f_0 = 50$ Hz. The wave impedance ratios of rock mass are $n = 0.4$ and 1.6, respectively.

Figure 5.23 illustrates that the amplitude of the tensile wave is influenced by the wave impedance ratio. The amplitude of the tensile wave is larger in value for a large wave impedance ratio. It is seen that there is an obvious difference between the amplitudes of the tensile wave if the wave impedance ratio is considered. When the wave impedance ratio $n = 1.6$, the amplitude of the tensile wave is larger than that of the tensile wave in the rock mass with the same wave impedances on two sides of the joint ($n = 1.0$), while the amplitude of the tensile wave is smaller than that of the tensile wave in the rock mass with the same wave impedances on two sides of the joint ($n = 1.0$) when the wave impedance ratio $n = 0.4$.

In traditional studies of stress wave propagation in the jointed rock mass, the wave impedances of rocks on two sides of the joint are generally assumed to be the same ($n = 1.0$), so the effect of wave impedance ratio on the amplitude of the tensile wave is not considered. In fact, the amplitude of the tensile wave is affected by the wave impedance ratio in the rock mass with different wave impedances on two sides of

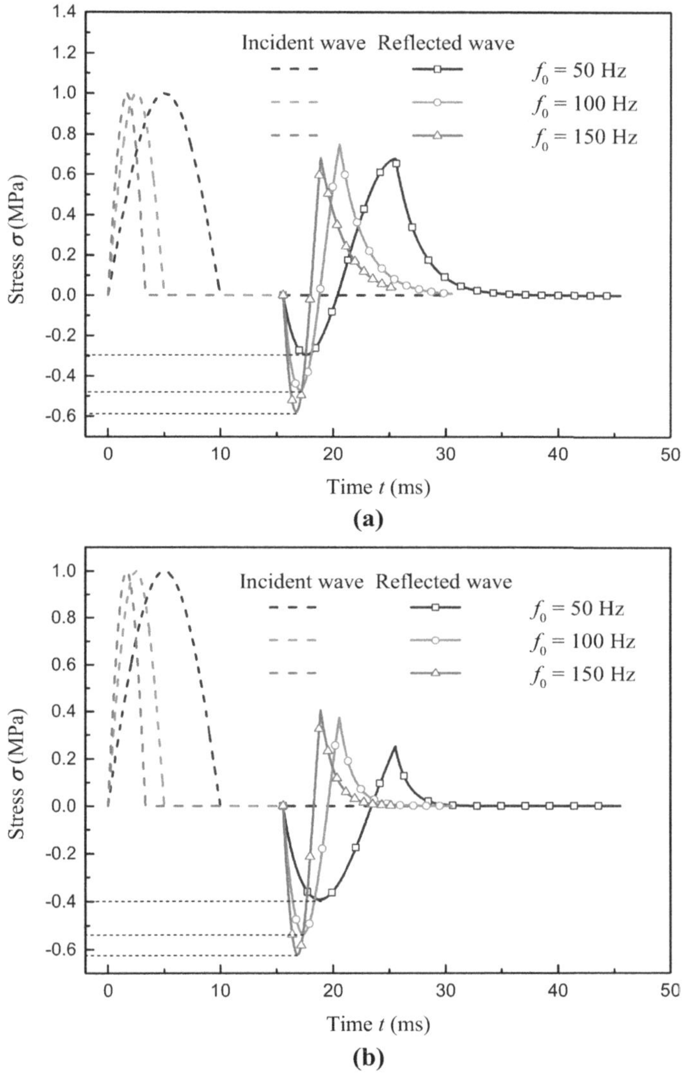

Fig. 5.21 Incident waves and the corresponding reflected waves at different frequencies in two kinds of rock mass. **a** $n = 0.4$. **b** $n = 1.6$

the joint, as shown in Fig. 5.23. Therefore, the effect of the wave impedance ratio should be taken into consideration to predict the reflected tensile wave accurately.

In order to evaluate the error induced by the wave impedance ratio, the difference of reflection coefficients was defined as

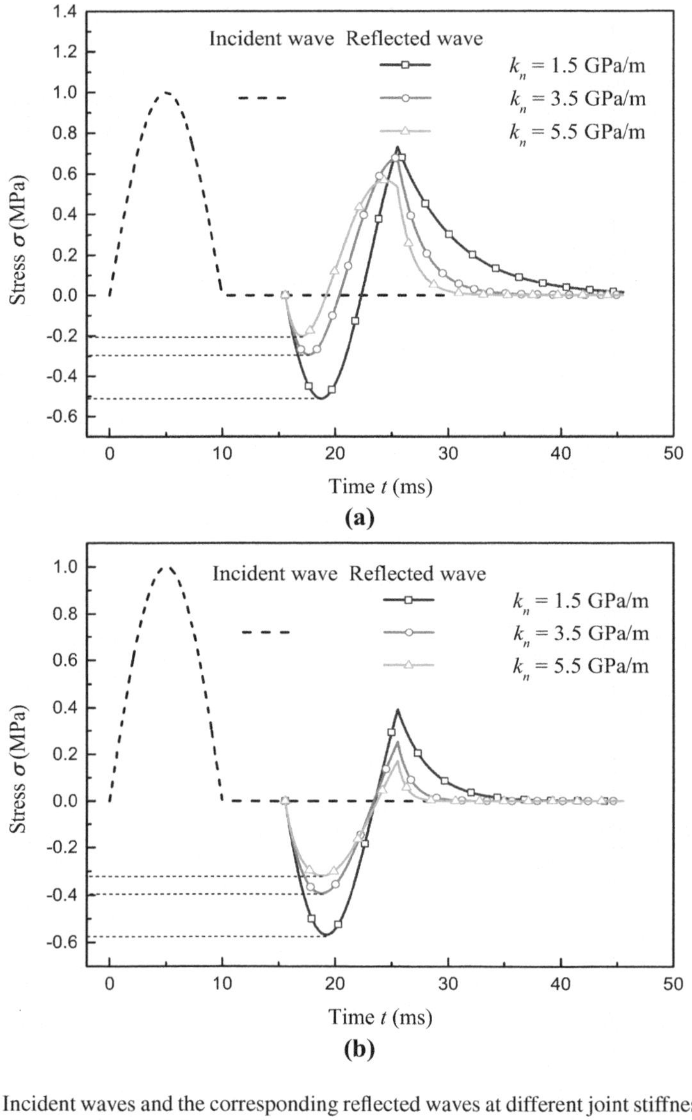

Fig. 5.22 Incident waves and the corresponding reflected waves at different joint stiffnesses in two kinds of rock mass. **a** $n = 0.4$. **b** $n = 1.6$

$$\Delta R = \frac{\left|R_{n\neq1.0} - R_{n=1.0}\right|}{R_{n\neq1.0}} \times 100\% \tag{5.59}$$

where $R_{n\neq1.0}$ is the reflection coefficient of rock mass with different wave impedances on two sides of the joint, and $R_{n=1.0}$ is the reflection coefficient of rock mass with the

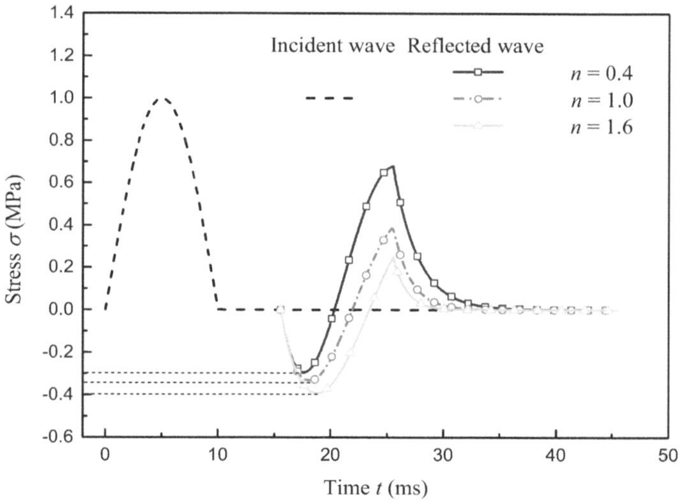

Fig. 5.23 Incident waves and the corresponding reflected waves at different wave impedance ratios

same wave impedances on two sides of the joint. The difference of reflection coefficients presents the influencing degree of wave impedance ratio on wave reflection. It can be used to effectively evaluate the reflected wave in the rock mass with different wave impedance ratios on two sides of the joint.

Based on Eq. (5.59), the difference of reflection coefficients are 14.13% and 14.33% for the wave impedance ratio $n = 0.4$ and $n = 1.6$ in Fig. 5.23, respectively. It should be noted that the reflection coefficient for the case of $n < 1.0$ is always smaller than that in the rock mass with the same wave impedances on two sides of the joint, while the reflection coefficient for the case of $n > 1.0$ is always larger than that in the rock mass with the same wave impedances on two sides of the joint.

5.3.3 Discussion

(1) Effect of incident frequency on the reflection coefficient

Figure 5.24a, b shows the effect of incident frequency on the reflection coefficient and the difference of reflection coefficients, respectively. The rock density and wave velocity in front of the joint are $\rho_{01} = 2400$ kg/m^3 and $C_{01} = 4500$ m/s, respectively. The joint stiffness is $k_n = 1.5$ GPa/m and the wave impedance ratios are $n = 0.2$, 0.4, 0.6, 1.4, 1.6 and 2.0, respectively.

In Fig. 5.24a, it is seen that the reflection coefficient R increases as the incident frequency increases. When the incident frequencies are relatively small, the reflection coefficients R in the cases of $n = 1.4$, $n = 1.6$ and $n = 2.0$ remain constants, while all of the reflection coefficients R in the cases of $n = 0.2$, $n = 0.4$ and $n = 0.6$ approach

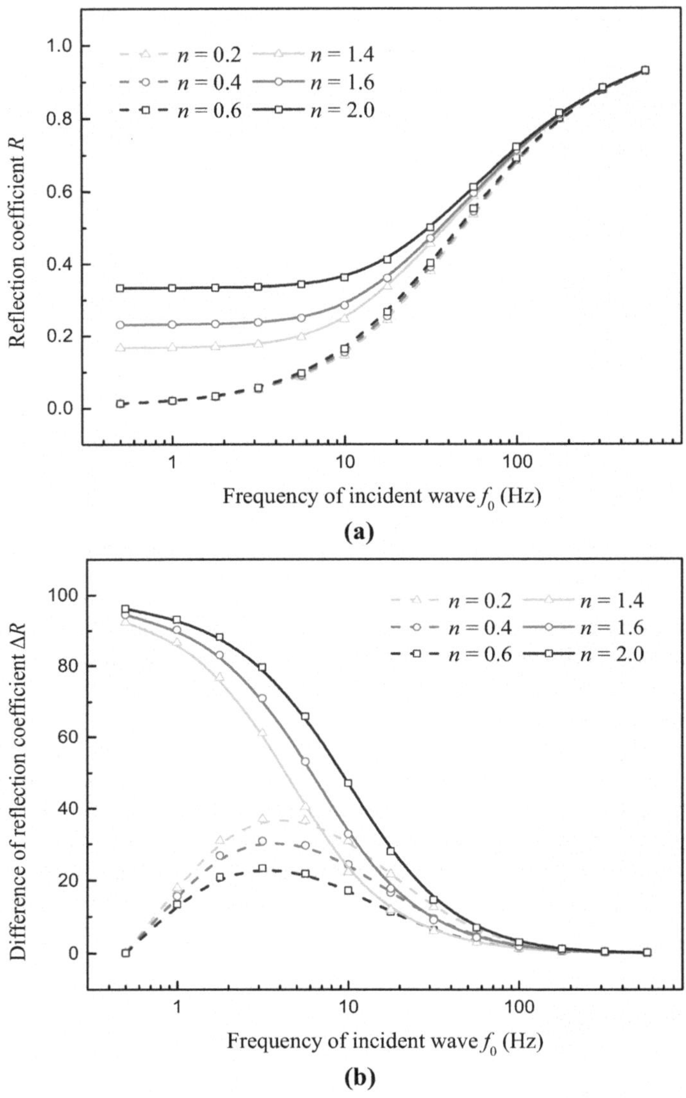

Fig. 5.24 Effect of frequency of incident wave on wave reflection. **a** Reflection coefficient. **b** difference of reflection coefficient

zero. The phenomenon illustrates that the tensile wave only exists in the rock mass where the wave impedance in front of the joint is larger than that behind the joint at small incident frequencies. Then, the reflection coefficients R increase sharply with the increase of incident frequency. In the cases of $n = 0.2$, $n = 0.4$ and $n = 0.6$, the reflection coefficients R are very close. When the incident frequencies become sufficiently large, all of the reflection coefficients R approach 1.0, which means that all of the waves are reflected at the joint and all of the reflected waves are tensile waves. At this time, the amplitude of the tensile wave is the largest. The joint acts as a small-pass filter and the incident wave with large frequencies is more likely to be reflected at the joint, which results in a large reflection coefficient as shown in Fig. 5.24a. Moreover, it can be also seen that the reflection coefficient R with a larger wave impedance ratio is larger in value than that with a smaller wave impedance ratio for a constant incident frequency.

Figure 5.24b shows the variation in the difference of reflection coefficients ΔR with respect to incident frequency. In Fig. 5.24b, the difference of reflection coefficients ΔR increases firstly and then decreases with the increases of incident frequency in the cases of $n = 0.2$, $n = 0.4$ and $n = 0.6$, while the difference of reflection coefficients ΔR always decreases as the incident frequency increases in the cases of $n = 1.4$, $n = 1.6$ and $n = 2.0$. When the wave impedance ratio $n < 1.0$, the difference of reflection coefficients ΔR approaches zero at sufficiently small incident frequencies, then increases with the increase of incident frequency and around the frequency of 5.6 Hz in Fig. 5.24b, it begins to decrease and finally approaches zero at sufficiently large incident frequencies. When the wave impedance ratio $n > 1.0$, the difference of reflection coefficients ΔR approaches 100% at sufficiently small incident frequencies, then decreases sharply with the increase of incident frequency and finally approaches zero at large incident frequencies. Therefore, the effect of the wave impedance ratio can be neglected at relatively small or large incident frequencies in the cases of $n < 1.0$, but it cannot be neglected at small incident frequencies in the cases of $n > 1.0$. In addition, for a constant incident frequency, the difference of reflection coefficients ΔR with a larger wave impedance ratio is larger in value when the wave impedance ratio $n > 1.0$, but the difference of reflection coefficients ΔR with a smaller wave impedance ratio is larger in value when the wave impedance ratio $n < 1.0$.

(2) Effect of joint stiffness on reflection coefficient

Figure 5.25a, b shows the effect of joint stiffness on the reflection coefficient and the difference of reflection coefficient, respectively. The rock density and wave velocity in front of the joint are $\rho_{01} = 2400$ kg/m^3 and $C_{01} = 4500$ m/s, respectively. The incident frequency is $f_0 = 50$ Hz and wave impedance ratios are $n = 0.2$, 0.4, 0.6, 1.4, 1.6 and 2.0, respectively.

It can be observed from Fig. 5.25a that the reflection coefficient R decreases as the joint stiffness increases. For sufficiently small joint stiffness, all the reflection coefficients R in the rock mass with different wave impedances on two sides of the joint approach 1.0. Under such conditions, the joint is weak, so all of the waves are reflected. Since all of the reflection coefficients R approach 1.0, only reflected tensile waves are produced, and the amplitude of the tensile wave is larger when the joint is weak. Then, the reflection coefficients R decrease sharply with the increase of joint stiffness and finally approach constant. If the joint stiffness is large enough, the reflection coefficients R in the cases of $n = 1.4$, $n = 1.6$ and $n = 2.0$ approach their minimum values of 0.17, 0.23 and 0.33 which are equal to the results of stress wave

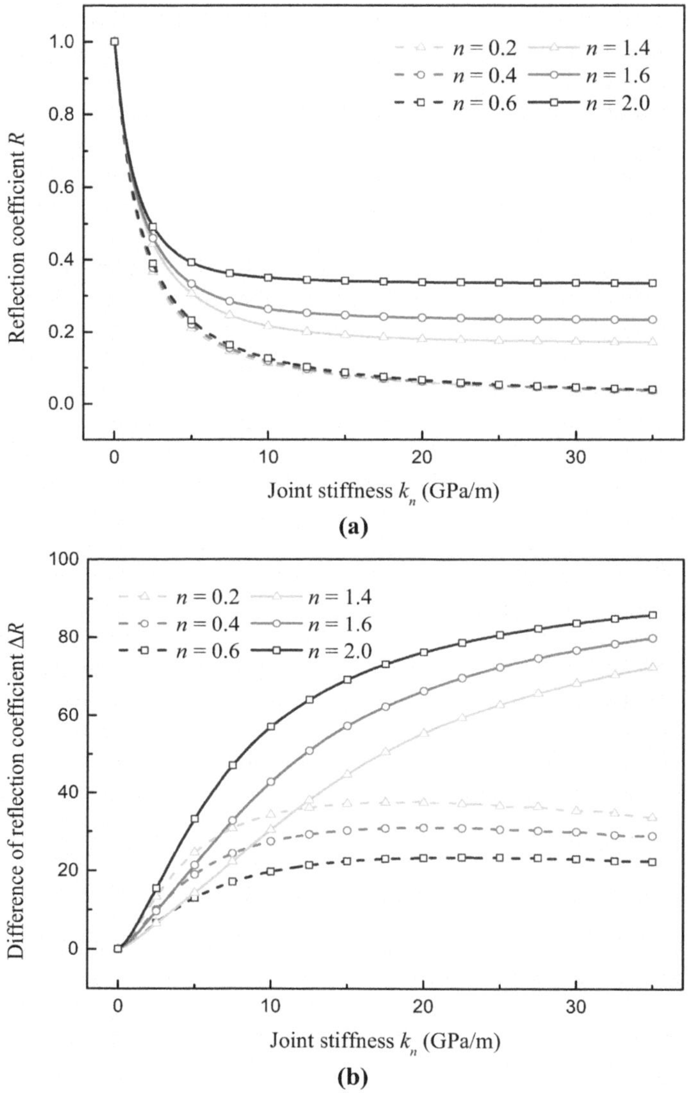

Fig. 5.25 Effect of joint stiffness on wave reflection. **a** Reflection coefficient. **b** Difference of reflection coefficient

propagating across the interface of different materials [24]. In the cases of $n = 0.2$, $n = 0.4$ and $n = 0.6$, the reflection coefficients R approach zero if the joint stiffness is large enough, which means that no tensile waves are produced. For the joint with large stiffness, the joint acts as a welded interface so there is no reflected wave caused by the joint. The reflected wave is only caused by the interface of different materials.

The reflection coefficients R in the cases of $n = 0.2$, $n = 0.4$ and $n = 0.6$ are very close in Fig. 5.25a. It can be also observed that the rock mass with a larger wave impedance ratio results in a larger reflection coefficient R once the joint stiffness is fixed.

Figure 5.25b shows the variation in the difference of reflection coefficients ΔR with respect to the joint stiffness. It can be observed from Fig. 5.25b that the difference of reflection coefficients ΔR increases firstly and then decreases with the increase of joint stiffness in the cases of $n = 0.2$, $n = 0.4$ and $n = 0.6$, while the difference of reflection coefficients ΔR always increases as the joint stiffness increases in the cases of $n = 1.4$, $n = 1.6$ and $n = 2.0$. When the wave impedance ratio $n < 1.0$, the difference of reflection coefficients ΔR approaches zero at sufficiently small joint stiffness, then increases with the increase of joint stiffness. After reaching the peak values, they begin to decrease slowly and finally approach zero if the joint stiffness is large enough. When the wave impedance ratio $n > 1.0$, the difference of reflection coefficients ΔR approaches zero at sufficiently small joint stiffness, then increases sharply with the increase of joint stiffness, and finally approaches 100% if the joint stiffness is large enough. It can be concluded that the effect of the wave impedance ratio can be neglected at relatively small or large joint stiffness in the cases of $n < 1.0$, but it cannot be neglected at large joint stiffness in the cases of $n > 1.0$. Similar to Fig. 5.24b, for a constant joint stiffness, the difference of reflection coefficients ΔR with a larger wave impedance ratio is larger in value when the wave impedance ratio $n > 1.0$, but the difference of reflection coefficients ΔR with a smaller wave impedance ratio is larger in value when the wave impedance ratio $n < 1.0$.

(3) Effect of wave impedance ratio on reflection coefficient

Figure 5.26a, b shows the effect of wave impedance ratio on the reflection coefficient and the difference of reflection coefficient, respectively. The rock density and wave velocity in front of the joint are $\rho_{01} = 2400$ kg/m^3 and $C_{01} = 4500$ m/s, respectively. The incident frequency is $f_0 = 50.0$ Hz and the joint stiffness is $k_n = 1.5$, 2.0, 2.5, 3.5 and 5.5 GPa/m, respectively.

In Fig. 5.26a, it illustrates that the reflection coefficient R increases as the wave impedance ratio increases. The reflection coefficient R approaches its minimum at sufficiently small wave impedance ratios. Subsequently, the reflection coefficient R increases sharply with the increase of wave impedance ratio. Finally, the reflection coefficient R approaches 1.0 at relatively large wave impedance ratios, and all of the waves are reflected off and only tensile waves are produced. When the wave impedance ratio is sufficiently large, the nonwelded interface acts as a free surface, so the waves are completely reflected. In addition, the rock mass with a small joint stiffness results in a large reflection coefficient R which is consistent with Fig. 5.25a.

In Fig. 5.26b, it is seen that the difference of reflection coefficients ΔR decreases firstly and then increases as the wave impedance ratio increases. When the wave impedance ratio $n < 1.0$, the difference of reflection coefficients ΔR decreases sharply and equals zero at $n = 1.0$. When the wave impedance ratio $n > 1.0$, the difference of reflection coefficients ΔR increases sharply at first, then increases slowly as the wave impedance ratio further increases, and finally approaches its maximum. For the

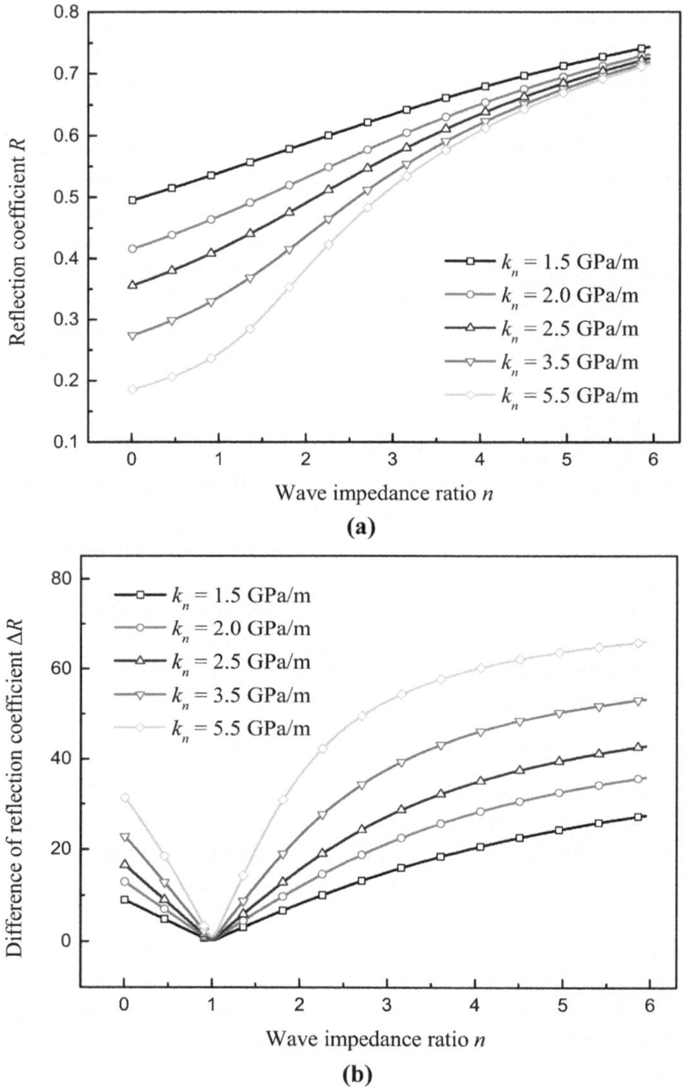

Fig. 5.26 Effect of wave impedance ratio on wave reflection. **a** Reflection coefficient. **b** Difference of reflection coefficient

same joint stiffness, the maximum value of the difference of reflection coefficients ΔR when the wave impedance $n < 1.0$ is much smaller than that of the difference of reflection coefficients ΔR when the wave impedance $n > 1.0$. In addition, the rock mass with a larger joint stiffness results in a larger difference of reflection coefficients

ΔR when the wave impedance ratio $n > 1.0$ which is consistent with the results in Fig. 5.25b.

5.4 Conclusions

(1) A modified characteristic method with four sets of characteristic lines is proposed to investigate the wave transmission through linearly jointed complex rock mass. The effect of the frequency, joint stiffness and wave impedance ratio on the particle velocity transmission ratio, stress transmission ratio and effective velocity are analyzed.

Wave propagation through conventional jointed rock mass is a special case of wave propagation through complex rock mass when the wave impedance ratio equals 1.0. The modified characteristic method with four sets of characteristic lines can be used to calculate the wave propagation through the complex rock mass.

For the cases of wave transmission through "soft-to-hard" rock mass, the particle velocity transmission ratios are always smaller than 1.0, while the stress transmission ratio may be larger than 1.0 when the wave impedance ratio and frequency are sufficiently small. For the cases of wave transmission through "hard-to-soft" rock mass, the stress transmission ratios are always smaller than 1.0, while the particle velocity transmission ratio may be larger than 1.0 when the wave impedance ratio is large enough for waves with sufficiently small frequencies.

It is concluded that the present study extended traditional studies to wave propagation through complex rock mass, which can be used to predict wave transmission more accurately. Moreover, the present study can be used as a reference for the design of disaster prevention and control. For example, we can design a "soft" layer on a "hard" foundation to reduce the stress, or we can design a "hard" layer on a "soft" foundation to reduce the particle velocity.

(2) Wave propagation through a nonlinearly jointed complex rock mass commonly occurs for blast or seismic engineering. The conventional characteristic method was improved to study the nonlinear effect of the frequency and amplitude of the incident wave, initial joint stiffness and wave impedance ratio of the complex rock mass on the particle velocity transmission coefficient and energy transmission coefficient.

The characteristic method can be modified to efficiently study stress wave propagation through a nonlinearly jointed complex rock mass. The traditional characteristic method can be regarded as a special case of the present modified method when the mechanical properties of the rock in front of and behind the joint are the same.

The wave impedance ratio is an important parameter affecting the propagation of stress waves in complex rock mass. Traditional studies of stress wave propagation through a rock mass are special cases of the present study (when the wave impedance ratio equals 1.0). It is concluded that the particle velocity transmission coefficients are always smaller than 1.0 for the "soft-to-hard" rock mass. However, it may exceed 1.0 as the wave impendence ratio increases and the frequency of incident waves decreases for the "hard-to-soft" rock mass. On the other hand, the energy transmission coefficient is always ≤ 1.0 for both the "soft-to-hard" rock mass and the "hard-to-soft" rock mass.

(3) The amplitude attenuation of the reflected wave in a jointed rock mass with different wave impedances on two sides of the joint is analyzed. The equation of reflected wave considering the variation of wave impedance ratio is derived based on the displacement discontinuity model and the method of characteristics. The parametric studies show that the reflection coefficient and the difference of reflection coefficients are not only affected by the incident frequency and joint stiffness, but also are affected by the wave impedance ratio.

When the stress wave propagates from the rock with large wave impedance to the rock with a small wave impedance, the tensile wave exists for all of the incident frequencies, joint stiffness and wave impedance ratios. The reflection coefficient increases with the increase of incident frequency, while it decreases with the increase of joint stiffness and increases as the wave impedance ratio increases. However, the difference of reflection coefficients decreases with the increase of incident frequency, and increases with the increase of joint stiffness and wave impedance ratio. Therefore, the effect of wave impedance ratio on the wave reflection through the rock mass with different wave impedances on two sides of the joint becomes less significant at sufficiently large incident frequencies and sufficiently small joint stiffness.

When the stress wave propagates from rock with small wave impedance to large wave impedance, the reflection coefficient also increases with the increase of incident frequency, decreases with the increase of joint stiffness and increases as the wave impedance ratio increases, while the difference of reflection coefficients increases firstly and then decreases as the incident frequency and the joint stiffness increase, and decreases as the wave impedance ratio increases. Therefore, the effect of wave impedance ratio on wave reflection through rock mass with different wave impedances on two sides of the joint can be neglected at sufficiently small or large incident frequencies and sufficiently small or large joint stiffness.

References

1. Li J, Ma G. Analysis of blast wave interaction with a rock joint. Rock Mech Rock Eng. 2009;43(6):777–87.
2. Zhao J, Cai JG. Transmission of elastic p-waves across single fractures with a nonlinear normal deformational behavior. Rock Mech Rock Eng. 2001;34(1):3–22.
3. Zhao XB, Zhao J, Cai JG. P-wave transmission across fractures with nonlinear deformational behaviour. Int J Numer Anal Met. 2006;30(11):1097–112.

4. Zhao J, Zhao XB, Cai JG. A further study of p-wave attenuation across parallel fractures with linear deformational behaviour. Int J Rock Mech Min. 2006;43(5):776–88.
5. Ma GW, Fan LF, Li JC. Evaluation of equivalent medium methods for stress wave propagation in jointed rock mass. Int J Numer Anal Met. 2013;37(7):701–15.
6. Li JC, Ma GW, Zhao J. An equivalent viscoelastic model for rock mass with parallel joints. J Geophys Res-Solid Ea. 2010;115(B3):B03305.
7. Schoenberg M. Elastic wave behavior across linear slip interfaces. J Acoust Soc Am. 1980;68(5):1516–21.
8. Miller RK. An approximate method of analysis of the transmission of elastic waves through a frictional boundary. J Appl Mech-T Asme. 1977;44(4):652–6.
9. Li JC, Ma GW. Experimental study of stress wave propagation across a filled rock joint. Int J Rock Mech Min. 2009;46(3):471–8.
10. Kulhawy FH. Stress deformation properties of rock and rock discontinuities. Eng Geol. 1975;9(4):327–50.
11. Zhao G-F. Developing a four-dimensional lattice spring model for mechanical responses of solids. Comput Method Appl M. 2017; 315(MAR.1):881–95.
12. Zhu WC, Niu LL, Li SH, et al. Dynamic brazilian test of rock under intermediate strain rate: pendulum hammer-driven shpb test and numerical simulation. Rock Mech Rock Eng. 2014;48(5):1867–81.
13. Chen X, Li JC, Cai MF, et al. A further study on wave propagation across a single joint with different roughness. Rock Mech Rock Eng. 2016;49(7):2701–9.
14. Bandis SC, Lumsden AC, Barton NR. Fundamentals of rock joint deformation. Int J Rock Mech Min. 1983;20(6):249–68.
15. Li ZL, Li JC, Li HB, et al. Effects of a set of parallel joints with unequal close-open behavior on stress wave energy attenuation. Wave Random Complex. 2020;2:1–25.
16. Chai S, Tian W, Zhao J. Study of the cylindrical wave propagation across a single rock joint with nonlinear normal deformation. Wave Random Complex. 2019;3:1–18.
17. Cai JG, Zhao J. Effects of multiple parallel fractures on apparent attenuation of stress waves in rock masses. Int J Rock Mech Min. 2000;37(4):661–82.
18. Wang M, Wang F, Zhu Z, et al. Modelling of crack propagation in rocks under shpb impacts using a damage method. Fatigue Fract Eng M. 2019;42(8):1699–710.
19. Wang Z-L, Li Y-C, Shen RF. Numerical simulation of tensile damage and blast crater in brittle rock due to underground explosion. Int J Rock Mech Min. 2007;44(5):730–8.
20. Mindlin RD. Waves and vibrations in isotropic elastic plates. New York: Pergamon Press; 1960.
21. Jones JP, Whittier JS. Waves at a flexibly bonded interface. J Appl Mech-T Asme. 1967;34(4):905–9.
22. Schoenberg M. Reflection of elastic waves from periodically stratified media with interfacial slip. Geophys Prospect. 1983;31(2):265–92.
23. Fan LF, Wang LJ, Wang M, et al. Investigation of stress wave transmission across a nonlinearly jointed complex rock mass. Int J Rock Mech Min. 2020;136:104485.
24. Gao J, Fan L, Wan Z. Thermal cycling effects on the dynamic behavior of granite and microstructural observations. B Eng Geol Environ. 2021;80:8711–23.

Chapter 6
Two-Characteristic Method for the Wave Propagation Through Complex Stratus Rock Mass with Parallel Joints

6.1 Wave Transmission Through Layered Composite Rock Mass Containing Multiple Parallel Joints with Different Wave Impedances on Two Sides of Joints

Figure 6.1 shows a typical representative elementary volume (REV) of the layered composite rock mass. According to the previous study, the representative elementary volume is assumed with two joints and three different types of rocks. Three types of rocks are with different wave impedances while two joints are with the same joint stiffness as shown in Fig. 6.1. The joints are parallelly located with the same joint spacing as S. Different from the wave propagation through a single joint, it is seen from Fig. 6.1 that the wave transmission is not only influenced by mechanical properties of rock mass (such as wave impedance and joint stiffness), but also influenced by the multiple reflections between the joints. Therefore, it is more complex to investigate the wave propagation through layered composite rock mass containing multiple joints and different types of rocks.

In order to analyze the multiple reflections of waves in layered composite rock mass more clearly, the composition of the total transmitted wave is shown in Fig. 6.2. In Fig. 6.2, the incident right-running wave is generated when the left boundary of the representative elementary volume of the layered composite rock mass is impacted. The incident right-running wave propagates to the first joint ($x = x_1$) and yields a transmitted right-running wave and a reflected left-running wave. The transmitted right-running wave propagates to the second joint ($x = x_2$) and also produces a transmitted right-running wave and a reflected left-running wave. The transmitted right-running wave without reflection is called the direct transmission wave. It should be noted that the left-running wave generated on the left side of the second joint ($x = x_2$) propagates to the first joint ($x = x_1$), and engenders a right-running wave with two reflections and a transmitted left-running wave. When a right-running wave with two reflections propagates to the second joint ($x = x_2$), a transmitted wave

L. Fan et al., *The Method of Characteristics for Stress Wave Propagation in the Rock Mass*, https://doi.org/10.1007/978-981-97-1440-7_6

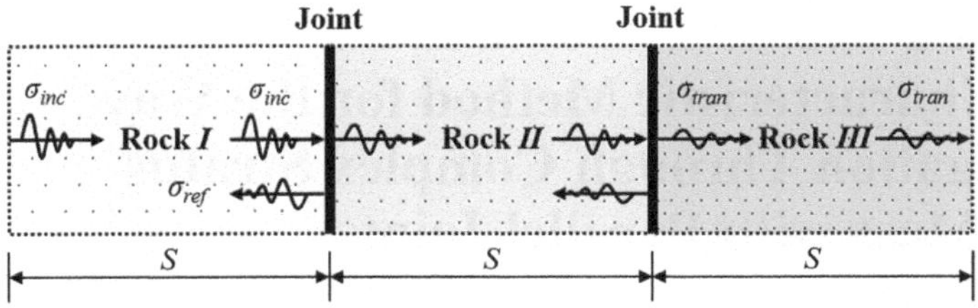

Fig. 6.1 Representative elementary volume of the layered composite rock mass

caused by two reflections and a reflected left-running wave are generated. The left-running wave is reflected again to form a right-running wave with four reflections. By similar superposition, the right-running wave obtained after six or eight times of reflection forms a series of transmitted waves. The superposition of these waves is called the indirectly transmitted wave caused by multiple reflections. Therefore, it can be observed from Fig. 6.2 that the total transmitted wave is composed of both the directly transmitted wave and the indirectly transmitted wave caused by multiple reflections between multiple joints.

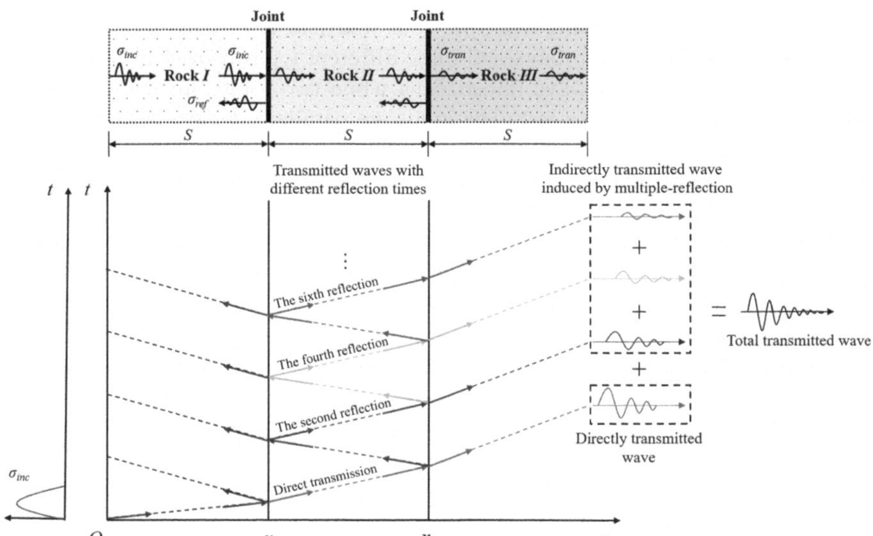

Fig. 6.2 Composition of the total transmitted wave

6.1.1 Modified Method of Characteristics in the Layered Composite Rock Mass

The displacement discontinuity model is commonly used to study wave propagation through the layered composite rock mass. Generally, the stress field is assumed to be continuous while the displacement field is discontinuous. Therefore, the displacement discontinuity model can be expressed as

$$\sigma^{-} = \sigma^{+} = \sigma \tag{6.1}$$

and

$$u^{-} - u^{+} = \frac{\sigma}{k_n} \tag{6.2}$$

where σ is the normal stress, u is the normal displacement and k_n is the joint stiffness; the superscripts $^{-}$ and $^{+}$ denote the parameters in front of and behind the joint, respectively.

After determining the boundary conditions of the joints in the layered composite rock mass, the wave propagation through such rock mass is studied. The method of characteristics is widely used to study wave propagation through the layered rock mass. According to the method of characteristics, the relations between the particle velocity v and stress σ can be derived as the left-running characteristic line $zv(x, t) - \sigma(x, t) = \text{constant}$ and the right-running characteristic line $zv(x, t) + \sigma(x, t) = \text{constant}$. The modified method of characteristics is proposed to study wave propagation through the layered composite rock mass with different wave impedances on two sides of joints, as shown in Fig. 6.3. It can be seen from Fig. 6.3 that when the wave propagates along the x-axis to joints in the layered composite rock mass, both the reflections and transmissions take place. The lines *ab*, *bc*, *cd*, *ce* and *ef* are the characteristic lines of waves in the x-t plane and the slopes of the characteristic lines are $-C_1$, C_1, $-C_2$, C_2 and $-C_3$, respectively. The slope of the characteristic line is affected by rock wave impedance in the layered composite rock mass. The rock wave impedance is proportion to the absolute value of the slope of the characteristic line.

The analytical solutions of transmission coefficient and reflection coefficient of internal wave propagation in layered composite rock mass are derived by using the modified method of characteristics and the displacement discontinuity model.

At the joint ($x = x_1$), the expression of the partial derivatives of Eq. (6.2) with respect to t is

$$v^{-}(x_1, t) - v^{+}(x_1, t) = \frac{1}{k_n}\frac{\partial \sigma(x_1, t)}{\partial t} \tag{6.3}$$

where $v^{-}(x_1, t)$ and $v^{+}(x_1, t)$ are the particle velocities in front of and behind the joint ($x = x_1$), respectively.

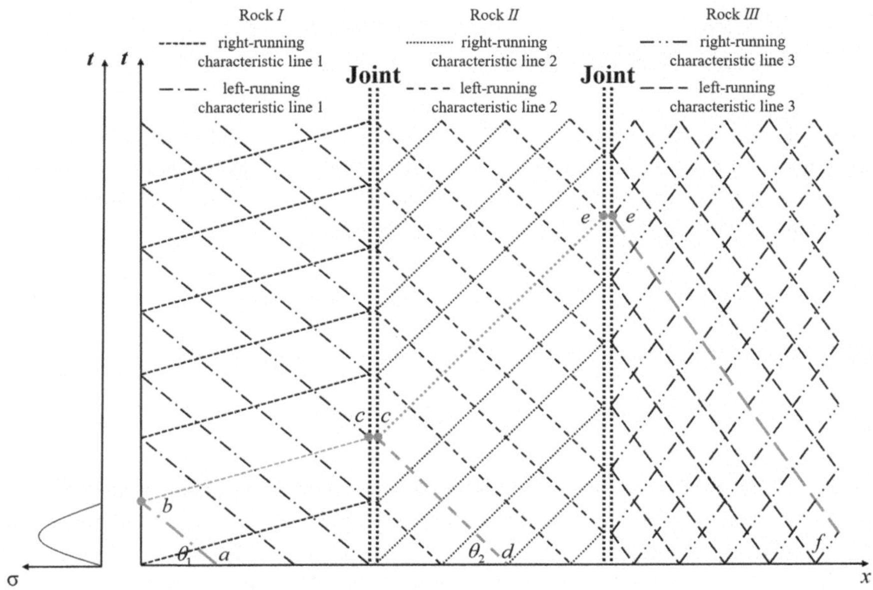

Fig. 6.3 Modified method of characteristics for layered composite rock mass

When $t = 0$, the rock mass is undisturbed and the initial conditions are $v(x, 0) = 0$ and $\sigma(x, 0) = 0$, so along the left-running characteristic line ab is

$$z_1 v_{\text{inc}}(0, t - x_1/C_1) - \sigma_{inc}(0, t - x_1/C_1) = 0 \tag{6.4}$$

where z_1 is the wave impedance of rock I, and $v_{inc}(0, t - x_1/C_1)$ and $\sigma_{inc}(0, t - x_1/C_1)$ are the particle velocities of the incident wave and the stress of the incident wave.

Along the right-running characteristic line bc in front of the joint ($x = x_1$) can be expressed as

$$z_1 v_{\text{inc}}(0, t - x_1/C_1) + \sigma_{\text{inc}}(0, t - x_1/C_1) = z_1 v^-(x_1, t) + \sigma(x_1, t) \tag{6.5}$$

Along the left-running characteristic line cd behind the joint ($x = x_1$) is

$$z_2 v^+(x_1, t) - \sigma(x_1, t) = 0 \tag{6.6}$$

where z_2 is the wave impedance of rock II.

Substituting Eq. (6.5) into Eq. (6.4) yields

$$z_1 v^-(x_1, t) + \sigma(x_1, t) = 2z_1 v_{\text{inc}}(0, t - x_1/C_1) \tag{6.7}$$

Substituting Eq. (6.7) into Eq. (6.6), the relationship of incident wave particle velocity, and the particle velocity in front of and behind the joint ($x = x_1$) can be obtained as

$$z_1 v^-(x_1, t) + z_2 v^+(x_1, t) = 2z_1 v_{\text{inc}}(0, t - x_1/C_1) \tag{6.8}$$

Combining Eqs. (6.3) and (6.6), the relationship between particle velocity in front of and behind the joint ($x = x_1$) can be obtained as

$$v^-(x_1, t) - v^+(x_1, t) = \frac{z_2 \partial v^+(x_1, t)}{k_n \partial t} \tag{6.9}$$

Substituting Eq. (6.8) into Eq. (6.9) gives

$$\frac{\partial v^+(x_1, t)}{\partial t} = \frac{2k_n}{z_2} v_{\text{inc}}(0, t - x_1/C_1) - k_n(1/z_1 + 1/z_2)v^+(x_1, t) \tag{6.10}$$

When Δt is small enough, the derivative in Eq. (6.10) can be replaced by a differential expression of

$$\frac{\partial v^+(x_1, t)}{\partial t} = \frac{v^+(x_1, t_{j+1}) - v^+(x_1, t_j)}{\Delta t} \tag{6.11}$$

Substituting Eq. (6.11) into Eq. (6.10) yields the iterative expression of transmitted wave particle velocity behind the joint ($x = x_1$) as

$$v_{\text{tran}} = v^+(x_1, t_{j+1}) = \frac{2k_n \Delta t}{z_2} v_{\text{inc}}(0, t - x_1/C_1) + [1 - k_n \Delta t(1/z_1 + 1/z_2)]v^+(x_1, t_j) \tag{6.12}$$

where v_{tran} is the particle velocity of the transmitted wave.

Simplified by Eq. (6.8), the particle velocity in front of the joint ($x = x_1$) can be obtained as

$$v^-(x_1, t_{j+1}) = 2v_{inc}(0, t - x_1/C_1) - v_{\text{tran}} \times z_2/z_1 \tag{6.13}$$

The particle velocity in front of the joint ($x = x_1$) is the sum of the particle velocity of the incident wave and the reflected wave, which can be expressed as

$$v^-(x_1, t_{j+1}) = v_{inc}(0, t - x_1/C_1) + v_{ref} \tag{6.14}$$

where v_{ref} is the particle velocity of the reflected wave.

Substituting Eq. (6.14) into Eq. (6.13), v_{ref} in front of the joint ($x = x_1$) can be obtained as

$$v_{\text{ref}} = v_{\text{inc}}(0, t - x_1/C_1) - v_{\text{tran}} \times z_2/z_1 \quad (6.15)$$

When the left particle velocity boundary condition $v_{inc}(0, t - x_1/C_1)$ and the initial particle velocity conditions $v^{+}(x_1, 0)$ are specified, Eqs. (6.12) and (6.15) determine v_{tran} behind the joint ($x = x_1$) and v_{ref} in front of the joint ($x = x_1$) in an iterative computing process for all values of x and t. Multiple reflections occur between the joint ($x = x_1$) and the joint ($x = x_2$). Therefore, multiple transmitted waves v_{tran} behind the joint ($x = x_2$) generate and superimpose to obtain the particle velocity of the total transmitted wave.

The above is an iterative expression for particle velocities of the transmitted wave and reflected wave; the iterative expressions of the stress transmitted wave and the stress reflected wave can also be derivated by using the displacement discontinuity model and the modified method of characteristics.

Substituting Eq. (6.5) into Eq. (6.4) gives

$$z_1 v^{-}(x_1, t) + \sigma(x_1, t) = 2\sigma_{inc}(0, t - x_1/C_1) \quad (6.16)$$

Substituting Eq. (6.6) into Eq. (6.3) yields

$$v^{-}(x_1, t) - \frac{\sigma(x_1, t)}{z_2} = \frac{\partial \sigma(x_1, t)}{k_n \partial t} \quad (6.17)$$

Substituting Eq. (6.16) into Eq. (6.17), the relationship between the incident stress and the transmitted stress can be obtained as

$$\frac{\partial \sigma(x_1, t)}{\partial t} = k_n \left[\frac{2\sigma_{\text{tnc}}(0, t - x_1/C_1) - \sigma(x_1, t)}{z_1} - \frac{\sigma(x_1, t)}{z_2} \right] \quad (6.18)$$

When Δt is sufficiently small, the derivative in Eq. (6.18) can be replaced by a differential expression of

$$\frac{\partial \sigma(x_1, t)}{\partial t} = \frac{\sigma(x_1, t_{j+1}) - \sigma(x_1, t_j)}{\Delta t} \quad (6.19)$$

Substituting Eq. (6.19) into Eq. (6.18), the iterative expression of the transmitted stress wave behind the joint ($x = x_1$) can be obtained as

$$\sigma_{\text{tran}} = \sigma(x_1, t_{j+1}) = \frac{2k_n \Delta t}{z_1} \sigma_{\text{inc}}(0, t - x_1/C_1) + [1 - k_n \Delta t(1/z_1 + 1/z_2)] \sigma(x_1, t_j) \quad (6.20)$$

where σ_{tran} is the stress of the transmitted wave.

The stress on the two sides of the joint is equal, which can be expressed as

$$\sigma_{inc}(0, t - x_1/C_1) + \sigma_{\text{ref}} = \sigma_{\text{tran}} \quad (6.21)$$

where σ_{ref} is the stress of the reflected wave.

According to Eq. (6.21), σ_{ref} in front of the joint ($x = x_1$) is obtained as

$$\sigma_{\text{ref}} = \sigma_{\text{tran}} - \sigma_{\text{inc}}\,(0, t - x_1/C_1) \tag{6.22}$$

When the corresponding parameters such as $\sigma_{inc}(0, t - x_1/C_1)$ and $\sigma(x_1, 0)$ are given, σ_{tran} behind the joint ($x = x_1$) and σ_{ref} in front of the joint ($x = x_1$) can be obtained in an iterative computing process for all values of x and t from Eqs. (6.20) and (6.22). Multiple reflections occur between the joint ($x = x_1$) and the joint ($x = x_2$). Therefore, multiple transmitted waves σ_{tran} behind the joint ($x = x_2$) generate and superimpose to obtain the stress of the total transmitted wave.

The iterative calculation of the above stress and particle velocity, and the superposition calculation of the directly transmitted wave and the indirectly transmitted wave have been implemented through computer programming in this study.

The ratio of the rock wave impedance in front of and behind any joints can be described by the wave impedance ratio n, which is defined as

$$n = \frac{z_f}{z_b} \tag{6.23}$$

where z_f is the wave impedance of the rock in front of the joint and z_b is the wave impedance of the rock behind the joint.

The particle velocity transmission coefficient T_v and the stress transmission coefficient T_σ are used to evaluate the particle velocity attenuation of the transmitted wave and the stress attenuation of the transmitted wave. The particle velocity transmission coefficient T_v and the stress transmission coefficient T_σ are defined as

$$T_v = \frac{\max(v_{tran})}{\max(v_{inc})} \tag{6.24}$$

and

$$T_\sigma = \frac{\max(\sigma_{tran})}{\max(\sigma_{inc})} \tag{6.25}$$

6.1.2 Transmitted Wave Waveform Though Different Layered Composite Rock Mass

Figure 6.4 shows the particle velocity waveforms and the stress waveforms of the wave transmission through layered composite rock mass with different wave impedance ratios. Figure 6.4a shows the variations of particle velocity as the wave propagates through layered composite rock mass with wave impedance ratios of n

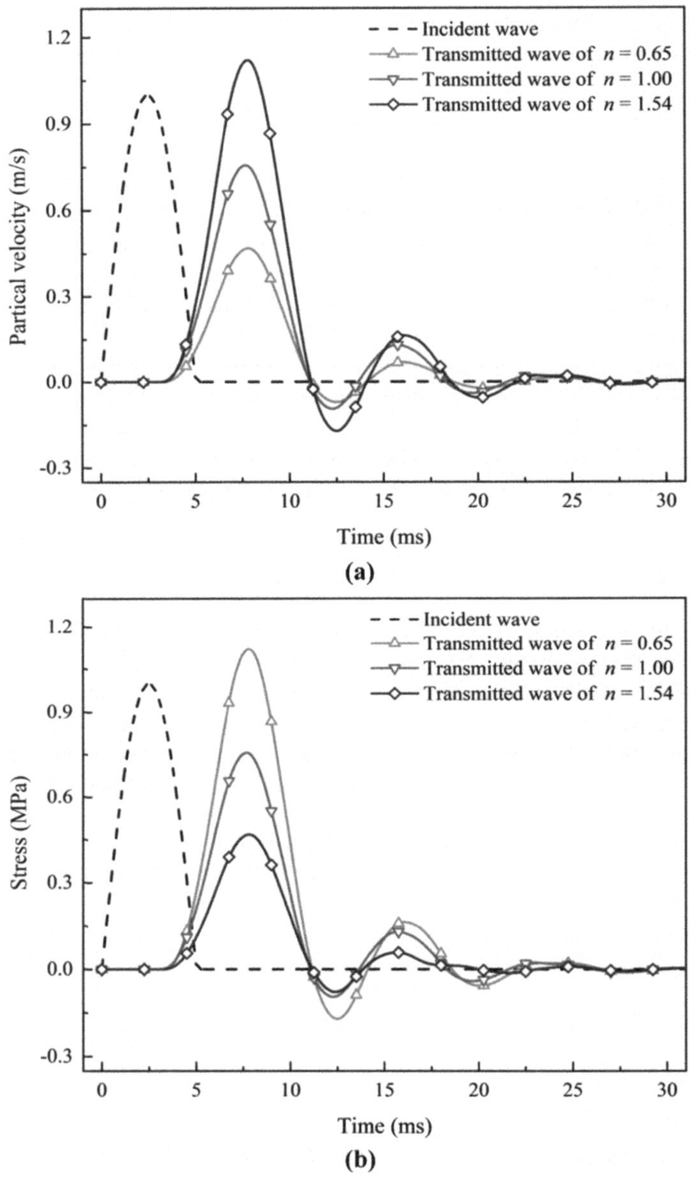

Fig. 6.4 Transmitted wave through layered composite rock mass with different wave impedance ratios. **a** Particle velocity transmission coefficient. **b** Stress transmission coefficient

= 0.65, 1.00 and 1.54. In Fig. 6.4a, when the wave impedance ratio is $n = 0.65$, the particle velocity transmission coefficient is $T_v = 0.47$. When the wave impedance ratio is $n = 1.00$, the particle velocity transmission coefficient is $T_v = 0.76$, and the effect of joints causes the amplitude of the wave to decrease. When the wave impedance ratio is $n = 1.54$, the particle velocity transmission coefficient is $T_v = 1.12$, and the amplitude of the wave is larger than the amplitude of the incident wave. In addition, the particle velocity transmission coefficient T_v increases as the wave impedance ratio increases.

Figure 6.4b shows the stress variations as the wave propagates through layered composite rock mass with wave impedance ratios of $n = 0.65$, 1.00 and 1.54. It can be seen from Fig. 6.4b that the stress transmission coefficient decreases as the wave impedance ratio increases. When wave impedance ratios are $n = 0.65$, 1.00 and 1.54, the stress transmission coefficient T_σ respectively are 1.12, 0.76 and 0.47. In addition, the amplitude of the transmitted wave is larger than the amplitude of the incident wave when the wave impedance ratio is $n = 0.65$.

6.1.3 *Effect of Incident Wave Frequency, Joint Stiffness, Joint Number and Wave Impedance Ratio on Particle Velocity and Stress Transmission Coefficient*

(1) Effect of incident wave frequency on particle velocity and stress transmission coefficient

Figure 6.5 shows the effect of the incident wave frequency on wave transmission through three groups of rock mass with different wave impedance ratios. The joint stiffness of all joints is $k_n = 3.5$ GPa/m and the joint spacing is $S = 4.0$ m.

Figure 6.5a shows the variations in the particle velocity transmission coefficients T_v of the total transmitted wave and directly transmitted wave with respect to frequency f_0 with wave impedance ratios of $n = 0.65$, 1.00 and 1.54. It is seen from Fig. 6.5a that particle velocity transmission coefficients of the total transmitted wave and directly transmitted wave decrease as the frequency increases whether it is total transmission or direct transmission. In Fig. 6.5a, when the frequency is $f_0 < 10$ Hz, the particle velocity transmission coefficients of the total transmitted wave and the directly transmitted wave remain stable and close, and both approach those of wave propagation through the welded interface. When the frequency f_0 increases from 10^1 to 10^2 Hz, the particle velocity transmission coefficient of the total transmitted wave firstly decreases and then increases, while the particle velocity transmission coefficient of the directly transmitted wave decreases slowly. The former is gradually larger than the latter and the difference between them gradually decreases until it reaches zero and then gradually increases. When the frequency f_0 increases from 10^2 to 10^4 Hz, the particle velocity transmission coefficients of the total transmitted wave and the directly transmitted wave both decrease rapidly. The former is always larger than the latter and the difference between them is getting smaller. When the

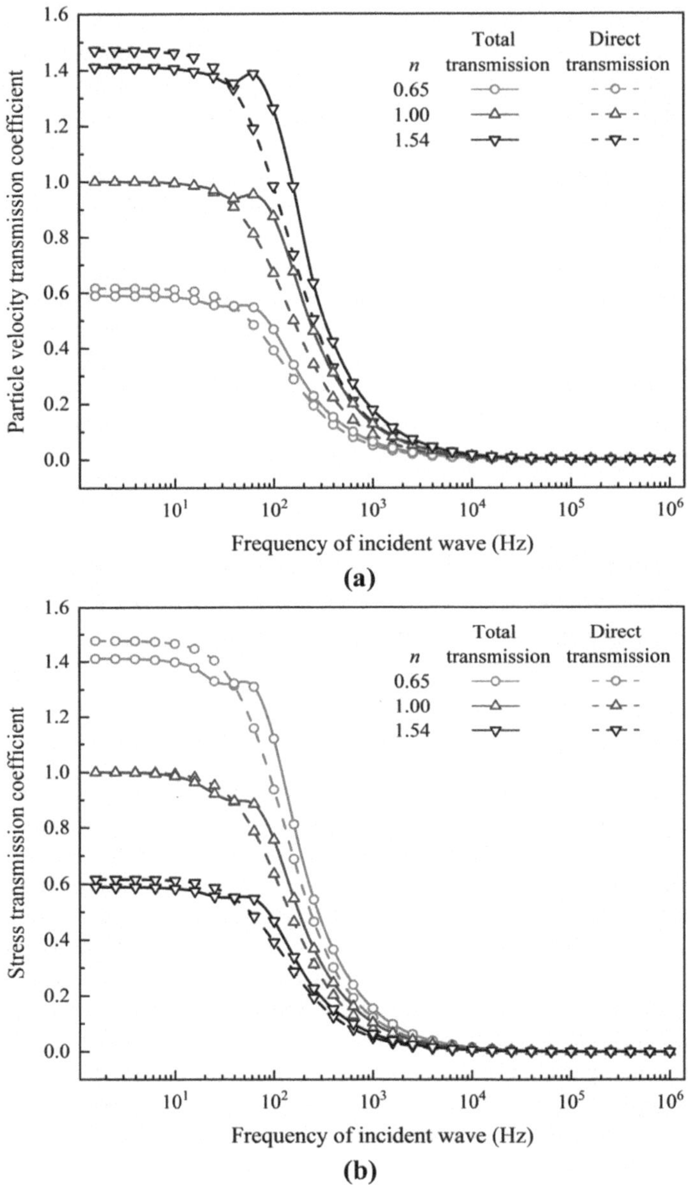

Fig. 6.5 Effect of incident wave frequency on wave transmission. **a** Particle velocity transmission coefficient. **b** Stress transmission coefficient

frequency is $f_0 > 10^4$ Hz, the particle velocity transmission coefficients of the total transmitted wave and the directly transmitted wave both remain stable and approach zero. The difference between them gradually approaches zero. In this case, the effect of indirect transmission caused by multiple reflections can be ignored. In addition, for a fixed incident wave frequency, whether it is total transmission or direct transmission, the particle velocity transmission coefficients increase as the wave impedance ratio increases.

Figure 6.5b shows the variations in the stress transmission coefficients T_σ of the total transmitted wave and directly transmitted wave with respect to frequency f_0 with wave impedance ratios of $n = 0.65$, 1.00 and 1.54. It can be observed from Fig. 6.5b that the stress transmission coefficient of the total transmitted wave firstly decreases from a constant value and next increases, then rapidly decreases until it approaches zero with the increase of frequency. The stress transmission coefficient of the directly transmitted wave rapidly decreases from a constant to zero. On the whole, the difference between the total transmitted wave and the directly transmitted wave decreases as the frequency increases. In other words, the effect of the indirectly transmitted wave induced by multiple reflections gradually decreases. In addition, whether it is the total transmitted wave or the directly transmitted wave, the stress transmission coefficients decrease with increasing wave impedance ratio for the fixed incident wave frequency.

(2) Effect of joint stiffness on particle velocity and stress transmission coefficient

Figure 6.6 shows the effect of joint stiffness k_n on wave transmission through three groups of rock mass with different wave impedance ratios. The incident wave frequency is $f_0 = 100.0$ Hz and the joint spacing is $S = 4.0$ m.

Figure 6.6a shows the variations in the particle velocity transmission coefficients T_v of the total transmitted wave and the directly transmitted wave with respect to joint stiffness k_n with wave impedance ratios of $n = 0.65$, 1.00 and 1.54. It has been proved from Fig. 6.6a that the particle velocity transmission coefficients of the total transmitted wave and the directly transmitted wave increase as the joint stiffness increases. At the same time, their rates of increase decrease. Therefore, when the joint stiffness is sufficiently large, the particle velocity transmission coefficients of the total transmitted wave and the directly transmitted wave are close to constant and equal. The interface of joints approaches the welded interface in this case. Figure 6.6a also shows the particle velocity transmission coefficient of the total transmitted wave is gradually larger than the particle velocity transmission coefficient of the direct transmission wave. The difference between them firstly increases and then decreases as the joint stiffness increases. The effect of the indirectly transmitted wave induced by multiple reflections firstly increases and then decreases. In addition, the particle velocity transmission coefficients of the total transmitted wave and the directly transmitted wave increase with increasing wave impedance ratio for the fixed joint stiffness.

Figure 6.6b shows the variations in the stress transmission coefficients T_σ of the total transmitted wave and the directly transmitted wave with respect to joint stiffness k_n with wave impedance ratios of $n = 0.65$, 1.00 and 1.54. Similar to the

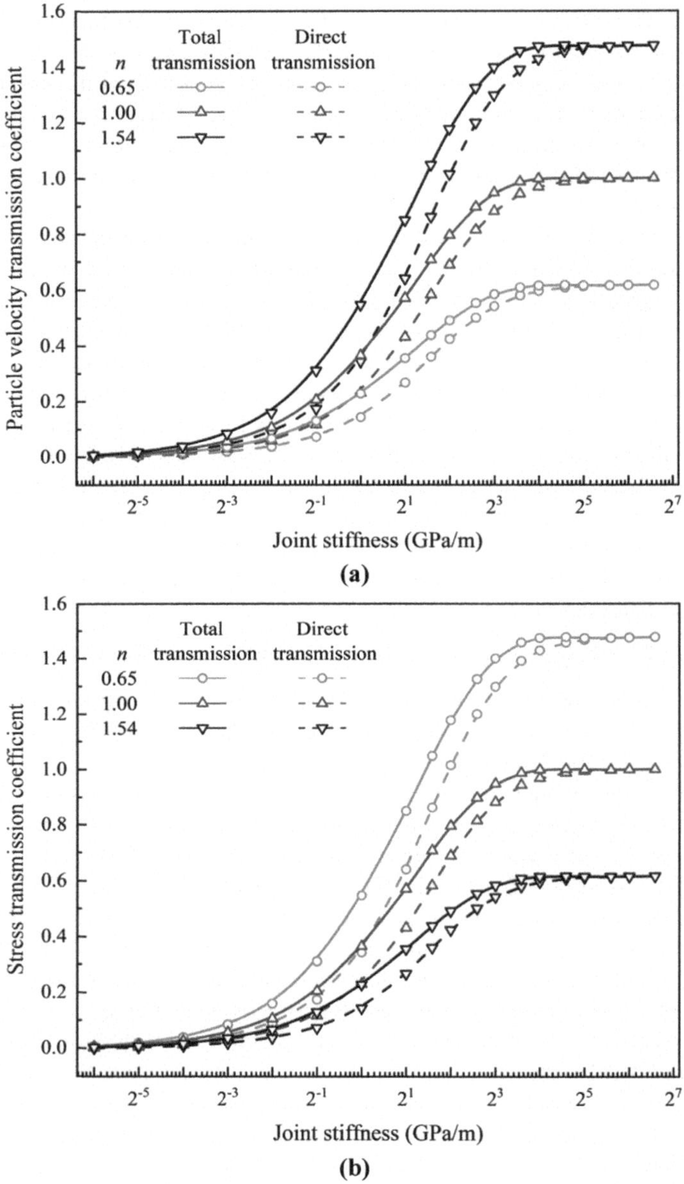

Fig. 6.6 Effect of joint stiffness on wave transmission. **a** Particle velocity transmission coefficient. **b** Stress transmission coefficient

variation of particle velocity transmission coefficient, whether it is total transmission or direct transmission, the stress transmission coefficients gradually increase to a constant as the joint stiffness increases. The stress transmission coefficient of the total transmitted wave is always larger than that of the direct transmission wave. The effect of the indirectly transmitted wave induced by multiple reflections on the stress transmission coefficient firstly increases and then decreases. In addition, under the fixed joint stiffness, the stress transmission coefficients of the total transmitted wave and the directly transmitted wave decrease as the wave impedance ratio increases.

(3) Effect of joint number on particle velocity and stress transmission coefficient

The more the joint number, the more complex the effect of indirectly transmitted waves induced by multiple reflection [1]. Figure 6.7 shows the effect of the joint number on the wave transmission through three groups of rock mass with different wave impedance ratios. The incident wave frequency is $f_0 = 100.0$ Hz. The joint stiffness of all joints is $k_n = 3.5$ GPa/m and the joint spacing is $S = 4.0$ m.

Figure 6.7a shows the variations in the particle velocity transmission coefficients T_v of the total transmitted wave and directly transmitted wave with respect to joint number with wave impedance ratios of $n = 0.65$, 1.00 and 1.54. It can be seen from Fig. 6.7a that when the joint number is 1, the particle velocity transmission coefficients of the total transmitted wave and the directly transmitted wave are equal. With the increase of the joint number, the particle velocity transmission coefficient of the directly transmitted wave decreases greatly. The particle transmission coefficient of the total transmitted wave has different trends with the variation of the wave impedance ratio. When the wave impedance ratio is 0.65, the particle velocity transmission coefficient of the total transmitted wave greatly decreases. When the wave impedance ratio is 1.00, the particle velocity transmission coefficient of the total transmitted wave decreases slightly. When the wave impedance ratio is 1.54, the particle velocity transmission coefficient of the total transmitted wave increases. Figure 6.7a also shows that the particle velocity transmission coefficient of the total transmitted wave is gradually larger than that of the directly transmitted wave as the joint number increases. The difference between them keeps increasing. In addition, for the fixed joint number, as the wave impedance ratio increases, the particle velocity transmission coefficients of total transmission and direct transmission increase. The former is gradually larger than the latter and the difference between them maintains increasing.

Figure 6.7b shows the variations in the stress transmission coefficients T_σ of the total transmitted wave and directly transmitted wave with respect to joint number with wave impedance ratios of $n = 0.65$, 1.00 and 1.54. In Fig. 6.7b, when the wave impedance ratio is $n = 0.65$, the stress transmission coefficient of the total transmitted wave gradually increases as the joint number increases. When the wave impedance ratio is $n = 1.00$, the stress transmission coefficient of the total transmitted wave drops slightly as the joint number increases. When the wave impedance ratio is $n = 1.54$, the stress transmission coefficient of the total transmitted wave drastically decreases as the joint number increases. However, for wave impedance ratios $n =$ 0.65, 1.00 and 1.54, the stress transmission coefficient of the directly transmitted

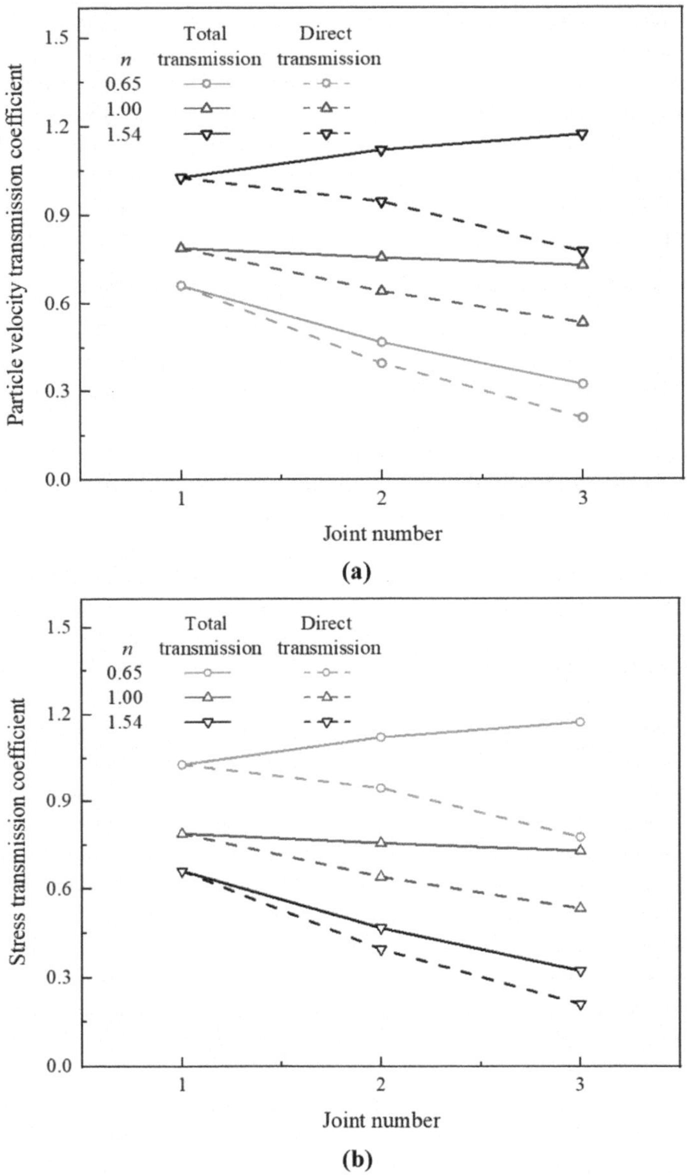

Fig. 6.7 Effect of joint number on wave transmission. **a** Particle velocity transmission coefficient. **b** Stress transmission coefficient

wave decreases with the increasing joint number. In addition, for the fixed joint number, as the wave impedance ratio increases, the particle velocity transmission coefficients of total transmission and direct transmission decrease. The former is gradually larger than the latter and the difference between them decreasing.

(4) Effect of wave impedance ratio on particle velocity and stress transmission coefficient

Figure 6.8 shows the effect of wave impedance ratio on wave transmission under conditions of different incident wave frequencies. The joint stiffness of all joints is $k_n = 3.5$ GPa/m and the joint spacing is $S = 4.0$ m.

Figure 6.8a shows the variations in the particle velocity transmission coefficients T_v of total transmission and direct transmission with respect to the wave impedance ratio n when the incident wave frequencies are 100, 250, 500, 1000 and 5000 Hz. It is revealed from Fig. 6.8a that the particle velocity transmission coefficients of the total transmitted wave and the directly transmitted wave increase as the wave impedance ratio increases under different frequency conditions. The former is always larger than the latter and the difference between them increases continuously. It can be seen that the multiple-reflection-induced indirect transmission has an increasing effect on the particle velocity transmission coefficient. Figure 6.8a also shows the particle velocity transmission coefficient must be smaller than 1.0 when the wave impedance ratio is smaller than 1.0. The particle velocity transmission coefficient may be larger than 1.0 when the wave impedance ratio is larger than 1.0. In Fig. 6.8a, for a fixed wave impedance ratio, the particle velocity transmission coefficients of the total transmitted wave and the directly transmitted wave decrease with the increase of the frequency. The particle velocity transmission coefficients of them finally are close to zero when the frequency is sufficiently large. The difference of particle velocity transmission coefficients between them gradually decreases to zero as the frequency increases, and these results are in accordance with those obtained in Fig. 6.5a.

Figure 6.8b shows the variations of the stress transmission coefficients T_σ of total transmission and direct transmission with respect to the wave impedance ratio n when the incident wave frequencies are 100, 250, 500, 1000 and 5000 Hz. It can be observed from Fig. 6.8b that the stress transmission coefficients of the total transmitted wave and the direct transmitted wave decrease as the wave impedance ratio increases. The former is always larger than the latter and the difference between them keeps decreasing. The effect of multiple-reflection-induced indirect transmission on the stress transmission coefficient decreases. Figure 6.8b also shows the stress transmission coefficient may be larger than 1.0 when the wave impedance ratio is smaller than 1.0. The stress transmission coefficient must be smaller than 1.0 when the wave impedance ratio is larger than 1.0. Moreover, for a fixed wave impedance ratio, the stress transmission coefficients of the total transmitted wave and the directly transmitted wave gradually decrease toward zero as the frequency increases. The former is larger than the latter and the difference between them gradually approaches zero, and the results are in accordance with those obtained in Fig. 6.5b.

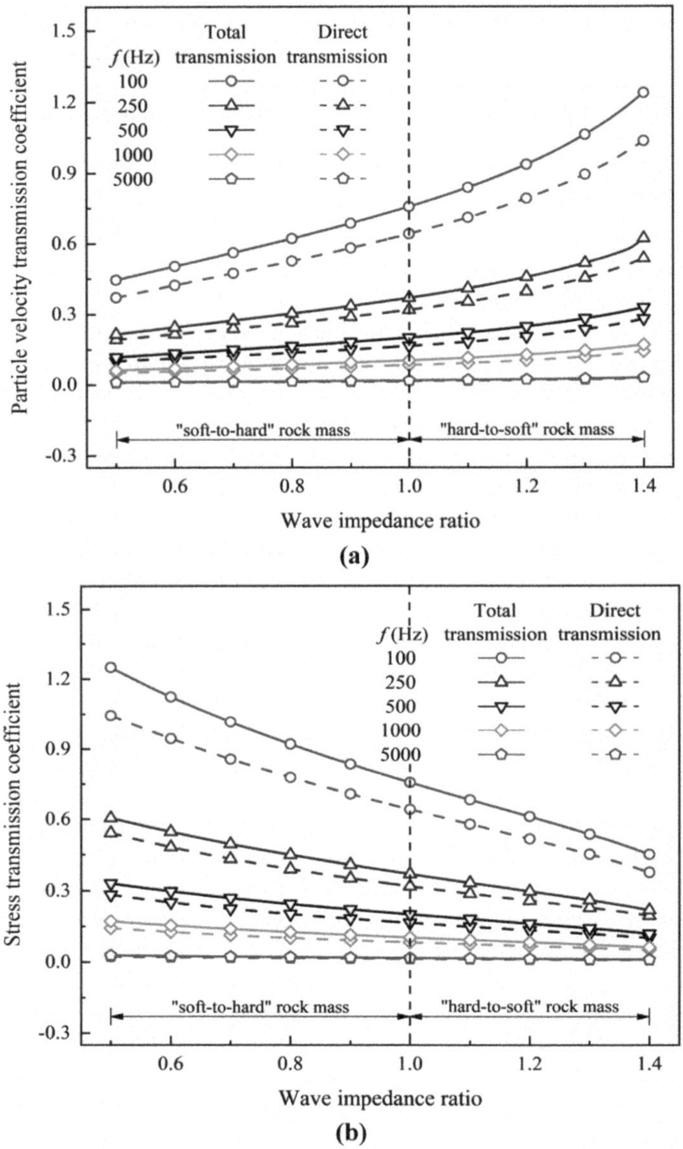

Fig. 6.8 Effect of wave impedance ratio on wave transmission. **a** Particle velocity transmission coefficient. **b** Stress transmission coefficient

6.2 The Effect of Joint Spacing on Transmission Characteristics of Stress Wave Through Layered Composite Rock Mass

In order to more meticulously discuss the effect of multiple reflections on the wave propagation characteristics, the wave propagation through rock mass with joint spacing S and nondimensional joint spacing ξ are distinguished. The nondimensional joint spacing ξ was usually defined as the ratio of the joint spacing and the wavelength.

Figure 6.9 shows the representative elementary volumes of layered rock mass. In all representative elementary volumes, liner joints are parallelly located. Figure 6.9a shows the representative elementary volume of layered homogeneous rock mass with joint spacing S. In Fig. 6.9a, the mediums have the same wave impedance and the joints have the same joint stiffness [2]. According to traditional research, the joint spacing S is defined as $\xi \cdot \lambda_1$, where λ_1 is the wavelength that the incident wave propagates through rock *I*. The lengths of rock *I* and rock *II* are equal to the joint spacing.

Different from the layered homogeneous rock mass, three types of mediums have different wave impedances in the layered composite rock mass, as shown in Fig. 6.9b and c. Figure 6.9b shows the representative elementary volume of layered composite rock mass with joint spacing S. Both the length of rocks and the joint spacing are S. Figure 6.9c shows the representative elementary volume of layered composite rock mass with nondimensional joint spacing ξ. Due to the wavelength of the wave propagating through different types of rocks varies, the joint spacing is $\xi \cdot \lambda_2$, where λ_2 is the wavelength that the incident wave propagates through rock *II*. The lengths

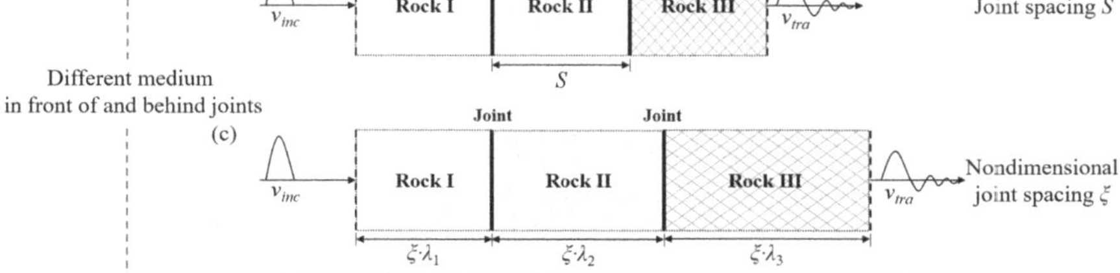

Fig. 6.9 Representative elementary volumes. **a** Layered homogeneous rock mass with joint spacing S. **b** Layered composite rock mass with joint spacing S. **c** Layered composite rock mass with nondimensional joint spacing ξ

of rock *I* and rock *III* are $\xi \cdot \lambda_1$ and $\xi \cdot \lambda_3$, where λ_3 is the wavelength that the incident wave propagates through rock *III*.

6.2.1 Modified Method of Characteristics for Layered Composite Rock Mass with Different Joint Spacings

The displacement discontinuity model is commonly used to determine the boundary conditions of joints in jointed rock mass. The model assumes the stress field is continuous when the wave propagates the joint while the displacement field is discontinuous on account of the joint deformation. Therefore, the stress relationship and displacement relationship in front of and behind joints in Fig. 6.9 are

$$\sigma_f = \sigma_b = \sigma \tag{6.26}$$

and

$$u_f - u_b = \frac{\sigma}{k_n} \tag{6.27}$$

where σ is the stress, u is the displacement and k_n is the joint stiffness. The superscripts "*f*" and "*b*" denote the parameters in front of and behind the joint, respectively.

Combined with the displacement discontinuity model, the method of characteristics is invariably applied to investigate wave propagation through jointed rock mass. The methods of characteristics for the layered homogeneous rock mass and the layered composite rock mass are shown in Fig. 6.10. Figure 6.10a shows the method of characteristics for layered homogeneous rock mass with the same rock in front of and behind joints. The relationship between the particle velocity and the stress can be established as $zv(x, t) \pm \sigma(x, t) =$ constant along the left-running and the right-running characteristic lines, where z is the wave impedance of rock. However, for the layered composite rock mass with different mediums in front of and behind joints, the previous method of characteristics is no longer applicable. Therefore, the modified method of characteristics for wave propagation through layered composite rock mass with joint spacing S and nondimensional joint spacing ξ is proposed, as shown in Fig. 6.10b and c. It can be observed from Fig. 6.10b and c that the calculation lengths of the layered composite rock mass with joint spacing S and nondimensional joint spacing ξ are different. In the modified method of characteristics, the absolute value of the characteristic line slope increases with the decrease of wave impedance.

In Fig. 6.10c, solve the derivative of Eq. (6.2) with respect to time to obtain the relationship of particle velocities in front of and behind the joint ($x = x_1$) as

$$v_f(x_1, t) - v_b(x_1, t) = \frac{1}{k_{n1}} \frac{\partial \sigma(x_1, t)}{\partial t} \tag{6.28}$$

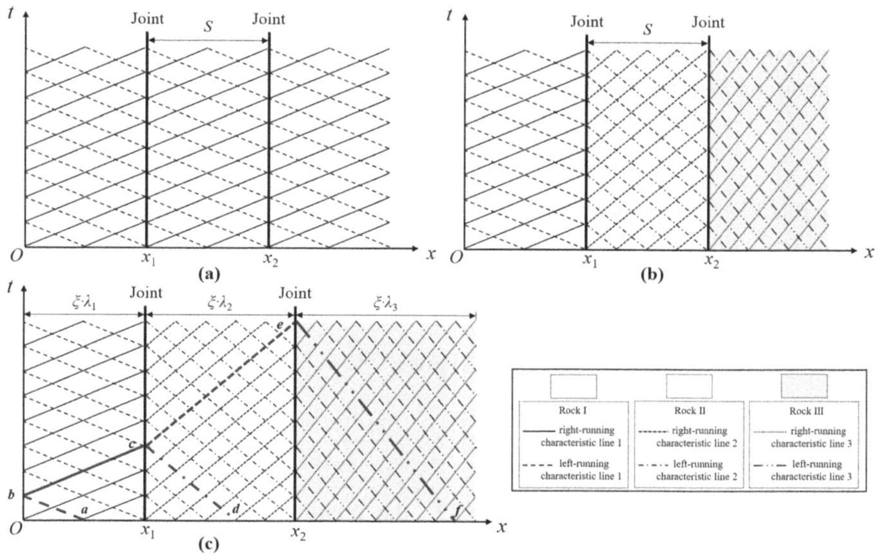

Fig. 6.10 Comparison of methods of characteristics. **a** Layered homogeneous rock mass with joint spacing S. **b** Layered composite rock mass with joint spacing S. **c** Layered composite rock mass with nondimensional joint spacing ξ

where $v_f(x_1, t)$ is the particle velocity in front of the joint ($x = x_1$), and $v_b(x_1, t)$ is the particle velocity behind the joint ($x = x_1$). $\sigma(x_1, t)$ is the stress at joint ($x = x_1$), and k_{n1} is the joint stiffness of joint ($x = x_1$).

Both of left-running characteristic lines and right-running characteristic lines in rock *I*, rock *II* and rock *III* are expressed as

$$\begin{cases} z_1 v \pm \sigma = \text{constant} \\ z_2 v \pm \sigma = \text{constant} \\ z_3 v \pm \sigma = \text{constant} \end{cases} \tag{6.29}$$

where z_1, z_2 and z_3 are wave impedances in rock *I*, rock *II* and rock *III*, respectively.

The rock mass is undisturbed when time is zero. The particle velocity and stress at arbitrary points on the x-axis are equal to zero. The intersection of the left-running characteristic line *ab* and right-running characteristic line *bc* is $(0,\ t - x_1/C_1)$. Combining Eq. (6.29), the relationship between the particle velocity and the stress along ab and *bc* can be expressed as

$$z_1 v_{\text{inc}}\,(0, t - x_1/C_1) - \sigma_{\text{inc}}\,(0, t - x_1/C_1) = 0 \tag{6.30}$$

and

$$z_1 v_{\text{inc}}\,(0, t - x_1/C_1) + \sigma_{\text{inc}}\,(0, t - x_1/C_1) = z_1 v_f\,(x_1, t) + \sigma\,(x_1, t) \tag{6.31}$$

where $v_{inc}(0, t - x_1/C_1)$ and $\sigma_{inc}(0, t - x_1/C_1)$ are the particle velocity of the incident wave and the stress of the incident wave.

Combining Eqs. (6.30) and (6.31) yields

$$z_1 v_f(x_1, t) + \sigma(x_1, t) = 2z_1 v_{\text{inc}}(0, t - x_1/C_1) \tag{6.32}$$

The relationship between the particle velocity and the stress along the left-running characteristic line *cd* is

$$z_2 v_b(x_1, t) - \sigma(x_1, t) = 0 \tag{6.33}$$

Substituting Eq. (6.33) into Eq. (6.28) gives

$$v_f(x_1, t) - v_b(x_1, t) = \frac{z_2 \partial v_f(x_1, t)}{k_{n1} \partial t} \tag{6.34}$$

Substituting Eq. (6.33) into Eq. (6.32), the relationship between the particle velocities is obtained as

$$z_1 v_f(x_1, t) + z_2 v_b(x_1, t) = 2z_1 v_{\text{inc}}(0, t - x_1/C_1) \tag{6.35}$$

Combining Eqs. (6.34) and (6.35), the particle velocity behind the joint ($x = x_1$) is obtained as

$$\frac{\partial v_b(x_1, t)}{\partial t} = \frac{2k_{n1}}{z_2} v_{inc}(0, t - x_1/C_1) - k_{n1}(1/z_1 + 1/z_2) v_b(x_1, t) \tag{6.36}$$

The derivative $\frac{\partial v_b(x_1, t)}{\partial t}$ can be replaced by $\frac{v_b(x_1, t_{j+1}) - v_f(x_1, t_j)}{\Delta t}$ when the time interval is small enough. Substituting it into Eq. (6.36), the particle velocity iterative expression of the transmitted wave is obtained as

$$v_{\text{tra}} = v_b(x_1, t_{j+1}) = \frac{2k_{n1}\Delta t}{z_2} v_{\text{inc}}(0, t - x_1/C_1) + [1 - k_{n1}\Delta t(1/z_1 + 1/z_2)] v_b(x_1, t_j) \tag{6.37}$$

where v_{tra} is the particle velocity of the transmitted wave.

The transmitted wave v_{tra} at the joint ($x = x_1$) is obtained in iterative calculation by Eq. (6.12) under the condition that the joint stiffness, wave impedance, incident wave and initial velocity are determined. The reflected wave can be further solved by combining the particle velocity in front of the joint ($x = x_1$).

Simplifying Eq. (6.35), the particle velocity in front of the joint ($x = x_1$) can be expressed as

$$v_f(x_1, t) = 2v_{inc}(0, t - x_1/C_1) - v_{tra} \times z_2/z_1 \tag{6.38}$$

In addition, the particle velocity in front of the joint ($x = x_1$) can also be expressed by the sum of incident wave particle velocity and reflected wave particle velocity:

$$v_f(x_1, t) = v_{inc}(0, t - x_1/C_1) + v_{ref} \tag{6.39}$$

where v_{ref} is the particle velocity of the reflected wave.

Substituting Eq. (6.14) into Eq. (6.13), the particle velocity iterative expression of the reflected wave in front of the joint ($x = x_1$) yields

$$v_{ref} = v_{inc}(0, t - x_1/C_1) - v_{tra} \times z_2/z_1 \tag{6.40}$$

The reflected wave particle velocity v_{ref} in front of the joint ($x = x_1$) can be acquired according to Eq. (6.15) when the transmitted wave particle velocity v_{tra} is determined.

The transmitted wave behind the joint ($x = x_1$) is regarded as the incident wave in front of the joint ($x = x_2$). Therefore, the particle velocity of the reflected wave in front of the joint ($x = x_2$) and the particle velocity of the directly transmitted wave behind the joint ($x = x_2$) are obtained. The reflected wave in front of the joint ($x = x_2$) generates multiple reflections between joints. Multiple transmitted waves are formed behind the joint ($x = x_2$). The particle velocity of the transmitted wave caused by multiple reflections is obtained. Therefore, the total transmitted wave can be considered as the superposition of a directly transmitted wave and transmitted wave caused by multiple reflections. In the solution for the particle velocity of the transmitted wave in the layered composite rock mass, both the iteration of the particle velocity with time and the superposition of the directly transmitted wave and the transmitted wave caused by multiple reflections are achieved.

The wave impedance ratio is the ratio with respect to wave impedances of the mediums in front of and behind the joint, which is expressed as

$$n = \frac{z_f}{z_h} \tag{6.41}$$

The wave impedance ratio is 1.0 for layered homogeneous rock mass with the same medium in front of and behind joints. The wave impedance ratio is not equal to 1.0 for layered composite rock mass with different mediums in front of and behind joints. Ulteriorly, the wave impedance ratio smaller than 1.0 indicates the layered composite rock mass with gradually increasing wave impedance of the medium. The wave impedance ratio larger than 1.0 indicates the layered composite rock mass with gradually decreasing wave impedance of the medium.

Figure 6.11 shows the effect of joint spacing on transmitted waves in layered composite rock mass. In Fig. 6.11, the fifth transmitted wave shows that both the directly transmitted wave and the transmitted wave caused by multiple reflections constitute the total transmitted wave when the joint spacing is adequately large. The amplitude of the total transmitted wave is the peak of the directly transmitted wave.

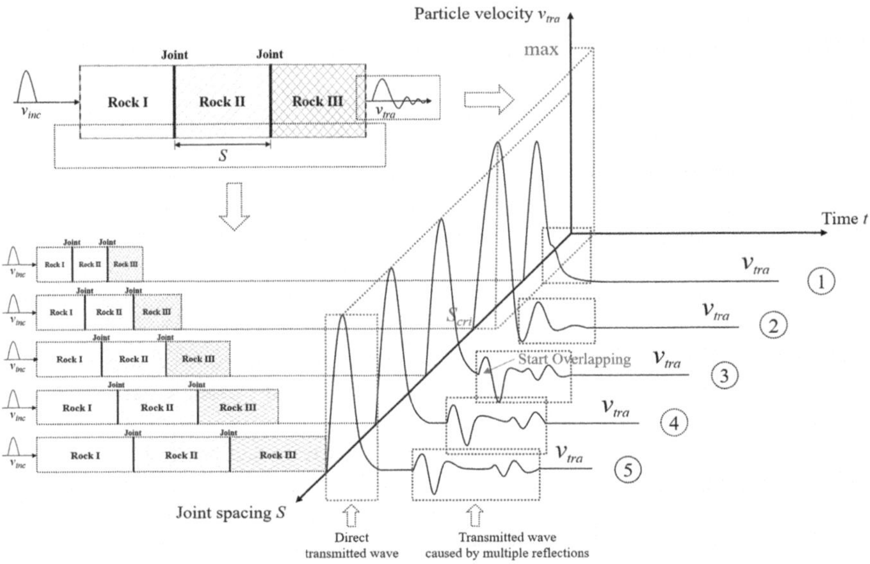

Fig. 6.11 Transmitted waves of layered composite rock mass under different joint spacings

Multiple transmissions and reflections cannot impact the amplitude of the total transmitted wave at this time. The transmitted wave caused by multiple reflections gradually catches up with the directly transmitted wave with the decrease of joint spacing. The peaks of the directly transmitted wave and transmitted wave caused by multiple reflections overlap as shown in the second transmitted wave. The joint spacing at this time is called the critical joint spacing S_{cri}. The amplitude of the total transmitted wave reaches maximum. Therefore, the amplitude of the transmitted wave for layered composite rock mass gradually decreases as the joint spacing decreases.

6.2.2 Wave Propagation Through Layered Composite Rock Mass with Joint Spacing S

Figure 6.12 shows the variations of the transmitted waves in the layered composite rock mass and the layered homogeneous rock mass under joint spacing of *S*. The wave impedance ratio of the layered composite rock mass is 0.8 while the wave impedance ratio of the layered homogeneous rock mass is 1.0. The joint spacing $S = 0.4\lambda$.

Figure 6.12a shows the total transmitted waves of layered composite rock mass and layered homogeneous rock mass. It can be indicated from Fig. 6.12a that the differences of total transmitted waves between layered composite rock mass and layered homogeneous rock mass are mainly presented in the onset time and the

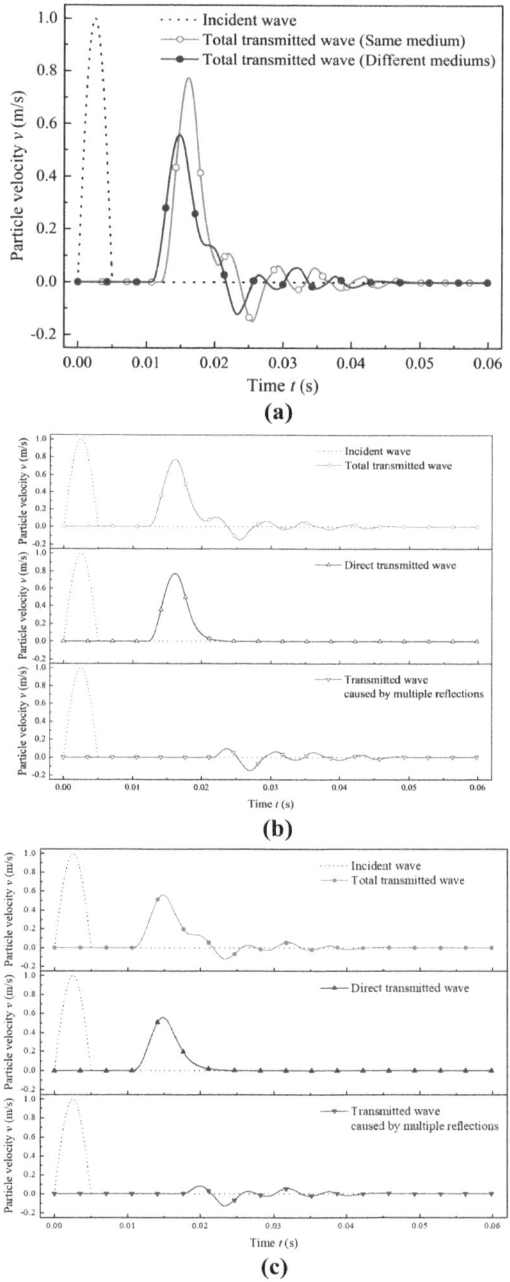

Fig. 6.12 Transmitted waves in different layered rock mass. **a** Total transmitted waves. **b** Composition of total transmitted wave in layered homogeneous rock mass. **c** Composition of total transmitted wave in layered composite rock mass

amplitude. In terms of the onset time, the onset time of the total transmitted waves in the layered composite rock mass is smaller than that of the layered homogeneous rock mass. The onset time of the total transmitted waves in the layered composite rock mass is 10.48 ms, while the onset time of the total transmitted waves in the layered homogeneous rock mass is 12.00 ms. In terms of amplitude, the amplitude of the total transmitted waves in layered composite rock mass is smaller than that in layered homogeneous rock mass. The amplitude of the total transmitted waves in layered composite rock mass is 0.56 m/s, while the amplitude of the total transmitted waves in layered homogeneous rock mass is 0.77 m/s. In addition, the variations of tail waveforms in the layered composite rock mass and the layered homogeneous rock mass show different trends.

In order to analyze the variations of the onset time, the amplitude and the tail waveform, the compositions of total transmitted waves in the layered composite rock mass and layered homogeneous rock mass are compared. Figure 6.12b shows the composition of the total transmitted waves in the layered homogeneous rock mass. In Fig. 6.12b, the onset time of the directly transmitted waves is 12.00 ms, while the onset time of the transmitted waves caused by multiple reflections is 20.00 ms. The onset times of the directly transmitted waves and the transmitted waves caused by multiple reflections are affected by the joint spacing. In addition, Fig. 6.12b also shows that the amplitude of the direct transmitted waves is 0.77 m/s and the first peak of the transmitted waves caused by multiple reflections is 0.10 m/s. The amplitudes of transmitted waves in the layered homogeneous rock mass are only affected by joint stiffness.

Figure 6.12c shows the compositions of the total transmitted waved in the layered composite rock mass. In Fig. 6.12c, the onset times of directly transmitted waves and transmitted waves caused by multiple reflections are 10.5 ms and 17.4 ms, respectively. It can be indicated from Fig. 6.12b and c that all onset times of transmitted waves in layered composite rock mass are smaller than those in layered homogeneous rock mass which leads to differences of tail waves. In layered composite rock mass, the onset times of directly transmitted waves and transmitted waves caused by multiple reflections are not only impacted by joint spacing, but also by wave impedance ratio. Figure 6.12c also shows that the amplitude of the direct transmitted waves is 0.56 m/s and the first peak of the transmitted waves caused by multiple reflections is 0.08 m/s. The amplitudes of transmitted waves in layered composite rock mass are affected by joint stiffness and wave impedance ratio.

Figure 6.13 shows the effect of joint spacing S on the particle velocity transmission coefficient under different wave impedance ratios. It can be seen from Fig. 6.13 that the critical of joint spacing S_{cri} and the threshold of joint spacing S_{thr} under different wave impedance ratios are determined. The particle velocity transmission coefficient curve is divided into the small spacing area ($S \leq S_{cri}$), the transition area ($S_{cri} < S < S_{thr}$) and the large spacing area ($S \geq S_{thr}$). In the small spacing area, the particle velocity transmission coefficient rapidly increases to the peak as the joint spacing increases. In the transition area, the particle velocity transmission coefficient gradually decreases to be constant as joint spacing increases. When joint spacing $S \leq S_{thr}$, the particle velocity transmission coefficient is affected by multiple reflections between parallel

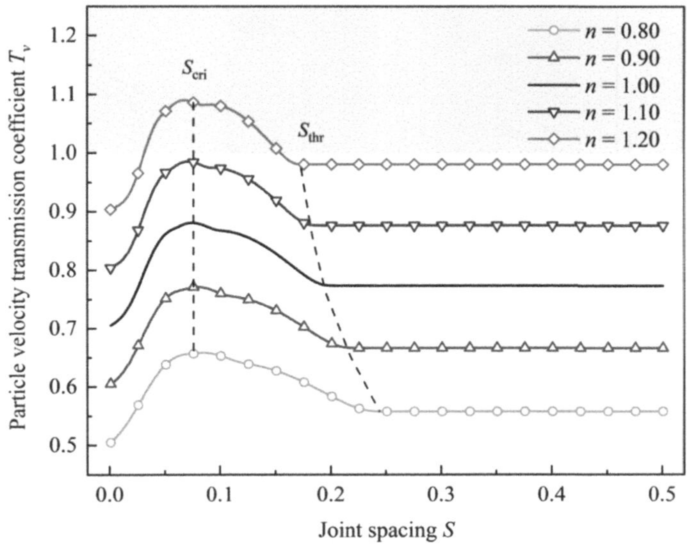

Fig. 6.13 Effect of joint spacing on particle velocity transmission coefficient

joints. In the large spacing area, the particle velocity transmission coefficient remains constant as joint spacing increases. It means that the particle velocity transmission coefficient at this area is not affected by multiple reflections between parallel joints. It is worth noting that the variation of the critical of joint spacing S_{cri} and threshold of joint spacing S_{thr} are shown by the dash lines in Fig. 6.13. The critical of joint spacing S_{cri} is almost invariant, while the threshold of joint spacing S_{thr} gradually decreases as the wave impedance ratio increases. Besides, the particle velocity transmission coefficient increases as the wave impedance ratio increases under the situation of fixed joint spacing. The particle velocity transmission coefficient is gradually larger than 1.0 as the wave impedance ratio increases larger than 1.0.

Figure 6.14 shows the effect of wave impedance ratio n on the particle velocity transmission coefficient under different joint spacings. It can be seen from Fig. 6.14 that the particle velocity transmission coefficient increases linearly with the increase of wave impedance ratio. The particle velocity transmission coefficient is always smaller than 1.0 when the wave impedance ratio is smaller than 1.0. The particle velocity transmission coefficient gradually becomes larger than 1.0 as the wave impedance ratio increases to larger than 1.0. It is worth noting that when the wave impedance ratio is 1.3, the particle velocity transmission coefficients of layered composite rock mass containing arbitrary joint spacings are larger than 1.0. In addition, the particle velocity transmission coefficient firstly increases and then decreases as the joint spacing increases under the situation of fixed wave impedance ratio. This is consistent with the conclusion shown in Fig. 6.13.

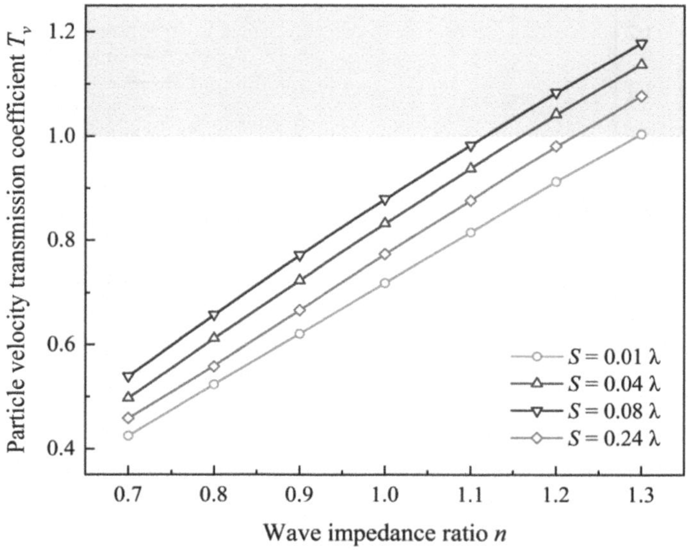

Fig. 6.14 Effect of wave impedance ratio on particle velocity transmission coefficient

6.2.3 Wave Propagation Through Layered Composite Rock Mass with Nondimensional Joint Spacing ξ

Figure 6.15 shows the variations of the transmitted waves in layered composite rock mass and layered homogeneous rock mass under nondimensional joint spacing ξ. Similarly, the wave impedance ratios of the layered composite rock mass and the layered homogeneous rock mass are 0.8 and 1.0, respectively. The nondimensional joint spacing $\xi = 0.4$.

Figure 6.15a shows the total transmitted waves of layered composite rock mass and layered homogeneous rock mass. In Fig. 6.15a, there is almost no difference in the onset time of the total transmitted wave between the layered composite rock mass and the layered homogeneous rock mass, but there is an obvious difference in the amplitude. With respect to the onset time, the onset time of the total transmitted wave in the layered composite rock mass is close to that in the layered homogeneous rock mass, which is 12.00 ms. With respect to the amplitude, the amplitude of the total transmitted wave in the layered composite rock mass is smaller than that in the layered homogeneous rock mass. The amplitude of the total transmitted wave in layered composite rock mass is 0.56 m/s, while the amplitude of the total transmitted wave in layered homogeneous rock mass is 0.77 m/s. Moreover, the tail waveforms of the layered homogeneous rock mass and the layered composite rock mass show similar trends.

By comparing the compositions of total transmitted waves in two types of rock mass, the reasons for the variation of the onset time, amplitude and tail waveform

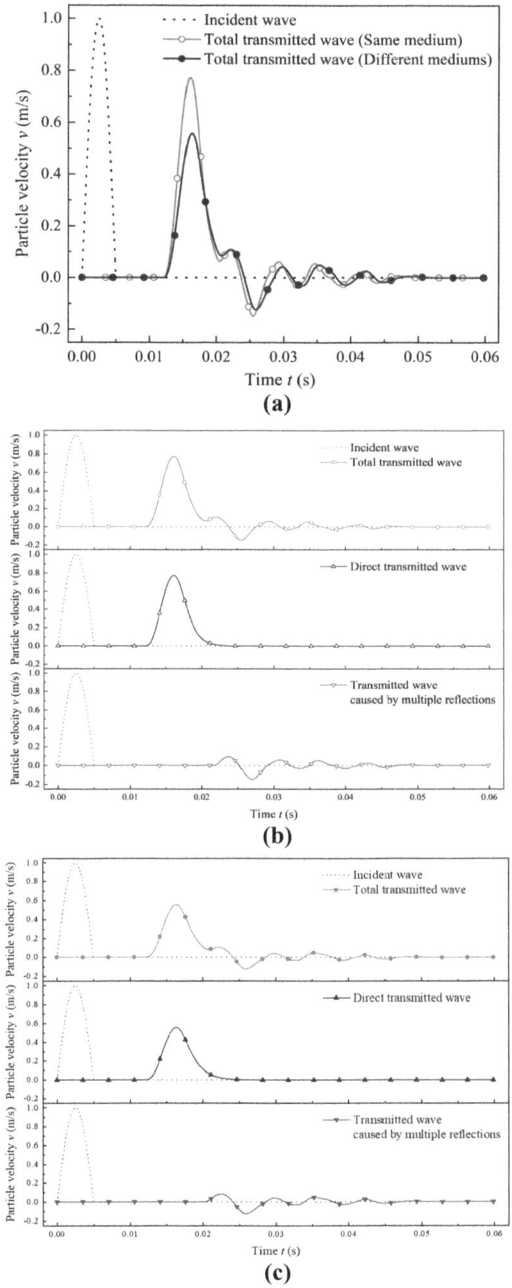

Fig. 6.15 Transmitted waves in different rock mass. **a** Total transmitted waves. **b** Composition of total transmitted wave in layered homogeneous rock mass. **c** Composition of total transmitted wave in layered composite rock mass

are further analyzed. Figure 6.15b shows the compositions of the total transmitted wave in the layered homogeneous rock mass with nondimensional joint spacing ξ. In Fig. 6.15b, the onset times of the directly transmitted wave and the transmitted wave caused by multiple reflections are 12.00 ms and 20.00 ms, respectively. In layered homogeneous rock mass, the onset times are only affected by joint spacing. Figure 6.15b also shows that the amplitude of the directly transmitted wave and the first peak of the transmitted wave caused by multiple reflections are 0.77 m/s and 0.10 m/s, respectively. The amplitudes of all transmitted waves in the layered homogeneous rock mass are only affected by joint stiffness.

Figure 6.15c shows the compositions of the transmitted wave in the layered composite rock mass. It can be indicated from Fig. 6.15c that the onset times of directly transmitted wave and transmitted wave caused by multiple reflections are the same as that in Fig. 6.15b. In the layered composite rock mass with nondimensional joint spacing, the onset time is not only affected by nondimensional joint spacing, but also by wave impedance ratio. The multiple transmitted waves in the layered homogeneous rock mass and multiple transmitted waves in the layered composite rock mass superimpose at the same time is caused by the nondimensional joint spacing.

Therefore, the various trends of the tail waves of total transmitted waves are similar. In addition, Fig. 6.15c also shows that the amplitude of direct transmitted wave is 0.56 m/s and the first peak of transmitted wave caused by multiple reflections is 0.08 m/s. They are smaller than the amplitude of the transmitted wave in the layered homogeneous rock mass with nondimensional joint spacing. The amplitude of the transmitted wave in layered composite rock mass is affected by the joint stiffness and the wave impedance ratio.

Figure 6.16 shows the effect of nondimensional joint spacing ξ on the particle velocity transmission coefficient under different wave impedance ratios. It can be observed from Fig. 6.16 that the curves of particle velocity transmission coefficient under different wave impedance ratios are divided into three areas by the critical of nondimensional joint spacing ξ_{cri} and the threshold of nondimensional joint spacing ξ_{thr}. Three areas are the small spacing areas ($\xi \leq \xi_{cri}$), the transition area ($\xi_{cri} < \xi < \xi_{thr}$) and the large spacing area ($\xi \geq \xi_{thr}$), respectively. As the nondimensional joint spacing increases, the particle velocity transmission coefficient increases to the peak in the small spacing area ($\xi \leq \xi_{cri}$) while that decreases to be constant in the transition area ($\xi_{cri} < \xi < \xi_{thr}$). Moreover, the particle velocity transmission coefficient in the large spacing area ($\xi \geq \xi_{thr}$) remains constant. In nondimensional joint spacing $\xi < \xi_{thr}$, the particle velocity transmission coefficient is affected by multiple reflections between joints. In other nondimensional joint spacing, the particle velocity transmission coefficient is not affected by multiple reflections between joints. In addition, it can be predicted from the two dash lines in Fig. 6.16 that the critical of nondimensional joint spacing ξ_{cri} and threshold of nondimensional joint spacing ξ_{thr} are nearly unaffected by wave impendence ratio. This is different from the conclusion from Fig. 6.13. For a given nondimensional joint spacing ξ, the particle velocity transmission coefficient increases to larger than 1.0 as the wave impedance ratio increases.

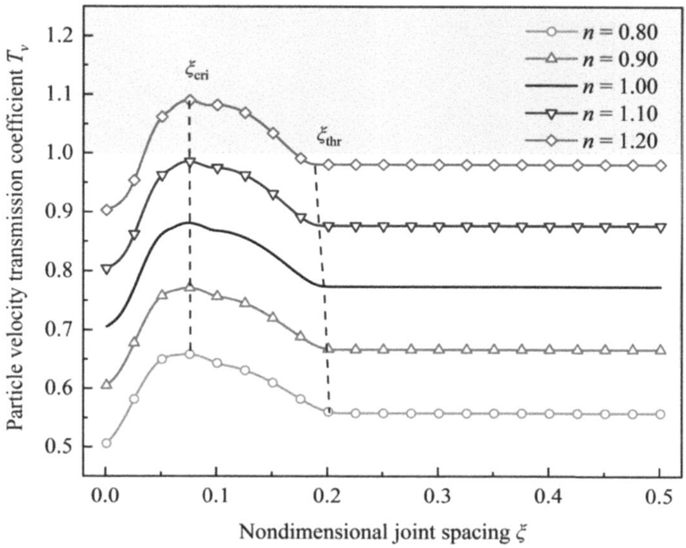

Fig. 6.16 Effect of nondimensional joint spacing on particle velocity transmission coefficient

Figure 6.17 shows the effect of wave impedance ratio n on the particle velocity transmission coefficient under different nondimensional joint spacings. It can be indicated from Fig. 6.17 that the particle velocity transmission coefficient increases with the increase of wave impedance ratio. The particle velocity transmission coefficient is always smaller than 1.0 when the wave impedance ratio is smaller than 1.0. The particle velocity transmission coefficient gradually becomes larger than 1.0 when the wave impedance ratio increases to larger than 1.0. It is noteworthy that when the wave impedance ratio of rock mass is $n = 1.3$, the particle velocity transmission coefficient of layered composite rock mass with arbitrary nondimensional joint spacings is larger than 1.0. This result is similar to that in Fig. 6.14. In addition, for a fixed wave impedance ratio, the particle velocity transmission coefficient firstly increases and then decreases as nondimensional joint spacing increases. This result is consistent with that shown in Fig. 6.16.

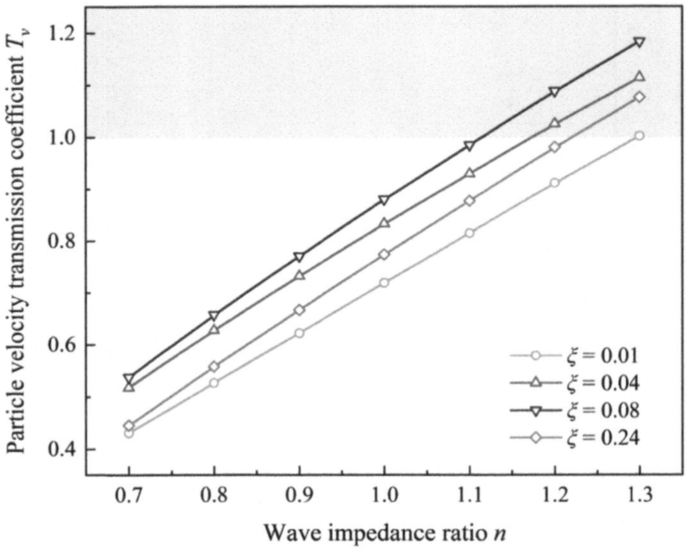

Fig. 6.17 Effect of wave impedance ratio on particle velocity transmission coefficient

References

1. Miller RK. An approximate method of analysis of the transmission of elastic waves through a frictional boundary[J]. J Appl Mech-T Asme. 1977;44(4):652–6.
2. Zhao J, Cai JG. Transmission of elastic p-waves across single fractures with a nonlinear normal deformational behavior[J]. Rock Mech Rock Eng. 2001;34(1):3–22.

Chapter 7
Three-Characteristic Method for the Wave Propagation Through Micro-Defected Rock Mass

7.1 Equivalent Viscoelastic Model

The attenuation and dispersion of stress wave propagation through micro-defected rock mass are similar to those of stress wave propagation through the viscoelastic medium. Therefore, a micro-defected rock mass is equivalently considered to be a viscoelastic medium to investigate stress wave propagation through a micro-defected rock mass. The equivalent viscoelastic behavior of the rock mass is obtained by pendulum impact experiments, and a modified three-element model is proposed to describe stress wave attenuation and dispersion.

7.1.1 Pendulum Impact Test

Figure 7.1 shows the granite bars used in this pendulum impact test. Some of the specimens are shown in Fig. 7.1a. To satisfy the one-dimensional wave propagation requirement and minimize the dispersion caused by lateral inertia, granite specimens are cylindrical with a cross-section of 4.5 cm and a length of 120.0 cm. The diameter of the cross-section varied within the range of $\pm$ 0.5 mm along the radial direction. The two ends of the granite bars are polished by a water-cooled diamond grinding wheel to ensure flatness, as shown in Fig. 7.1b. Two free ends are parallel to each other and perpendicular to the axis of the granite bar, as shown in Fig. 7.1c.

Figure 7.2a shows the pendulum impact setup used to study the equivalent viscoelastic behavior of granite. Figure 7.2b shows the steel pendulum hammer used for impact. A stress wave is generated in the cylindrical granite bar under the impact of the steel pendulum hammer. The amplitude of the stress wave could be adjusted by changing the swinging angle of the pendulum hammer. The same pendulum angle is used for the pendulum impact test on granite bars after different temperature treatments. The strain gauge resistance is 120 Ω, and the sensitivity coefficient is 2.0 $\pm$

L. Fan et al., *The Method of Characteristics for Stress Wave Propagation in the Rock Mass*, https://doi.org/10.1007/978-981-97-1440-7_7

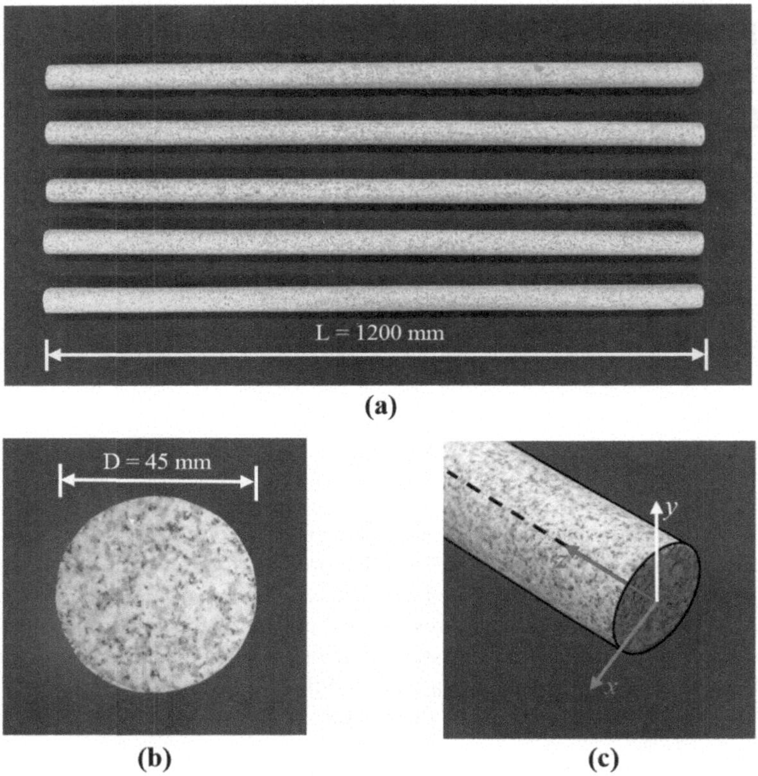

Fig. 7.1 Granite bars for dynamic test. **a** Geometric dimension. **b** Cross-section. **c** End face of specimen

1%. The longitudinal strain wave is measured by pasting four strain gauges in the middle of the granite bar, as shown in Fig. 7.2c. The longitudinal strain is measured by the full bridge method. Two strain gauges are symmetrically pasted, which are parallel to the axis to reduce the error caused by the angle between the strain gauge and the axis. The remaining two strain gauges are symmetrically pasted, which are perpendicular to the axis to reduce the error caused by radial strain. The longitudinal strain history is recorded at a time rate of $1 \times 10^7\ s^{-1}$, which could provide sufficient data points to perform a discrete Fourier transform (DFT). A pair of rollers is designed to ensure that the granite specimen can move freely under a stress wave.

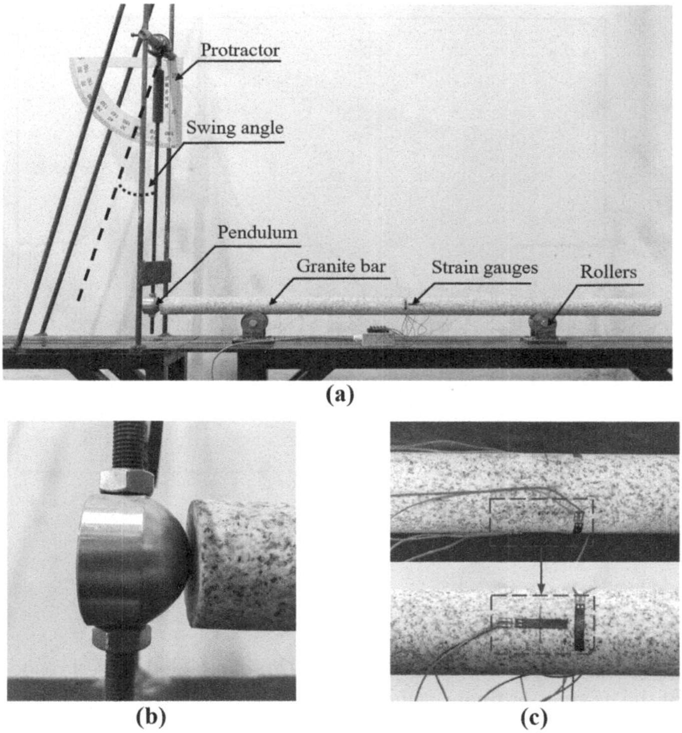

Fig. 7.2 Pendulum impact setup for dynamic test. **a** Experimental setup. **b** Pendulum. **c** Strain gauges

7.1.2 Strain History in Granite

Figure 7.3 shows the strain wave history of granite in the pendulum impact test. The strain wave history can be identified as $\varepsilon_1(t)$ and $\varepsilon_2(t)$. $\varepsilon_1(t)$ is the first measured strain pulse in the increasing x direction produced by the impact of the pendulum. $\varepsilon_2(t)$ is the second measured strain pulse in the decreasing x direction, which is the reflection of the strain pulse $\varepsilon_1(t)$ on the free surface. Obvious wave attenuation can also be observed in Fig. 7.3. Figure 7.3 shows that the amplitude of $\varepsilon_2(t)$ is 7.07×10^{-5}, which is smaller by 5.80% than the amplitude of $\varepsilon_1(t)$.

7.1.3 Propagation Coefficients of Granite

According to the stress wave propagation theory, the attenuation coefficient $\alpha(\omega)$ and wave number $k(\omega)$ can be obtained from the measured strain pulses of $\varepsilon_1(t)$ and

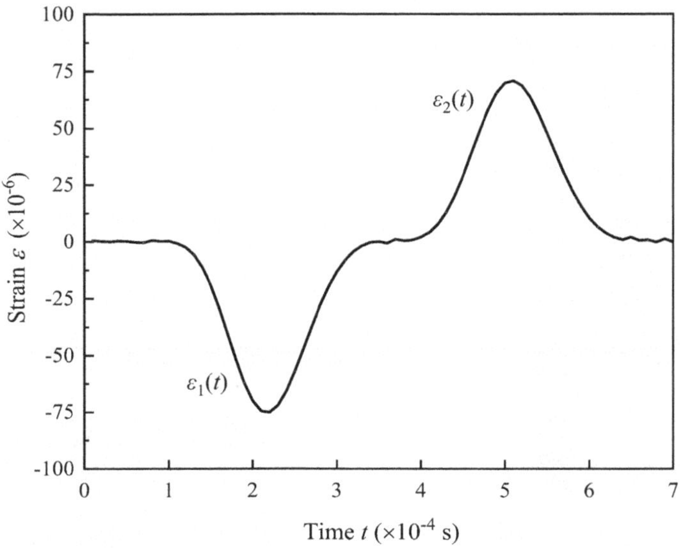

Fig. 7.3 Strain history in granite

$\varepsilon_2(t)$ by the Fourier transform. The $\alpha(\omega)$ and $k(\omega)$ can be expressed respectively as equations of the axial coordinate x and angular frequency ω as

$$\alpha(\omega) = -Re\left[ln\left(\frac{F|\varepsilon_2|}{F|\varepsilon_1|}\right)\Big/ l\right] \tag{7.1}$$

and

$$k(\omega) = -Im\left[ln\left(\frac{F|\varepsilon_2|}{F|\varepsilon_1|}\right)\Big/ l\right] \tag{7.2}$$

where ω is the angular frequency of a harmonic component wave after Fourier transformation, ε_1 is the first measured strain pulse in the increasing x direction produced by the impact of the pendulum, ε_2 is the second measured strain pulse in the decreasing x direction, which is the reflection of the strain pulse ε_1 on the free surface, l is the propagation distance between ε_1 and ε_2, $\mathbb{R}$is the Fourier transform, and Re and Im are the real and imaginary parts of the complex expression, respectively.

Figure 7.4 shows the relationship between the attenuation coefficient of granite and the frequency of the harmonic component wave. It can be observed from Fig. 7.4 that the attenuation coefficient increases slowly when the frequency of the harmonic component wave is in the small range (e.g., $\omega < 1000 \times 2\pi$ Hz); then it rapidly increases as the frequency of the harmonic component wave further increases.

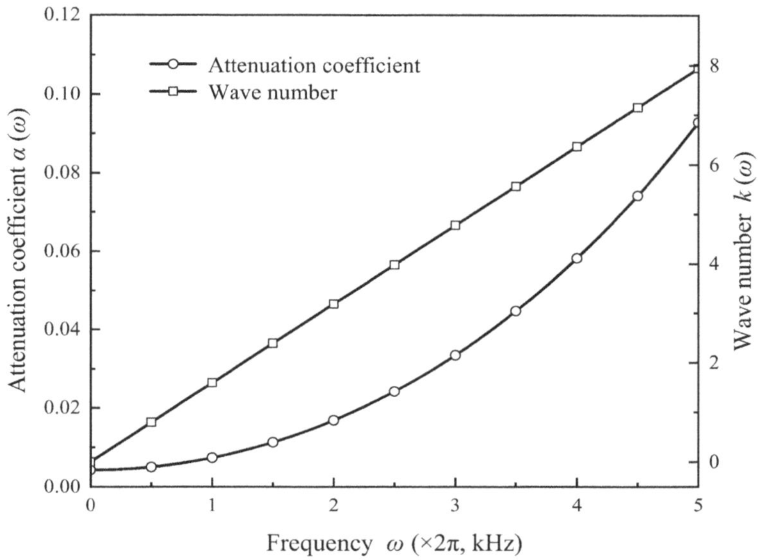

Fig. 7.4 Attenuation coefficient and wave number of stress wave in granite

Figure 7.4 shows the relationship between the wave number of granite and the frequency of the harmonic component wave. When the frequency of the harmonic component approaches zero, the wave number of granite approaches zero. It can be observed that the wave number increases approximately linearly as the frequency of the harmonic component wave increases.

7.1.4 Storage Modulus and Loss Modulus of Granite

Based on the attenuation coefficient and wave number determined respectively by Eqs. (7.1) and (7.2), the storage modulus and loss modulus of granite can be obtained respectively as

$$E'(\omega) = \frac{\rho\omega^2\left(k^2 - \alpha^2\right)}{\left(k^2 + \alpha^2\right)^2} \tag{7.3}$$

and

$$E''(\omega) = \frac{2\rho\omega^2 k\alpha}{\left(k^2 + \alpha^2\right)^2} \tag{7.4}$$

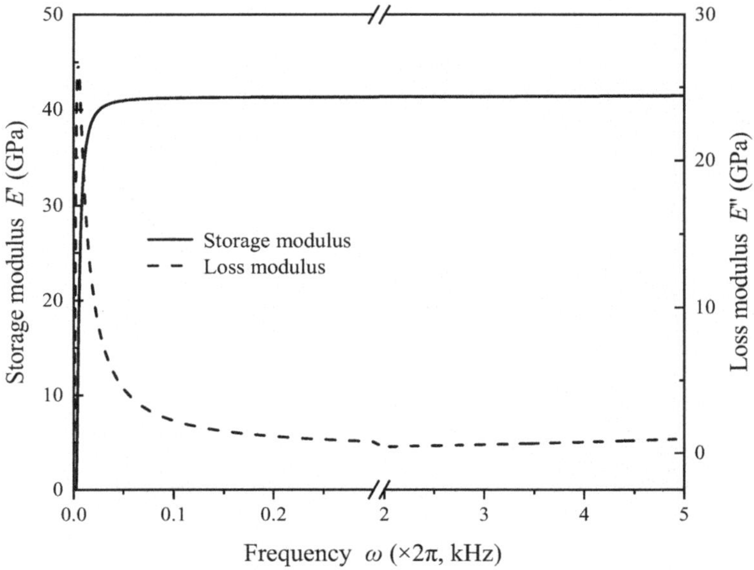

Fig. 7.5 Storage modulus and loss modulus of in granite

where E ' is the storage modulus, E '' is the loss modulus, and ρ is the density of granite.

Figure 7.5 shows the relationship between the storage modulus of granite and the frequency of harmonic component wave. It can be observed from Fig. 7.5 that the storage modulus increases sharply as the frequency of the harmonic component wave increases initially, and then the storage modulus tends to stabilize as the frequency of the harmonic component wave further increases.

Figure 7.5 shows the relationship between the loss modulus of granite and the frequency of the harmonic component wave. It can be observed from Fig. 7.5 that the loss modulus increases rapidly to its maximum value as the frequency of the harmonic component wave increases initially, and then the loss modulus decreases as the frequency of the harmonic component wave increases. It can also be observed from Fig. 7.5 that the loss modulus approaches a constant value as the frequency of the harmonic component wave increases.

7.1.5 An Equivalent Viscoelastic Model

An equivalent viscoelastic model is established to describe the storage modulus and loss modulus of granite under stress waves, as shown in Fig. 7.6. The proposed model consists of a nonlinear spring in parallel with a Maxwell rheological element.

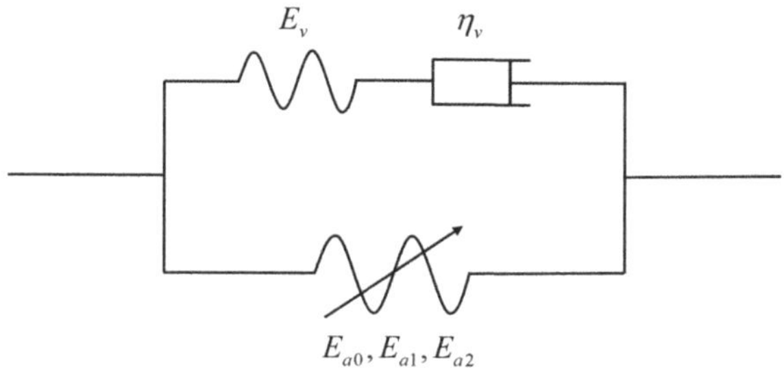

Fig. 7.6 Equivalent viscoelastic model

The stress–strain relationship of the nonlinear spring can be expressed by the following polynomial equation as

$$\sigma = E_{a0}\varepsilon + E_{a1}\varepsilon^2 + E_{a2}\varepsilon^3\sigma \tag{7.5}$$

where σ is the stress, ε is the strain, and E_{a0}, E_{a1} and E_{a2} are the material parameters which are obtained from the quasi-static uniaxial compression test.

Therefore, the constitutive relationship shown in Fig. 7.6 can be expressed as

$$\sigma = E_{a0}\varepsilon + E_{a1}\varepsilon^2 + E_{a2}\varepsilon^3 + E_v \int_0^t \dot{\varepsilon}(\zeta) exp\left[-\frac{t-\zeta}{\eta_v/E_v}\right]d\zeta \tag{7.6}$$

where $\dot{\varepsilon}(\zeta)$ is the strain rate, ζ is the integral variable, and E_v and η_v are the elastic modulus and viscosity coefficient of the modified Maxwell element, respectively.

In Eq. (7.6), the absolute values of E_{a1} are 2 to 3 orders of magnitude larger than those of E_{a0}, and E_{a2} is 3 to 5 orders of magnitude larger than that of E_{a0}. However, the strain amplitude is on the order of 10^{-4} when the stress wave propagates through granite after different high-temperature treatments. Therefore, the effect of the second and third terms in Eq. (7.6) can be ignored.

The viscoelastic constitutive relationship when the stress wave propagates through granite after different high-temperature treatments can be expressed as

$$\sigma = E_{a0}\varepsilon + E_v \int_0^t \dot{\varepsilon}(\zeta) \exp\left[-\frac{t-\zeta}{\eta_v/E_v}\right]d\zeta \tag{7.7}$$

Equation (7.7) can be rewritten in the differential form as

$$E_v\sigma + \eta_v\dot{\sigma} = E_v E_{a0}\varepsilon + (E_{a0} + E_v)\eta_v\dot{\varepsilon} \tag{7.8}$$

where $\dot{\sigma}$ is the stress rate.

Fourier transform is applied to Eq. (7.8), the complex dynamic modulus of granite after different high-temperature treatments can be expressed as

$$E^* = \frac{\tilde{\sigma}}{\tilde{\varepsilon}} = \frac{E_v E_{a0} + i\omega(E_{a0} + E_v)\eta_v}{E_v + i\omega\eta_v} \tag{7.9}$$

where E^* is the complex dynamic modulus, and $\tilde{\sigma}$ and $\tilde{\varepsilon}$ are the stress and strain after Fourier transform, respectively.

The storage modulus E' and loss modulus E'' of granite, which are defined as the real and imaginary parts of the complex dynamic modulus respectively, can be obtained from Eq. (7.9) as

$$E' = \frac{E_{a0}E_v^2 + (E_{a0} + E_v)\eta_v^2\omega^2}{E_v^2 + \eta_v^2\omega^2} \tag{7.10}$$

and

$$E'' = \frac{E_v^2\eta_v\omega}{E_v^2 + \eta_v^2\omega^2} \tag{7.11}$$

The three material parameters E_{a0}, elastic modulus E_v and viscosity coefficient η_v, as shown in Eqs. (7.10) and (7.11) are determined by the experimental results of the quasi-static test and the pendulum impact test.

Figure 7.7a shows the relationship between the storage modulus determined by the proposed model and the pendulum impact test. It can be seen from Fig. 7.7a that the storage modulus determined by the proposed model agrees well with the storage modulus determined by the pendulum impact test. Figure 7.7b shows the loss modulus determined by the proposed model, which is compared with the loss modulus determined by the pendulum impact test. It can also be seen from Fig. 7.7b that the loss modulus determined by the proposed model agrees well with the loss modulus determined by the pendulum impact test, especially in the range of large frequency of the harmonic component wave. Therefore, the present model can be used to describe the dynamic behavior of granite.

7.2 Three-Characteristic Method

The micro-defected rock mass is equivalent to a viscoelastic medium as shown in Fig. 7.8, which is composed of a Maxwell element and a spring in parallel. The constitutive equation of the viscoelastic medium is given as

$$\frac{\partial\varepsilon}{\partial t} - \frac{1}{E_a + E_M}\frac{\partial\sigma}{\partial t} + \frac{E_a E_M \varepsilon}{(E_a + E_M)\eta_M} - \frac{E_M\sigma}{(E_a + E_M)\eta_M} = 0 \tag{7.12}$$

where σ, ε and t are the stress, strain and time, respectively. E_M and η_M are the elastic modulus of spring and viscous coefficient of the dashpot in the Maxwell element, respectively. E_a is the elastic modulus of spring in parallel with the Maxwell element.

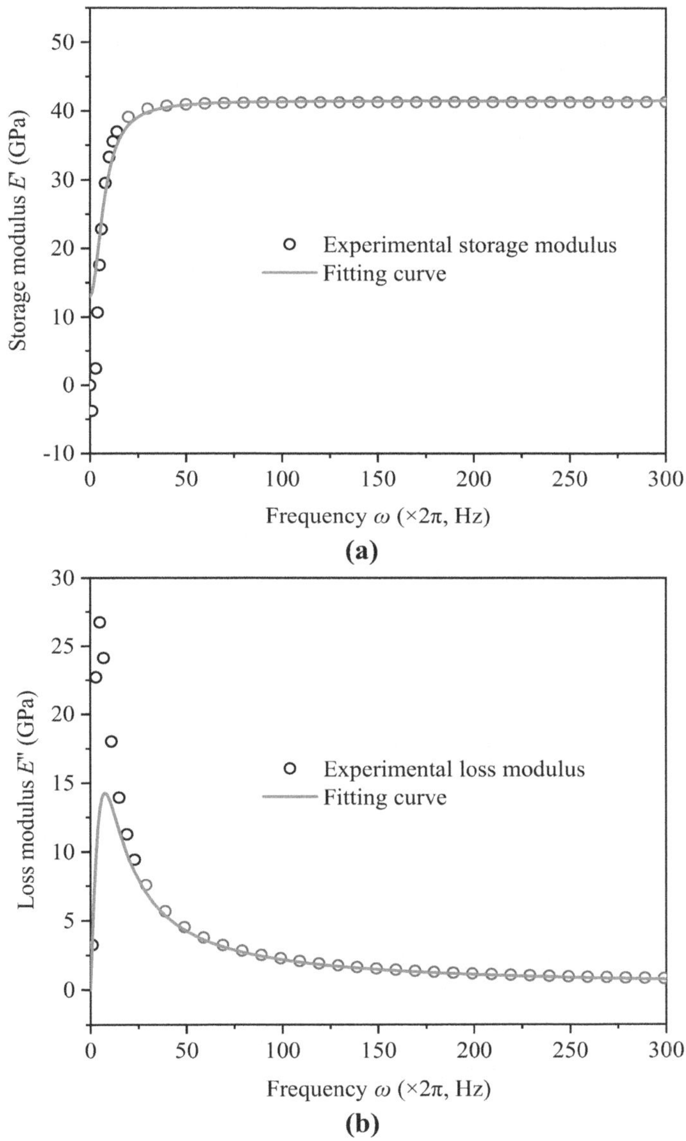

Fig. 7.7 Experiment and fitting equivalent viscoelastic modulus. **a** Storage modulus. **b** Loss modulus

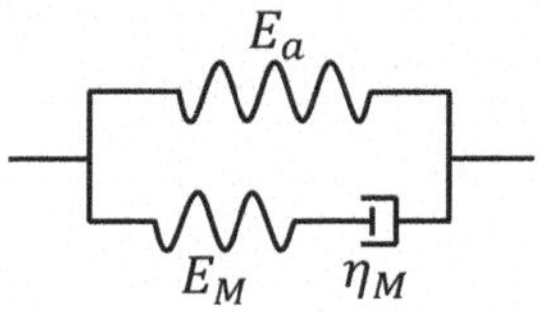

Fig. 7.8 Equivalent viscoelastic model of micro-defected rock mass

The motion equation and the continuity equation of one-dimensional longitudinal stress wave propagation are respectively governed by

$$\rho_0 \frac{\partial v}{\partial t} - \frac{\partial \sigma}{\partial x} = 0 \tag{7.13}$$

and

$$\frac{\partial \varepsilon}{\partial t} - \frac{\partial v}{\partial x} = 0 \tag{7.14}$$

where ρ_0 is the density, x is Cartesian coordinate along the path of stress wave propagation and v is the particle velocity.

Based on the governing equations of Eqs. (7.12) to (7.14), three sets of characteristic lines and the corresponding compatibility conditions can be obtained.

The first and the second sets of characteristic lines are determined as

$$\frac{dx}{dt} = \pm\sqrt{\frac{E_a + E_M}{\rho_0}} = \pm C_V \tag{7.15}$$

where C_V is the velocity of the stress wave.

The corresponding two sets of compatibility conditions along the characteristic lines of Eq. (7.15) are

$$dv = \pm\frac{1}{\rho_0 C_V} d\sigma + \left[\frac{\sigma - E_a\varepsilon}{(E_a + E_M)\theta_M}\right] dx \tag{7.16}$$

where θ_M is the relaxation time and is equal to the ratio of η_M to E_M.

The third set of characteristic lines is determined as

$$dx = 0 \tag{7.17}$$

The corresponding compatibility condition along the characteristic line of Eq. (7–17) is

$$d\varepsilon - \frac{d\sigma}{E_a + E_M} - \frac{\sigma - E_a\varepsilon}{(E_a + E_M)\theta_M} dt = 0 \tag{7.18}$$

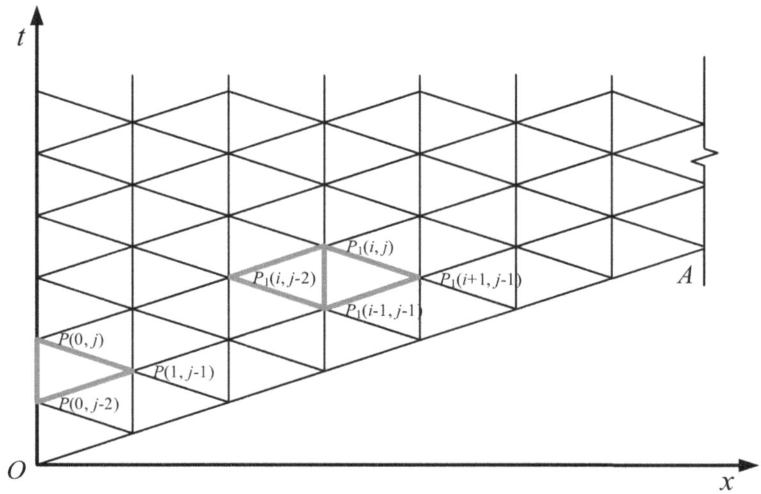

Fig. 7.9 Three-characteristic method

It can be known from Eqs. (7.15) and (7.17) that the wave propagation in the micro-defected rock mass can be calculated by the three sets of characteristic lines in the x-t plane. As shown in Fig. 7.9, the x-t plane is discreted by the three sets of characteristic lines with differential steps of Δx and Δt, where Δx equals $C_V \cdot \Delta t$ according to Eq. (7.15).

7.2.1 *Wave Propagation in Continuous Media on Characteristics OA*

Assume that the initial condition of rock mass is $\sigma_0(x, 0) = \varepsilon_0(x, 0) = v_0(x, 0) = 0$ at $t = 0$, it means that it is in a static undisturbed state, and then suddenly undergoes a half-sinusoidal impact loading $\sigma^*(0, t)$ at its end at $x = 0$, and the P-wave will propagate in the rock bar. The problem of P-wave propagation in the bar is to solve three unknown quantities σ, ε and v at any point on the x-t plane in Figs. 7.7, 7.8, 7.9 according to the known initial and boundary conditions. The AOx region is a constant region. According to the Hugoniot relations and compatibility conditions along the characteristic line OA, the three unknowns, $\sigma(x, t)$, $\varepsilon(x, t)$ and $v(x, t)$, in front of the macro-joint along the characteristic line OA are expressed as

$$\sigma(x, t) = \sigma^*(0, 0)\exp\left[-\frac{\rho_0 C_V}{2\eta_M (1 + E_a/E_M)^2} x\right] \tag{7.19}$$

$$v(x,t) = -\frac{\sigma^*(0,0)}{\rho_0 C_V}\exp\left[-\frac{\rho_0 C_V}{2\eta_M(1+E_a/E_M)^2}x\right] \tag{7.20}$$

and

$$\varepsilon(x,t) = \frac{\sigma^*(0,0)}{E_a+E_M}\exp\left[-\frac{\rho_0 C_V}{2\eta_M(1+E_a/E_M)^2}x\right] \tag{7.21}$$

7.2.2 Wave Propagation in Left Boundary Points

The stress σ at an arbitrary point of the left boundary characteristics Ot, as shown in Fig. 7.9, has been given by the boundary condition, while strain ε and velocity v at an arbitrary point along Ot are unknown, for example the point $P(0, j)$ in Fig. 7.9. According to Eqs. (7.16) and (7.18), it is known that the strain ε and velocity v are based on the compatibility conditions along characteristics $P(1, j\text{-}1)\text{-}P(0, j)$ and $P(0, j\text{-}2)\text{-}P(0, j)$; the forms of the solution are

$$\begin{aligned} v(0,j) &= v(1,j-1) - \tfrac{1}{\rho_0 C_V}\left[\sigma(0,j) - \sigma(1,j-1)\right] \\ &\quad + \tfrac{\sigma(1,j-1)-E_a\varepsilon(1,j-1)}{(E_a+E_M)\theta_M}\left[x(0,j) - x(1,j-1)\right] \end{aligned} \tag{7.22}$$

and

$$\begin{aligned} \varepsilon(0,j) &= \varepsilon(0,j-2) + \tfrac{1}{E_a+E_M}\left[\sigma(0,j) - \sigma(0,j-2)\right] \\ &\quad + \tfrac{\sigma(0,j-2)-E_a\varepsilon(0,j-2)}{(E_a+E_M)\theta_M}\left[t(0,j) - t(0,j-2)\right] \end{aligned} \tag{7.23}$$

Thus, the two unknown quantities, $\varepsilon(0, j)$ and $v(0, j)$, at an arbitrary left boundary point can be solved by Eqs. (7.22) and (7.23).

7.2.3 Wave Propagation in Ordinary Interior Points

When solving at an arbitrary ordinary interior point in the domain AOt, for example the point $P_1(i, j)$ in Fig. 7–9. Similarly, the solution is based on the compatibility conditions along characteristics $P_1(i\text{-}1, j\text{-}1)\text{-}P_1(i, j)$, $P_1(i+1, j\text{-}1)\text{-}P_1(i, j)$ and $P_1(i, j\text{-}2)\text{-}P_1(i, j)$, respectively, which, in the form of finite differences, are

$$\begin{aligned} v(i,j) - v(i-1,j-1) &= \tfrac{1}{\rho_0 C_V}\left[\sigma(i,j) - \sigma(i-1,j-1)\right] \\ &\quad + \tfrac{\sigma(i-1,j-1)-E_a\varepsilon(i-1,j-1)}{(E_a+E_M)\theta_M}\left[x(i,j) - x(i-1,j-1)\right] \end{aligned} \tag{7.24}$$

$$\begin{aligned} v(i,j) - v(i+1,j-1) = &-\tfrac{1}{\rho_0 C_V}\left[\sigma(i,j) - \sigma(i+1,j-1)\right] \\ &+ \tfrac{\sigma(i+1,j-1) - E_a \varepsilon(i+1,j-1)}{(E_a + E_M)\theta_M}\left[x(i,j) - x(i+1,j-1)\right] \end{aligned} \tag{7.25}$$

and

$$\begin{aligned} \varepsilon(i,j) - \varepsilon(i,j-2) = &\tfrac{1}{E_a + E_M}\left[\sigma(i,j) - \sigma(i,j-2)\right] \\ &+ \tfrac{\sigma(i,j-2) - E_a \varepsilon(i,j-2)}{(E_a + E_M)\theta_M}\left[t(i,j) - t(i,j-2)\right] \end{aligned} \tag{7.26}$$

Thus, the three unknown quantities, $\sigma(i, j)$, $\varepsilon(i, j)$ and $v(i, j)$, for an arbitrary ordinary interior point can be solved by Eqs. (7.24) to (7.26).

Thus, the three unknown quantities, $\sigma^+(i, j)$, $\varepsilon^+(i, j)$ and $v^+(i, j)$, of the wave propagating through the joint in domain *AOt* can be solved.

Through the above calculation, the stress $\sigma(i, j)$, strain $\varepsilon(i, j)$ and velocity $v(i, j)$ of a one-dimensional stress wave propagating across micro-defected rock mass to different positions can be obtained.

Chapter 8
A Split Three-Characteristic Method for Stress Wave Propagation Through Rock Mass with Double-Scale Discontinuities

8.1 A Split Three-Characteristic Method for Stress Wave Propagation Through Rock Mass with Double-Scale Discontinuities

Discontinuities ranging from micro-defect to macro-joint widely exist in the natural rock mass, which significantly influence the stress wave propagation [1–3]. The traditional characteristic method has been successfully applied for the stress wave propagation through single-scale discontinuity [4, 5]. It is still difficult to analyze the different mechanisms of double-scale discontinuities (macro-joint and micro-defect) on stress wave propagation through rock mass.

This section presents a split three-characteristic method combined with the displacement discontinuity method (DDM) to solve the stress wave propagation through rock mass with double-scale discontinuities. By comparing the effect of macro-joint and micro-defect on wave propagation, the differences between considering the micro-defect effect and without considering the micro-defect effect on wave propagation in macro-joint are further studied. In addition, the effect of joint stiffness and frequency on the transmission coefficient and effective velocity of rock mass with double-scale discontinuities are discussed in detail and compared with macro-joint.

8.1.1 Split Three-Characteristic Method

(1) Scheme of split three-characteristic method

Figure 8.1a and b shows the influence mechanism and calculation method of macro-joint and micro-defect on wave propagation, respectively. The macro-joint causes the reflection and transmission of stress waves as shown in Fig. 8.1a. The stress field in

L. Fan et al., *The Method of Characteristics for Stress Wave Propagation in the Rock Mass*, https://doi.org/10.1007/978-981-97-1440-7_8

front of the macro-joint is the superposition of the incident wave and reflected wave, and the stress field behind the macro-joint is the transmitted wave. For the macro-joint with a small thickness and large planar, the stress field is assumed continuous while the displacement field is discontinuous (Fig. 8.1a). The change of displacement field on two sides of the macro-joint equals the deformation of the macro-joint [6]. In addition, the practical rocks usually contain micro-defect (Fig. 8.1b) that lead to stress wave attenuation and dispersion, and experimental studies have shown that micro-defected rock mass can be equivalently described by a viscoelastic medium (Fig. 8.1b). In the process of wave propagation, the viscoelastic behavior leads to stress not only related to strain but also to strain rate, which can be effectively solved by the three-characteristic method (Fig. 8.1b) proposed by Wang et al. [7]. Besides theoretical and experimental studies, numerous numerical studies further revealed the effect of micro-defects and macro-joints in the rock mass on stress wave transmission [8, 9].

In existing studies, the effect of single-scale discontinuous rock mass on stress wave propagation is mainly investigated. However, the discontinuities in the rock mass from micro-defect to macro-joint greatly influence stress wave transmission. Micro-defect and macro-joint have shown a complex interaction in the process of stress wave propagation, and the effect of micro-defects and macro-joints in the rock mass on stress wave attenuation of velocity and amplitude is different in mechanism. Therefore, to solve the more general problem that stress wave propagation through double-scale discontinuities as shown in Fig. 8.1c. Combined with Fig. 8.1a

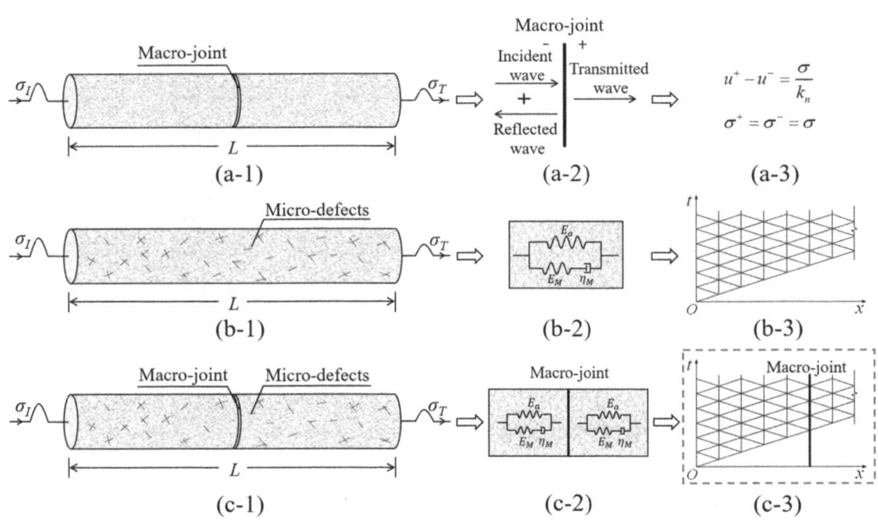

Fig. 8.1 Scheme of split three-characteristic method. **a** Reflection and transmission mechanism of wave propagation by a single macro-joint and calculation method of DDM. **b** Equivalent viscoelastic medium of micro-defected rock mass and three-characteristic method. **c** Homogeneous-discontinuous viscoelastic medium of double-scale discontinuous rock mass and split three-characteristic method

and b, the rock mass with double-scale discontinuities (Fig. 8.1c) is equivalent to a homogeneous-discontinuous viscoelastic medium (Fig. 8.1c). A split three-characteristic method combined with the displacement discontinuity method (DDM) is proposed in the present study (Fig. 8.1c).

To solve the stress wave propagation through double-scale discontinuous rock mass, the three-characteristic lines are split at $x = x_1$, where is assumed as the location of a single macro-joint. The split three-characteristic lines are divided into three basic elements, as shown in Fig. 8.2. Figure 8.3 shows in detail the three basic elements of the split three-characteristic method: triangle, diamond and separated diamond elements. Similar to the traditional three-characteristic method, the triangle and diamond elements are used to solve wave propagation at left boundary points $P(0, j)$ and ordinary interior points $P_1(i, j)$ of the rock mass. To further study the propagation of stress waves at the macro-joint, the separated diamond element is proposed. The displacement discontinuity method (DDM) is introduced into the separated diamond element to solve wave propagation at pre-interior points $P_2^-(i, j)$ and post-interior points $P_2^+(i, j)$ of the macro-joint.

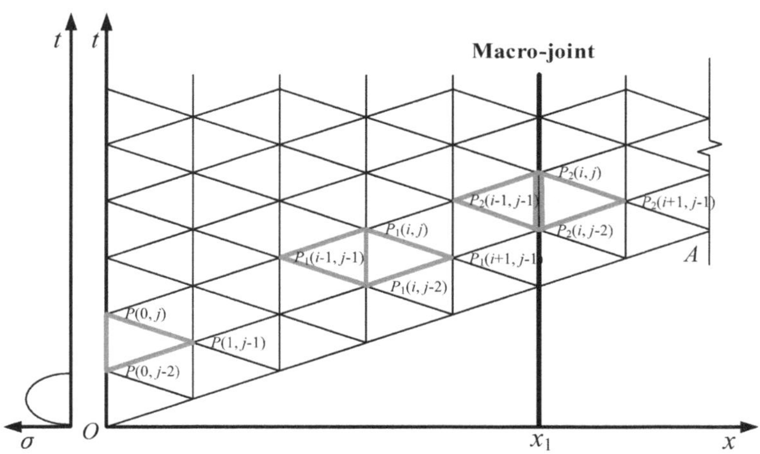

Fig. 8.2 The split three-characteristic method

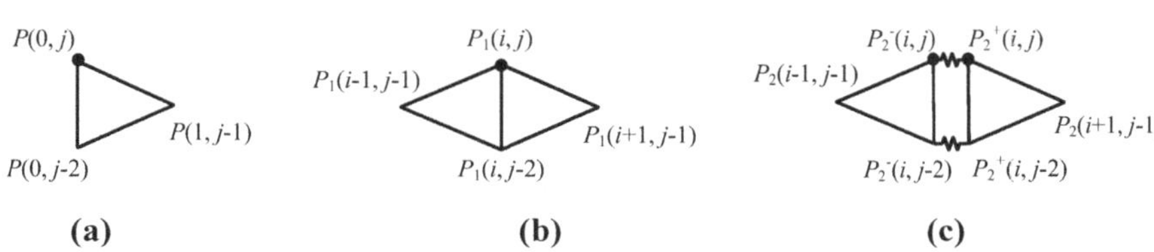

Fig. 8.3 Three basic elements of the split three-characteristic method. **a** Triangle element. **b** Diamond element. **c** Separated diamond element

(2) Split three-characteristic method combined with linear displacement discontinuity method

Assume the initial condition of double-scale discontinuous rock mass is $\sigma_0(x, 0) = 0$, $\varepsilon_0(x, 0) = 0$ and $v_0(x, 0) = 0$ at $t = 0$, which denotes a static undisturbed state. An impact loading $\sigma^*(0, t)$ is then applied at $x = 0$ resulting in the stress wave propagation through double-scale discontinuous rock mass. The stress wave propagation in the double-scale discontinuous rock mass can be solved by obtaining three unknowns, stress σ, strain ε and velocity v at any point on the x-t plane as in Fig. 8.2 according to the initial and boundary conditions. Region AOx is the undisturbed region.

When a load is applied at the end of the double-scale discontinuous rock mass, the three unknowns stress σ, strain ε and velocity v along the characteristic line OA in front of the macro-joint can be obtained first. The three unknowns along characteristic line OA shown in Fig. 8.2 satisfy the Hugoniot relations as [7]

$$v = -\frac{\sigma}{\rho_0 C_V} \tag{8.1}$$

and

$$v = -C_V \varepsilon \tag{8.2}$$

Moreover, the characteristic line OA coincides with the first set of characteristic lines, and the three unknowns stress σ, strain ε and velocity v should satisfy the corresponding compatibility condition as

$$dv = \frac{1}{\rho_0 C_V} d\sigma + \frac{\sigma - E_a \varepsilon}{\rho_0 C_V \theta_M} dt = \frac{1}{\rho_0 C_V} d\sigma + \left[\frac{\sigma - E_a \varepsilon}{(E_a + E_M)\theta_M} \right] dx \tag{8.3}$$

Together with Eqs. (8.1) and (8.2), an ordinary differential equation about σ is obtained:

$$\frac{d\sigma}{dx} = -\frac{\rho_0 C_V \sigma}{2\eta_M (1 + E_a/E_M)^2} \tag{8.4}$$

The general solution of the differential Eq. (8.4) is

$$\sigma = R \cdot \exp\left[-\frac{\rho_0 C_V x}{2\eta_M (1 + E_a/E_M)^2} \right] \tag{8.5}$$

According to the boundary condition, the integral constant R is equal to $\sigma^*(0, 0)$. Then, by substituting Eq. (8.5) into Eqs. (8.1) and (8.2), the three unknowns $\sigma(x, t)$, $\varepsilon(x, t)$ and $v(x, t)$ along OA in front of the macro-joint can be respectively obtained:

$$\sigma(x,t) = \sigma^*(0,0)\exp\left[-\frac{\rho_0 C_V}{2\eta_M(1+E_a/E_M)^2}x\right] \tag{8.6}$$

$$v(x,t) = -\frac{\sigma^*(0,0)}{\rho_0 C_V}\exp\left[-\frac{\rho_0 C_V}{2\eta_M(1+E_a/E_M)^2}x\right] \tag{8.7}$$

and

$$\varepsilon(x,t) = \frac{\sigma^*(0,0)}{E_a+E_M}\exp\left[-\frac{\rho_0 C_V}{2\eta_M(1+E_a/E_M)^2}x\right] \tag{8.8}$$

There are six unknowns at the two sides of the macro-joint along the characteristic line *OA*. According to Eqs. (8.6)–(8.8), the three unknown quantities, $\sigma^-(x_1,t)$, $\varepsilon^-(x_1, t)$ and $v^-(x_1,t)$, in front of the macro-joint along the characteristic line *OA* are $\sigma^-(x_1, t) = \sigma(x_1,t)$, $\varepsilon^-(x_1,t) = \varepsilon(x_1,t)$ and $v^-(x_1,t) = v(x_1,t)$, respectively. On the other hand, the three unknown quantities, $\sigma^+(x_1, t)$, $\varepsilon^+(x_1, t)$ and $v^+(x_1, t)$ behind the macro-joint along the characteristic line *OA* can be derived according to the DDM.

The displacement discontinuity boundary conditions of the macro-joint are described by the constitutive equation of joint normal deformation and the equation of stress continuity. The constitutive equation of joint normal deformation is

$$u^-(x,t) - u^+(x,t) = \frac{\sigma(x,t)}{k_n} \tag{8.9}$$

In addition, according to the stress continuity condition, the transmitted stress behind the macro-joint is equal to the summation of incident stress and reflected stress in front of the macro-joint. However, the value of reflected stress at the macro-joint on the characteristic line *OA* equals zero according to the boundary condition, which leads to the stress in front of the macro-joint equal to the incident wave. Therefore, the equation of stress continuity at the macro-joint can be described as

$$\sigma^-(x,t) = \sigma^+(x,t) = \sigma^*(0,0)\exp\left[-\frac{\rho_0 C_V}{2\eta_M(1+E_a/E_M)^2}x\right] \tag{8.10}$$

where k_n is stiffness of the macro-joint, and superscripts "–" and "+" refer to the wave fields in front of and behind the macro-joint, respectively.

The expressions of partial derivatives of t and x in Eq. (8.9) are

$$v^-(x,t) - v^+(x,t) = \frac{\partial\sigma(x,t)}{\partial t}\frac{1}{k_n} \tag{8.11}$$

and

$$\varepsilon^-(x,t) - \varepsilon^+(x,t) = \frac{\partial\sigma(x,t)}{\partial x}\frac{1}{k_n} \tag{8.12}$$

Substituting Eq. (8.10) into Eqs. (8.11) and (8.12) yields

$$v^{+}(x,t) = v^{-}(x,t) - C_V\sigma^{*}(0,0)\left(-\frac{\rho_0 C_V}{2\eta_M(1+E_a/E_M)^2}\right)\exp\left[-\frac{\rho_0 C_V}{2\eta_M(1+E_a/E_M)^2}x\right]\frac{1}{k_n} \quad (8.13)$$

and

$$\varepsilon^{+}(x,t) = \varepsilon^{-}(x,t) - \sigma^{*}(0,0)\left(-\frac{\rho_0 C_V}{2\eta_M(1+E_a/E_M)^2}\right)\exp\left[-\frac{\rho_0 C_V}{2\eta_M(1+E_a/E_M)^2}x\right]\frac{1}{k_n} \quad (8.14)$$

Substituting x equals x_1 into Eqs. (8.10), (8.13) and (8.14), the three unknown quantities, $\sigma^{+}(x_1,\ t)$, $\varepsilon^{+}(x_1,\ t)$ and $v^{+}(x_1,\ t)$ behind the macro-joint along the characteristic line *OA* can be solved.

Once the unknowns stress σ, strain ε and velocity v along the characteristic line *OA* are obtained, the remaining unknowns in the *AOt* domain can be obtained according to the triangle, diamond and separated diamond elements in Fig. 8.3.

The unknowns stress σ, strain ε and velocity v of left boundary points $P(0,j)$ can be obtained by triangle elements in Fig. 8.3a. The stress $\sigma(0,j)$ at an arbitrary point of the left boundary characteristic line *Ot* has been given by the boundary condition, while strain $\varepsilon(0,j)$ and velocity $v(0,j)$ at an arbitrary point along *Ot* are unknown. The strain $\varepsilon(0,j)$ and velocity $v(0,j)$ can be solved by the compatibility conditions along characteristic line $P(1,j-1) - P(0,j)$ and $P(0,j-2) - P(0,j)$ in triangle element as

$$\begin{aligned} v(0,j) &= v(1,j-1) - \frac{1}{\rho_0 C_V}\left[\sigma(0,j) - \sigma(1,j-1)\right] \\ &\quad + \frac{\sigma(1,j-1) - E_a\varepsilon(1,j-1)}{(E_a+E_M)\theta_M}\left[x(0,j) - x(1,j-1)\right] \end{aligned} \quad (8.15)$$

and

$$\begin{aligned} \varepsilon(0,j) &= \varepsilon(0,j-2) + \frac{1}{E_a+E_M}\left[\sigma(0,j) - \sigma(0,j-2)\right] \\ &\quad + \frac{\sigma(0,j-2) - E_a\varepsilon(0,j-2)}{(E_a+E_M)\theta_M}\left[t(0,j) - t(0,j-2)\right] \end{aligned} \quad (8.16)$$

Thus, the two unknown quantities, $\varepsilon(0,j)$ and $v(0,j)$, at arbitrary left boundary points can be solved by Eqs. (8.15) and (8.16).

The unknowns stress σ, strain ε and velocity v of ordinary interior points $P_1(i,j)$ can be obtained by the diamond elements in Fig. 8.3b. Similarly, the three unknowns can be solved by the compatibility conditions along characteristic line $P_1(I-1,j-1) - P_1(i,j)$, $P_1(i+1,j-1)$-$P_1(i,j)$ and $P_1(i,j-2)$-$P_1(i,j)$ in the diamond element, and the solutions are

$$v(i,j) - v(i-1,j-1) = \frac{1}{\rho_0 C_V}\left[\sigma(i,j) - \sigma(i-1,j-1)\right]$$

$$+\frac{\sigma(i-1,j-1)-E_a\varepsilon(i-1,j-1)}{(E_a+E_M)\theta_M}[x(i,j)-x(i-1,j-1)] \quad (8.17)$$

$$v(i,j)-v(i+1,j-1)=-\frac{1}{\rho_0 C_V}[\sigma(i,j)-\sigma(i+1,j-1)]$$
$$+\frac{\sigma(i+1,j-1)-E_a\varepsilon(i+1,j-1)}{(E_a+E_M)\theta_M}[x(i,j)-x(i+1,j-1)] \quad (8.18)$$

and

$$\varepsilon(i,j)-\varepsilon(i,j-2)=\frac{1}{E_a+E_M}[\sigma(i,j)-\sigma(i,j-2)]$$
$$+\frac{\sigma(i,j-2)-E_a\varepsilon(i,j-2)}{(E_a+E_M)\theta_M}[t(i,j)-t(i,j-2)] \quad (8.19)$$

Thus, the three unknown quantities, $\sigma(i,j)$, $\varepsilon(i,j)$ and $v(i,j)$, for arbitrary ordinary interior points can be solved by Eqs. (8.17)–(8.19).

The unknowns of interior points at macro-joint $P_2(i,j)$ can be obtained by separated diamond elements in Fig. 8.3c. From Fig. 8.3c, it can be observed that the interior points $P_2(i,j)$ include pre-interior points $P_2^-(i,j)$ and post-interior points $P_2^+(i,j)$ of the macro-joint, which leads to six unknowns, $\sigma^-(i,j)$, $\sigma^+(i,j)$, $v^-(i,j)$, $v^+(i,j)$, $\varepsilon^-(i,j)$ and $\varepsilon^+(i,j)$, at the points $P_2(i,j)$.

When stress wave propagates through the macro-joint, according to the DDM, the displacement discontinuity boundary conditions of the macro-joint are

$$u^+(x_1,t)-u^-(x_1,t)=\frac{\sigma(x_1,t)}{k_n} \quad (8.20)$$

and

$$\sigma^+(x_1,t)=\sigma^-(x_1,t)=\sigma(x_1,t) \quad (8.21)$$

The derivative expression of Eq. (8.20) is

$$v^+(x_1,t)-v^-(x_1,t)=\frac{1}{k_n}\frac{\partial\sigma(x_1,t)}{\partial t} \quad (8.22)$$

In order to obtain the numeric solution for $\sigma(x_1, t)$, the time interval [0, t] is divided into n equal time steps, namely [0, t] = [0, t_1, …t_j, t_{j+1}…, t]. Within each time step, a linear approximation of $\partial\sigma(x_1,t)/\partial t$ is adopted as

$$\frac{\partial\sigma(x_1,t_j)}{\partial t}=\frac{\sigma(x_1,t_j)-\sigma(x_1,t_{j-2})}{2dt} \quad (8.23)$$

Substituting Eqs. (8.23) into (8.22), it can be expressed as

$$v^{+}(x_1, t_j) - v^{-}(x_1, t_j) = \frac{1}{k_n}\frac{\sigma(x_1, t_j) - \sigma(x_1, t_{j-2})}{2dt} \tag{8.24}$$

Assuming coordinate (x_1, t_j) is the point $P_2(i,j)$ in Fig. 8.3c, Eqs. (8.24) and (8.21) are transformed to

$$v^{+}(i,j) - v^{-}(i,j) = \frac{1}{k_n}\frac{\sigma(i,j) - \sigma(i,j-2)}{2dt} \tag{8.25}$$

and

$$\sigma^{+}(i,j) = \sigma^{-}(i,j) = \sigma(i,j) \tag{8.26}$$

The six unknowns can be solved by the compatibility conditions along characteristic line $P_2(I-1, j-1) - P_2(i, j)$, $P_2(i+1, j-1) - P_2(i, j)$, $P_2^{+}(i, j) - P_2^{+}(i, j-2)$ and $P_2^{-}(i, j) - P_2^{-}(i, j-2)$ in the separated diamond element and the displacement discontinuity boundary conditions Eqs. (8.25) and (8.26). The four compatibility conditions of considering Eq. (8.26) can be expressed as

$$\begin{aligned} v^{-}(i,j) - v(i-1,j-1) = {} & \frac{1}{\rho_0 C_V}\left[\sigma(i,j) - \sigma(i-1,j-1)\right] \\ & + \frac{\sigma(i-1,j-1) - E_a\varepsilon(i-1,j-1)}{(E_a + E_M)\theta_M}dx \end{aligned} \tag{8.27}$$

$$\begin{aligned} v^{+}(i,j) - v(i+1,j-1) = {} & -\frac{1}{\rho_0 C_V}\left[\sigma(i,j) - \sigma(i+1,j-1)\right] \\ & - \frac{\sigma(i+1,j-1) - E_a\varepsilon(i+1,j-1)}{(E_a + E_M)\theta_M}dx \end{aligned} \tag{8.28}$$

$$\begin{aligned} \varepsilon^{-}(i,j) - \varepsilon^{-}(i,j-2) = {} & \frac{1}{E_a + E_M}\left[\sigma(i,j) - \sigma(i,j-2)\right] \\ & + \frac{\sigma(i,j-2) - E_a\varepsilon^{-}(i,j-2)}{(E_a + E_M)\theta_M}(2dt) \end{aligned} \tag{8.29}$$

and

$$\begin{aligned} \varepsilon^{+}(i,j) - \varepsilon^{+}(i,j-2) = {} & \frac{1}{E_a + E_M}\left[\sigma(i,j) - \sigma(i,j-2)\right] \\ & + \frac{\sigma(i,j-2) - E_a\varepsilon^{+}(i,j-2)}{(E_a + E_M)\theta_M}(2dt) \end{aligned} \tag{8.30}$$

It can be obtained by using Eqs. (8.25), (8.27) and (8.28):

$$\sigma(i,j) = \frac{2k_n\rho_0 C_V dt}{\rho_0 C_V + 4k_n dt}\left[\begin{array}{c} \frac{\sigma(i+1,j-1)+\sigma(i-1,j-1)}{\rho_0 C_V} - v(i-1,j-1) \\ +v(i+1,j-1) + \frac{1}{k_n 2dt}\sigma(i,j-2) \\ -\frac{\sigma(i+1,j-1)-E_a\varepsilon(i+1,j-1)}{(E_a+E_M)\theta_M}dx \\ -\frac{\sigma(i-1,j-1)-E_a\varepsilon(i-1,j-1)}{(E_a+E_M)\theta_M}dx \end{array}\right] \quad (8.31)$$

Substituting Eq. (8.31) into Eqs. (8.27) to (8.30), the four unknown quantities, $v^-(i,j)$, $v^+(i,j)$, $\varepsilon^-(i,j)$ and $\varepsilon^+(i,j)$ can be solved, respectively.

Therefore, the stress $\sigma(i, j)$, strain $\varepsilon(i, j)$ and velocity $v(i, j)$ of stress wave propagation through rock mass with double-scale discontinuities can be obtained.

8.1.2 Examples

The influence of macro-joint and micro-defect on stress wave propagation is studied by the illustrations of stress wave propagation in macro-jointed rock mass and rock mass with double-scale discontinuities. Table 8.1 shows the mechanical properties in the examples.

In the following study, the incident stress wave is assumed to be a half-sinusoidal stress wave in the form:

$$\sigma^*(0,t) = \begin{cases} A_0 sin(2\pi f_0 t) & \text{when} \quad 0 \le t \le 1/(2f_0) \\ 0 & \text{others} \end{cases} \quad (8.32)$$

where A_0 is the amplitude and f_0 is the frequency of the incident stress wave. In the present study, $A_0 = 0.02$ MPa and $f_0 = 50$ Hz.

(1) Stress wave propagation through macro-jointed rock mass (MJ) and micro-defected rock mass (MD)

Figures 8.4 and 8.5 show the effect of micro-defect and macro-joint on stress wave propagation, respectively.

Figure 8.4a shows the amplitude attenuation of transmitted stress wave of macro-jointed rock mass for different macro-joint stiffnesses of $k_n = 2.5$, 3.5 and 5.0 GPa/m, respectively. The three macro-joint stiffnesss are selected based on previous studies [5, 10, 11]. The micro-defect in the macro-jointed rock mass is not considered. The macro-jointed rock mass is denoted as "MJ" as in Fig. 8.1a. It can be observed from

Table 8.1 Mechanical parameters of rocks

Parameter	Value
Young's modulus E_a	25.0 GPa
Young's modulus E_M	15.0 GPa
Density ρ_0	2500.0 kg/m^3

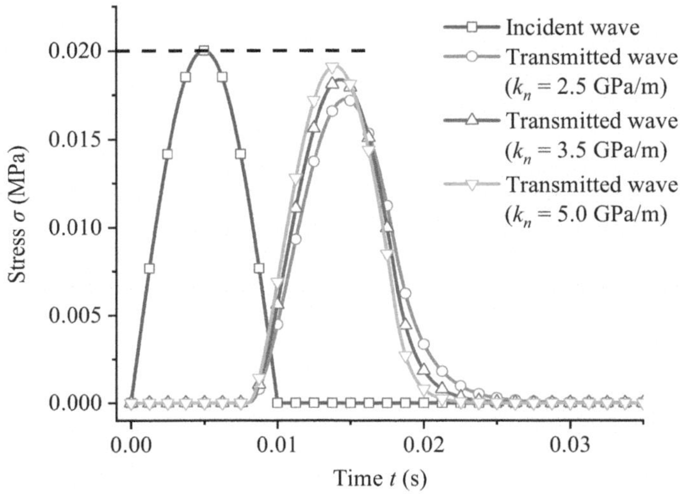

(a)

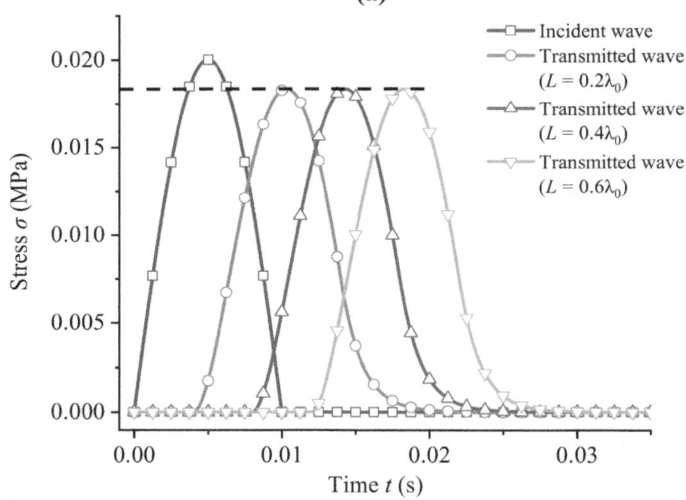

(b)

Fig. 8.4 Stress wave propagation in macro-jointed rock mass (micro-defect not considered). **a** The effect of macro-joint stiffness (L = 0.4λ0). **b** The effect of propagation distance (kn = 3.5 GPa/m)

Fig. 8.4a that the amplitude of the transmitted stress wave increases as the macro-joint stiffness increases. Figure 8.4b shows the relationship between the transmitted stress wave of MJ and propagation distances of $L = 0.2\lambda_0$, $0.4\lambda_0$ and $0.6\lambda_0$, where λ_0 is the wavelength of the incident stress wave. It can be seen from Fig. 8.4b that the amplitude of the transmitted stress wave of MJ remains constant as the propagation distance increases.

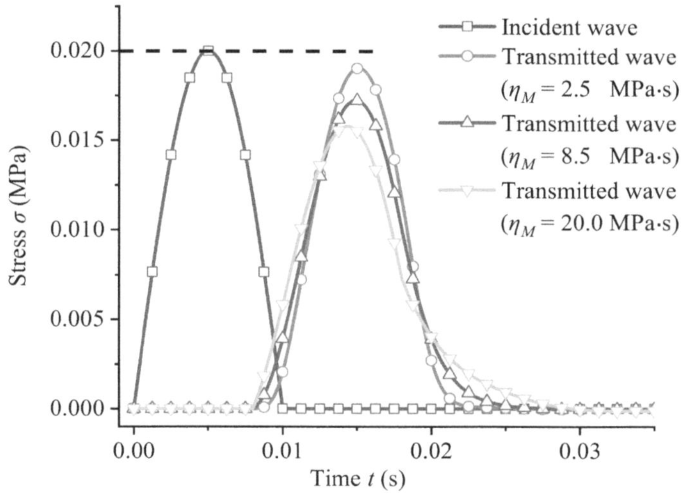

(a)

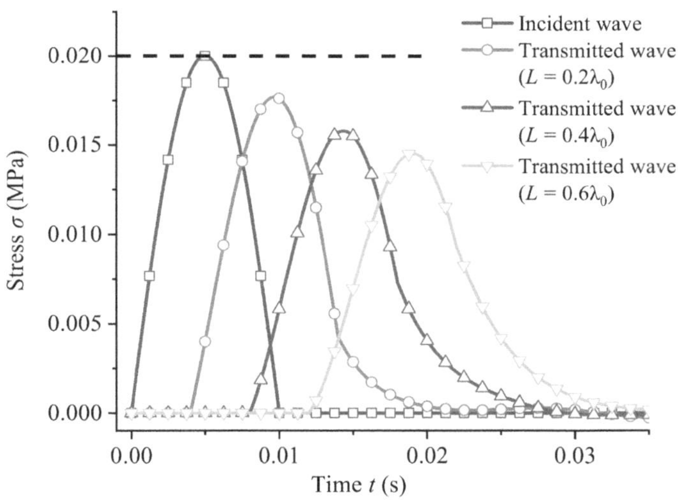

(b)

Fig. 8.5 Stress wave propagation in micro-defected rock mass (macro-joint not considered). **a** The effect of viscous coefficient (L = 0.4λ0). **b** The effect of propagation distance ($\eta_M = 20.0$ MPa · s)

Figure 8.5a shows the amplitude attenuation of the transmitted stress wave of micro-defected rock mass for different viscous coefficients of $\eta_M = 2.5$, 8.5 and 20.0 MPa · s, respectively. The macro-joint in the micro-defected rock mass is not considered. The micro-defected rock mass is denoted as "MD," the same as in Fig. 8.1b. It can be observed from Fig. 8.5a that the amplitude of the transmitted stress wave decreases as the viscous coefficient increases. Figure 8.5b shows the relationship between the transmitted stress wave of MD and propagation distances of $L = 0.2\lambda_0$, $0.4\lambda_0$ and $0.6\lambda_0$. It can be found from Fig. 8.5b that the amplitude of stress wave propagation in MD decreases with the increase of propagation distance.

Table 8.2 shows the transmission coefficients of stress wave propagation in the MJ and MD, in which the macro-joint stiffness is $k_n = 3.5$ GPa/m and viscosity coefficient is $\eta_M = 20.0$ MPa · s. It can be obtained from Table 8.2 that the transmission coefficients of MJ are independent of the propagation distances L, while the transmission coefficients of MD are obviously affected by the propagation distances L. When the propagation distance is sufficiently small, such as $L = 0.005\lambda_0$ in the present study, the transmission coefficient of MJ is 0.92 and that of MD is about 0.99, and the influence of micro-defect on transmitted stress wave can be neglected. On the other hand, as the propagation distance increases, such as $L = 0.6\lambda_0$, the transmission coefficient of MD decreases to 0.71, while that of MJ is 0.92. Therefore, the influence of the micro-defect on transmission stress waves should be considered with the increase of propagation distance.

(2) Stress wave propagation through macro-jointed rock mass (MJ) and double-scale discontinuous rock mass (MD-MJ)

Figure 8.6 compares the stress wave transmission between MJ and double-scale discontinuous rock mass, which is denoted as "MD-MJ" as in Fig. 8.1c. The propagation distance of $L = 0.2\lambda_0$, $0.4\lambda_0$ and $0.6\lambda_0$ and the viscous coefficient of $\eta_M = 2.5$ and 20.0 MPa · s are illustrated, respectively. It can be observed from Fig. 8.6 that both the stress waves disperse and attenuate when propagating through MJ and MD-MJ, while the dispersion degree and attenuation degree of the stress wave propagation through MD-MJ are larger than those of MJ.

In addition, Fig. 8.6a, c and e shows that the stress wave propagation at distances of $L = 0.2\lambda_0$, $0.4\lambda_0$ and $0.6\lambda_0$ for a given viscous coefficient $\eta_M = 2.5$ MPa · s, respectively. It is observed from Fig. 8.6a, c and e that the amplitude of the transmitted stress wave of MJ remains constant as the propagation distance increases from $0.2\lambda_0$ to $0.6\lambda_0$, similarly as shown in Fig. 8.4b. Moreover, it is also observed from Fig. 8.66a, c and e that the transmitted stress wave of MD-MJ disperses and attenuates more and more obviously with the increase of distance. A similar phenomenon can be observed from Fig. 8.6b, d and f that the stress wave propagation at distances of $L = 0.2\lambda_0$, $0.4\lambda_0$ and $0.6\lambda_0$ for a given viscous coefficient $\eta_M = 20.0$ MPa · s, respectively.

On the other hand, it is seen from Fig. 8.6 that, for a given propagation distance, such as $L = 0.2\lambda_0$ for Fig. 8.6a and b, $L = 0.4\lambda_0$ for Fig. 8.6c and d, and $L = 0.6\lambda_0$ for Fig. 8.6e and f, the amplitude of transmitted stress wave of MJ remains constant

Table 8.2 Comparisons of transmission coefficients for both of MJ and MD at different propagation distances

Rock mass type	$T_{(L = 0.005\lambda0)}$	$T_{(L = 0.05\lambda0)}$	$T_{(L = 0.2\lambda0)}$	$T_{(L = 0.4\lambda0)}$	$T_{(L = 0.6\lambda0)}$
MJ	0.92	0.92	0.92	0.92	0.92
MD	0.99	0.97	0.88	0.79	0.71

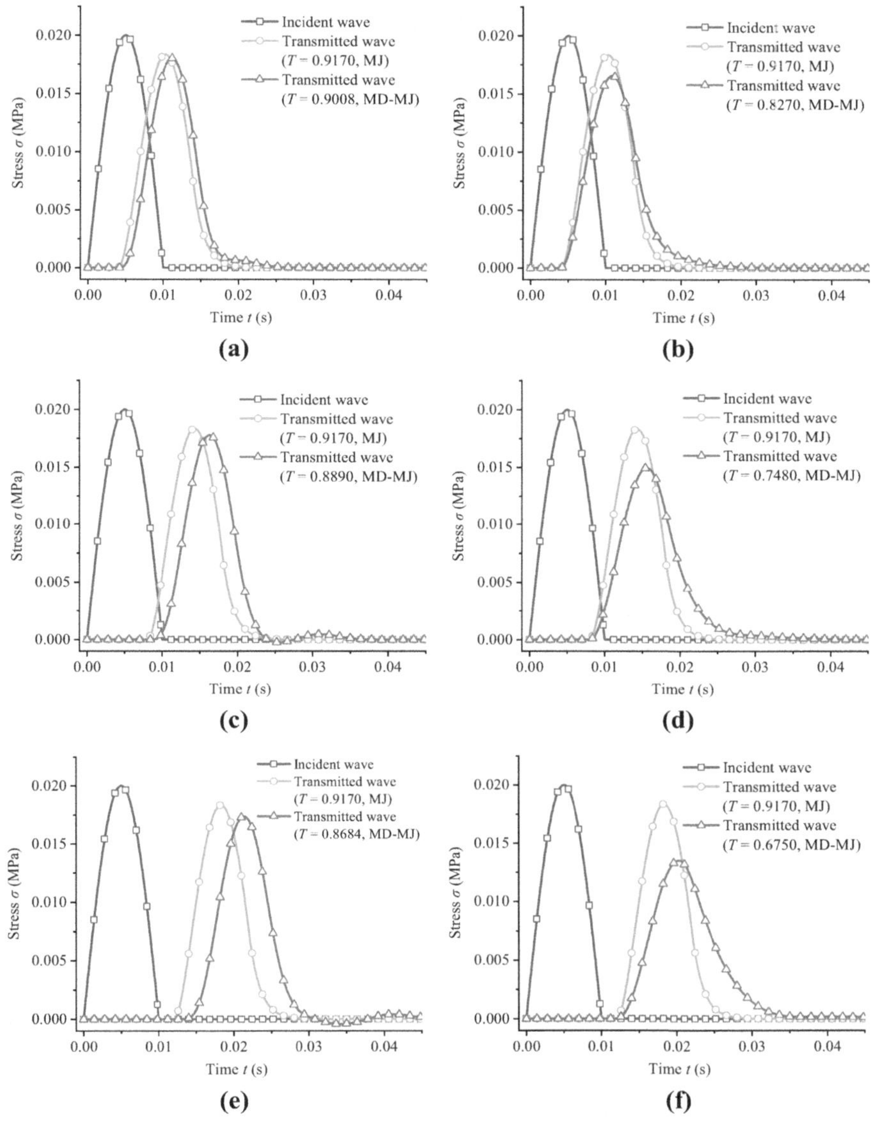

Fig. 8.6 Comparison of transmitted stress waves of MD-MJ and MJ for different propagation distances and viscous coefficients ($k_n = 3.5$ GPa/m). **a** $L = 0.2\lambda_0$ and $\eta_M = 2.5$ MPa·s. **b** $L = 0.2\lambda_0$ and $\eta_M = 20.0$ MPa · s. **c** $L = 0.4\lambda_0$ and $\eta_M = 2.5$ MPa · s. **d** $L = 0.4\lambda_0$ and $\eta_M = 20.0$ MPa · s. **e** $L = 0.6\lambda_0$ and $\eta_M = 2.5$ MPa · s. **f** $L = 0.6\lambda_0$ and $\eta_M = 20.0$ MPa · s

while the transmitted stress wave of MJ-MD attenuates as the viscous coefficient η_M increases from 2.5 to 20.0 MPa · s.

8.1.3 Transmission Coefficient and Effective Velocity of Stress Wave Propagation Through Double-Scale Rock Mass Discontinuities

The transmission coefficient T is used to evaluate the attenuation of the transmitted stress wave, which equals the ratio of the amplitude of the transmitted stress wave to the amplitude of the incident stress wave, i.e.,

$$T = \frac{\max(\sigma_T)}{\max(\sigma_I)} \tag{8.33}$$

where $\max(\sigma_T)$ and $\max(\sigma_I)$ are the peak stresses of transmitted and incident stress waves, respectively.

The effective velocity C_{eff} is the ratio of the length of the rock mass to the time difference between the peak values of the incident stress wave and transmitted stress wave, i.e.,

$$C_{eff} = \frac{L}{t_T - t_I} \tag{8.34}$$

where L is the distance of stress wave propagation, and t_T and t_I are the times for peak stress of incident stress wave σ_I and transmitted stress wave σ_T, respectively.

An incident stress wave in the form of Eq. (8.32) propagating across MD-MJ is investigated. The influences of incident stress wave frequency and macro-joint stiffness on transmission coefficient and effective velocity are calculated and compared with those obtained by incident stress wave propagating across MJ, as shown in Figs. 8.7 and 8.8.

(1) Effect of incident stress wave frequency on stress wave transmission and effective velocity

Generally, rock mass is affected by dynamic loads ranging from a few Hertz, such as earthquakes, to thousands of Hertz, such as explosions, so it is necessary to study the relationship between frequency and transmission stress wave, as shown in Fig. 8.7.

Figure 8.7a shows the effect of incident stress wave frequency on the transmission coefficient for macro-joint stiffness of $k_n = 3.5$, 11.5, and 26.5 GPa/m with a fixed propagation distance of $L = 0.05\lambda_0$ and the viscous coefficient of MD-MJ is 20.0 MPa·s. It is observed from Fig. 8.7a that the general decreasing trend of the transmission coefficient of MD-MJ with the increase of incident stress wave frequency is the same as MJ. The transmission coefficient is almost a constant of 1.0 in the small-frequency region of $f_0 < 10^1$ Hz, which decreases rapidly with the increase

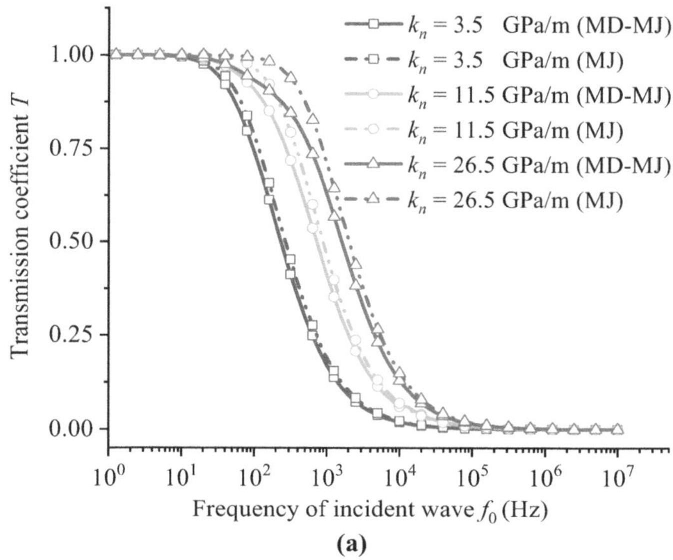

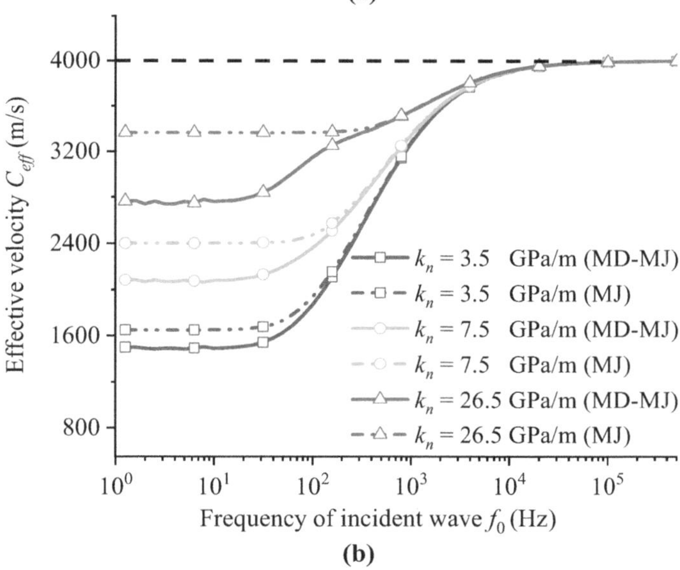

Fig. 8.7 Effect of incident stress wave frequency on stress wave transmission. **a** Transmission coefficient. **b** Effective velocity

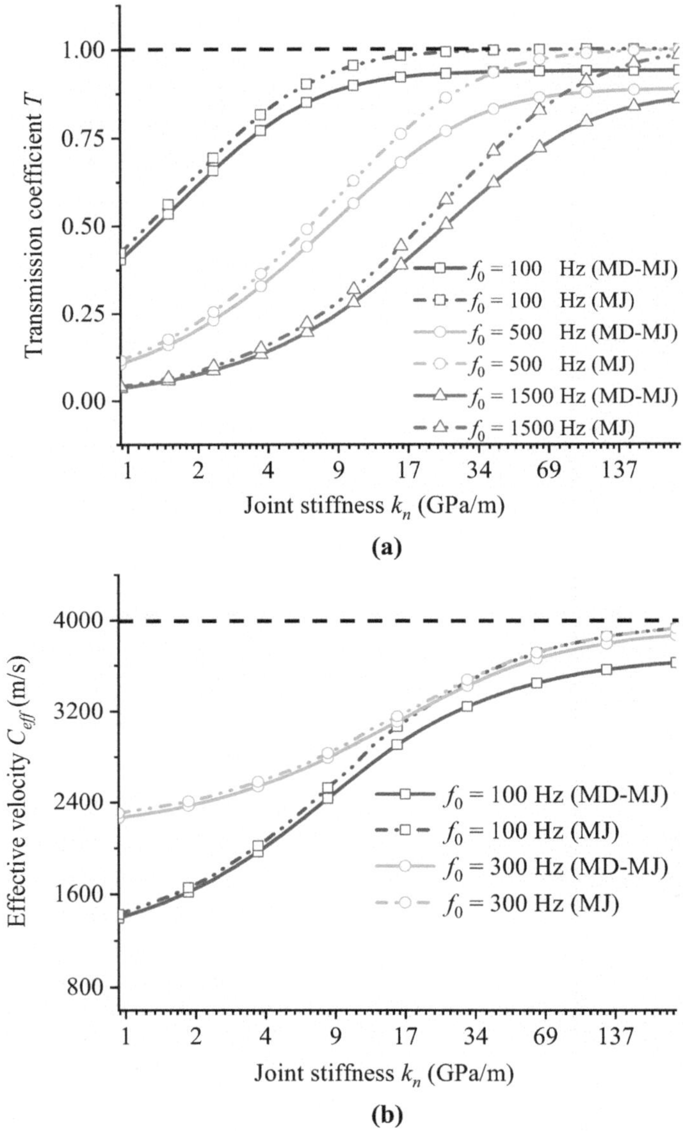

Fig. 8.8 Effect of macro-joint stiffness on stress wave transmission. **a** Transmission coefficient. **b** Effective velocity

of frequency in the medium-frequency region from 10^1 to 10^5 Hz and gradually becomes zero in the large-frequency region of $f_0 > 10^5$ Hz. This explains why small-frequency earthquake loads propagate farther than large-frequency explosive loads. However, it is found from Fig. 8.7a that the transmission coefficient of MJ is larger than that of MD-MJ in the medium-frequency region with respect to the same macro-joint stiffness, which shows that the influence of micro-defect on transmission stress wave increases in the medium-frequency region. Moreover, Fig. 8.7a also shows that the transmission coefficient increases with the increase of macro-joint stiffness in the medium-frequency region.

Figure 8.7b shows the relationship between incident stress wave frequency and effective velocity for macro-joint stiffness of $k_n = 3.5$, 7.5 and 26.5 GPa/m, with the same propagation distance and viscous coefficient as in Fig. 8.7a. The results show that the effective velocity stabilizes to a minimum at a small frequency $f_0 < 10^1$ Hz, then it increases rapidly with the increase of frequency in the medium-frequency region of $10^1 < f_0 < 10^5$ Hz. When the frequency is sufficiently large as $f_0 > 10^5$ Hz, the effective velocity approaches the velocity of the stress wave in intact rock, which is 4000 m/s. In addition, the effective velocity increases as macro-joint stiffness increases for a given incident stress wave frequency. However, the difference decreases with the increase in frequency. It is also observed from Fig. 8.7b that when the macro-joint stiffness of MJ and MD-MJ are the same, the effective velocity of MJ is larger than that of MD-MJ in the small-frequency region and the difference increase with the increase of macro-joint stiffness.

(2) Effect of macro-joint stiffness on stress wave transmission and effective velocity

Figure 8.8 shows the effect of macro-joint stiffness on stress wave transmission. The propagation distance in MJ and MD-MJ, and the viscous coefficient of MD-MJ are the same as in Sect. 4.1.

Figure 8.8a shows the relationship between macro-joint stiffness and transmission coefficient for different frequencies of $f_0 = 100$, 500 and 1500 Hz. From Fig. 8.8a, it is noticed that the transmission coefficient increases with the increase of macro-joint stiffness. When the macro-joint stiffness is sufficiently large, the transmission coefficient of MJ approaches 1.0, similar to the results of wave propagation across MJ in the previous research [8], while that of MD-MJ with micro-defect is smaller than 1.0. The transmission coefficient in Fig. 8.8a increases as the frequency decreases for a given macro-joint stiffness, which can be also observed in Fig. 8.7a. Moreover, the transmission coefficient of MJ is always larger than that of MD-MJ at the same frequency, and the difference increases as the macro-joint stiffness increases. It also shows that the influence of the micro-defect in MD-MJ on transmission stress wave increases with the increase of macro-joint stiffness.

Figure 8.8b shows the relationship between the macro-joint stiffness and the effective velocity with respect to different frequencies of $f_0 = 100$ and 300 Hz. It is observed from Fig. 8.8b that the effective velocity with the increase of macro-joint stiffness increases. When the macro-joint stiffness is large enough, the effective velocity of MJ approaches the stress wave velocity, which is 4000 m/s, while that of MD-MJ approaches a value smaller than 4000 m/s due to the consideration of the micro-defect. For the same macro-joint stiffness, the effective velocity increases with the increase of incident stress wave frequency. In addition, it is seen from Fig. 8.8b that when the frequency is small enough, the effective velocity of MD-MJ with micro-defect is smaller than that of MJ, and the difference increases as the macro-joint stiffness increases. However, the difference decreases with the increase in frequency, which can be also observed in Fig. 8.7b.

8.2 Effect of Nonlinearly Deformational Macro-Joint on Stress Wave Propagation Through a Double-Scale Discontinuous Rock Mass

The study of stress wave propagation in natural rock mass is of great significance for the dynamic stability of rock engineering. The discontinuities of micro-defects and macro-joints that exist in the rock mass significantly influence the stress wave transmission. The micro-defect causes the viscoelastic behavior of the rock mass under dynamic loading [12, 13]. On the other hand, the completely deformational behavior of filled macro-joint is generally nonlinear [14–16], which makes the effective mechanical properties of a double-scale discontinuous rock mass difficult to investigate.

This section presents a study on stress wave propagation through a double-scale discontinuous rock mass with a nonlinearly deformational macro-joint by combining the split three-characteristic method with the piecewise linear DDM. The effect of the nonlinearly deformational macro-joint and micro-defect on stress wave propagation is revealed. In addition, the effect of amplitude, frequency and initial macro-joint stiffness on the transmission coefficient obtained based on the double-scale discontinuous rock mass with the nonlinearly deformational macro-joint is discussed in detail. Subsequently, the differences observed with and without considering the nonlinearly deformational macro-joint effect on the wave propagation through the double-scale discontinuous rock mass are further studied.

8.2.1 Split Three-Characteristic Method Based on the Piecewise Linear DDM

(1) Introduction of split three-characteristic method and piecewise linear DDM

In natural rock mass, micro-defects (e.g., pores and cracks) are uniformly distributed, and macro-joints are assumed to be parallel [17, 18]. The representative elementary volume (REV) of a double-scale discontinuous rock mass with a nonlinearly deformational macro-joint is shown in Fig. 8.9a. Figure 8.9b and d are magnified rectangular regions of Fig. 8.9a. The existence of micro-defects in the rock mass can be observed from the magnified rectangular region, as shown in Fig. 8.9b, which are too small to be seen by the naked eye in the rock mass. The incident wavelength is larger than the maximum geometric dimension of the micro-defect by an order of magnitude. The presence of micro-defects leads to the viscoelastic behavior of rock mass. Figure 8.9c shows the three-element viscoelastic model that describes the dynamic behavior of micro-defected rock. The three-characteristic method is especially convenient and effective for solving the problem of stress wave propagation through micro-defected rock mass with viscoelastic properties [19]. In addition, Fig. 8.9d shows the existence of filled macro-joint in the rock mass, and the deformational behavior of filled joints is usually nonlinear as shown in Fig. 8.9e. The macro-joint is considered as a large crack in size and thin in thickness related to the wavelength. The nonlinear DDM has been widely used in solving the problem of stress wave propagation through a nonlinearly deformational macro-joint.

However, in order to solve the problem of stress wave propagation through a double-scale discontinuous rock mass with a nonlinearly deformational macro-joint, it is necessary to propose a method combining the three characteristic lines with the nonlinear DDM. Therefore, the split three-characteristic method is proposed, as shown in Fig. 8.10a. It can be seen from Fig. 8.10a that the x-t plane is divided into equidistant grids Δx and Δt by the three sets of characteristic lines [20]. The three-characteristic lines are split at $x = x_1$, where is assumed as the location of a single macro-joint.

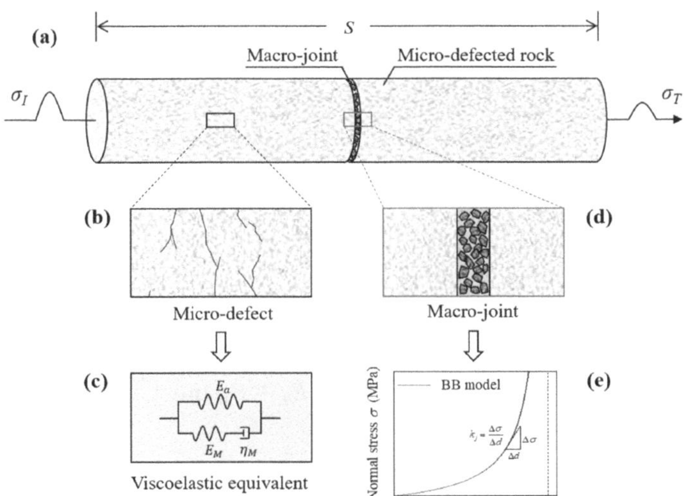

Fig. 8.9 Stress wave propagation through the double-scale discontinuous rock mass with nonlinearly deformational macro-joint. **a** Representative elementary volume (REV) of the double-scale discontinuous rock mass. **b** Micro-defect in the rock mass. **c** Three-element viscoelastic model of equivalent micro-defect. **d** Single filled macro-joint in the rock mass. **e** The nonlinearly deformational behavior of the macro-joint

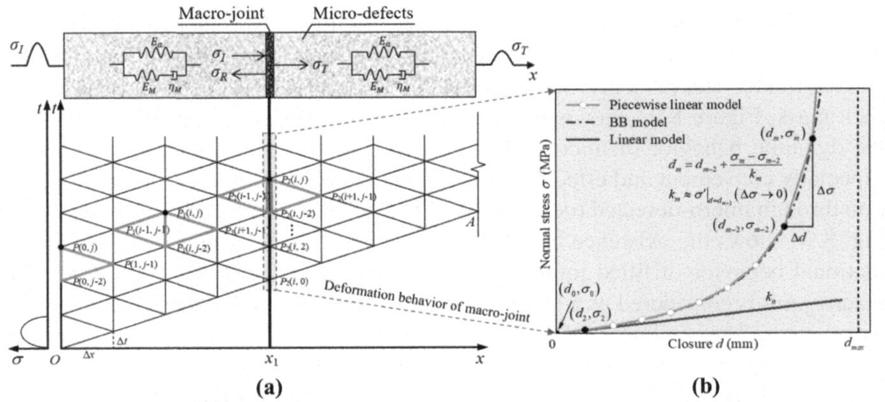

Fig. 8.10 Split three-characteristic combined with piecewise linear DDM. **a** The split three-characteristic method. **b** Piecewise linear model, BB model and linear model of the macro-joint

According to the three-element viscoelastic model in Fig. 8.9c, three sets of characteristics in Fig. 8.10a can be obtained as

$$\frac{dx}{dt} = \pm\sqrt{\frac{E_a + E_M}{\rho_0}} = \pm C_V \tag{8.35}$$

and

$$dx = 0 \tag{8.36}$$

and the corresponding three sets of characteristic compatibility relationships can be obtained as

$$\begin{aligned} dv &= \pm\frac{1}{\rho_0 C_V}d\sigma \pm \frac{(\sigma - E_a\varepsilon)E_M}{\rho_0 C_V \eta_M}dt \\ &= \pm\frac{1}{\rho_0 C_V}d\sigma + \left[\frac{(\sigma - E_a\varepsilon)E_M}{(E_a + E_M)\eta_M}\right]dx \end{aligned} \tag{8.37}$$

and

$$d\varepsilon - \frac{d\sigma}{E_a + E_M} - \frac{(\sigma - E_a\varepsilon)E_M}{(E_a + E_M)\eta_M}dt = 0 \tag{8.38}$$

where x and t are Cartesian coordinate and time, respectively. σ, ε and v are the particle stress, particle strain and particle velocity, respectively. ρ_0 is the density. C_V is the wave velocity. E_M and η_M are the elastic modulus of spring and viscous coefficient of the dashpot in the Maxwell element in Fig. 8.9c, respectively. E_a is the elastic modulus of spring in parallel with the Maxwell element in Fig. 8.9c.

Different from the linear model of macro-joint, the BB model involves macro-joint where the stiffness increases with the increasing stress as shown in Fig. 8.10b, which makes it very complicated to solve stress wave propagation through a double-scale discontinuous rock mass with a nonlinearly deformational macro-joint by combining the split three-characteristic method with nonlinear DDM. Therefore, a piecewise linear model is proposed to consider the nonlinearly deformational behavior of the macro-joint as shown in Fig. 8.10b, which approximates the BB model with enough accuracy when the closure difference Δd is small enough. Next, the closure-stress relationship and macro-joint stiffness of each segment in Fig. 8.10b can be obtained, which are combined with the stress continuity condition of each segment to form the piecewise linear DDM. The closure-stress relationship at the ends of each segment of the piecewise linear model satisfies the BB model as

$$d = \frac{\sigma}{k_n + \sigma/d_{\max}} \quad \text{or} \quad \sigma = \frac{dk_n}{1 - d/d_{\max}} \tag{8.39}$$

Based on Eq. (8.39), the macro-joint stiffness of any point in the BB model is

$$k = \frac{\partial \sigma}{\partial d} = \frac{k_n}{(1 - d/d_{max})^2} \tag{8.40}$$

where d is the macro-joint closure, σ is the normal stress, $d_{\max}$ is the maximum allowable closure of the macro-joint, k_n is the initial macro-joint stiffness, and k is the macro-joint stiffness of any point in the BB model.

For the piecewise linear model shown in Fig. 8.10b, $(d_0, \sigma_0) - (d_2, \sigma_2)$ is defined as segment 2, $(d_2, \sigma_2) - (d_4, \sigma_4)$ is defined as segment 4, and ... $(d_{m-2}, \sigma_{m-2}) - (d_m, \sigma_m)$ is defined as segment m, where m takes an even number. Assuming that the point (d_{m-2}, σ_{m-2}) at the left end of segment m $(d_{m-2}, \sigma_{m-2}) - (d_m, \sigma_m)$ is known and substituted into Eq. (8.40), the macro-joint stiffness of segment m can be expressed as

$$k_m = \frac{k_n}{(1 - d_{m-2}/d_{max})^2}, m = 2, 4, 6 \cdots\cdots \tag{8.41}$$

According to the point (d_{m-2}, σ_{m-2}) and macro-joint stiffness of segment m, the stress-closure relation of segment m (d_{m-2}, σ_{m-2})-(d_m, σ_m) can be described as.

$$d_m = d_{m-2} + \frac{\sigma_m - \sigma_{m-2}}{k_m}, m = 2, 4, 6 \cdots\cdots \tag{8.42}$$

when $m = 2$, $d_{m-2} = d_0 = 0$ and $k_m = k_2 = k_n$.

The closure-stress relationship of each point of the split three characteristic lines at $x = x_1$ in Fig. 8.10a can correspond to the piecewise linear model in Fig. 8.10b. We assume that the macro-joint closure $d(i, j–2)$ and normal stress $\sigma(i, j–2)$ of point $P_2(i, j–2)$ in Fig. 8.10a correspond to the point (d_{m-2}, σ_{m-2}) in Fig. 8.10b. According to

Eqs. (8.41) and (8.42), the macro-joint stiffness $k(i, j)$ and the stress-closure relation of point $P_2(i, j)$ in Fig. 8.10a can be obtained as

$$k(i,j) = \frac{k_n}{\left[1 - d(i,j-2)/d_{max}\right]^2}, \quad i = x_1/\Delta x, \quad j = 2, 4, 6 \cdots\cdots \tag{8.43}$$

and

$$d(i,j) = d(i,j-2) + \frac{\sigma(i,j) - \sigma(i,j-2)}{k(i,j)}, \quad i = x_1/\Delta x, \quad j = 2, 4, 6 \cdots\cdots \tag{8.44}$$

The displacement discontinuity boundary conditions of point $P_2(i,j)$ at the macro-joint in Fig. 8.10a can be obtained by Eq. (8.44) of the stress-closure relation and Eq. (8.45) of the stress continuity condition, which is the piecewise linear DDM:

$$\sigma^+(i,j) = \sigma^-(i,j) = \sigma(i,j), i = x_1/\Delta x, j = 2, 4, 6 \cdots\cdots \tag{8.45}$$

where $d(i,j) = u^+(i,j) - u^-(i,j)$, $u(i,j)$ denote the displacement normal to the macro-joint at point $P_2(i, j)$, and the signs "–" and "+" denote the wave fields in front of and behind the macro-joint, respectively. $i = x_1/\Delta x$ denotes the location of the macro-joint.

The derivative of Eq. (8.44) with respect to t is Eq. (8.46):

$$v^+(i,j) - v^-(i,j) = \frac{1}{k(i,j)} \frac{\partial \sigma(i,j)}{\partial t}, \quad i = x_1/\Delta x, \quad j = 2, 4, 6 \cdots\cdots \tag{8.46}$$

The derivative form of Eq. (8.46) is replaced by a differential form, so Eq. (8.46) is transformed into

$$v^+(i,j) - v^-(i,j) = \frac{1}{k(i,j)} \frac{\sigma(i,j) - \sigma(i,j-2)}{2\Delta t}, \quad i = x_1/\Delta x, \quad j = 2, 4, 6 \cdots\cdots \tag{8.47}$$

***

(2) Calculation process of stress wave propagation

The problem of stress wave propagation through a double-scale discontinuous rock mass with a nonlinearly deformational macro-joint is to solve three unknowns of stress σ, strain ε and velocity v at any point in Fig. 8.10a according to the initial and boundary conditions by combining the split three-characteristic method with the piecewise linear DDM. Similar to the calculation process of wave propagation through double-scale discontinuous rock mass, the three unknowns, $\sigma(x, t)$, $\varepsilon(x, t)$ and $v(x, t)$, along the characteristic line OA in Fig. 8.10a can be obtained first, and then the remaining unknowns of the left boundary points $P(0, j)$, the interior points $P_1(i,j)$ and the interior points $P_1(i,j)$ in the AOt domain can be further solved according to the triangle, diamond and separated diamond elements in Fig. 8.10a, respectively. However, in order to study the stress wave propagation at the nonlinearly deformational macro-joint, the piecewise linear DDM is introduced into the separated diamond element.

We assume that the rock mass is in an undisturbed state at $t = 0$ and then undergoes a half-sinusoidal impact loading at $x = 0$. The initial and boundary conditions can be expressed as

$$\begin{cases} \sigma_0(x,0) = 0 \\ \varepsilon_0(x,0) = 0 \ , \ 0 \le x \le S \\ v_0(x,0) = 0 \end{cases} \tag{8.48}$$

and

$$\sigma^*(0,t) = \begin{cases} A_0 sin(2\pi f_0 t) \\ 0 \end{cases}, \quad \text{when} \quad \begin{matrix} 0 \le t \le 1/(2f_0) \\ \text{others} \end{matrix} \tag{8.49}$$

where S is the length of representative elementary volume (REV), A_0 is the amplitude and f_0 is the frequency of incident stress wave.

Once the initial and boundary conditions of rock mass are determined, the stress wave propagating through the rock mass can be described by using Eqs. (8.37) to (8.38) of three sets of characteristic compatibility relationships. According to the Hugoniot relations and compatibility conditions along the characteristic line *OA*, the three unknowns, $\sigma^-(x,t)$, $\varepsilon^-(x,t)$ and $v^-(x,t)$, in front of the macro-joint along characteristic line *OA* are expressed as

$$\sigma^-(x,t) = \sigma^*(0,0)\exp\left[-\frac{\rho_0 C_V}{2\eta_M(1+E_a/E_M)^2}x\right] \tag{8.50}$$

$$v^-(x,t) = -\frac{\sigma^*(0,0)}{\rho_0 C_V}\exp\left[-\frac{\rho_0 C_V}{2\eta_M(1+E_a/E_M)^2}x\right] \tag{8.51}$$

and

$$\varepsilon^-(x,t) = \frac{\sigma^*(0,0)}{E_a+E_M}\exp\left[-\frac{\rho_0 C_V}{2\eta_M(1+E_a/E_M)^2}x\right] \tag{8.52}$$

The three unknowns, $\sigma^+(x,t)$, $\varepsilon^+(x,t)$ and $v^+(x,t)$, behind the macro-joint along characteristic line *OA* are derived to be

$$\sigma^-(x,t) = \sigma^-(x,t) \tag{8.53}$$

$$v^+(x,t) = v^-(x,t) - C_V\sigma^*(0,0)\left(-\frac{\rho_0 C_V}{2\eta_M(1+E_a/E_M)^2}\right)\exp\left[-\frac{\rho_0 C_V}{2\eta_M(1+E_a/E_M)^2}x\right]\frac{1}{k_n} \tag{8.54}$$

and

$$\varepsilon^{+}(x,t) = \varepsilon^{-}(x,t) - \sigma^{*}(0,0)\left(-\frac{\rho_0 C_V}{2\eta_M(1+E_a/E_M)^2}\right)\exp\left[-\frac{\rho_0 C_V}{2\eta_M(1+E_a/E_M)^2}x\right]\frac{1}{k_n} \tag{8.55}$$

where x is equal to $C_V \times t$ along the characteristic line OA.

Then, according to the compatibility conditions corresponding to characteristic line $P(1,j-1)-P(0,j)$ and $P(0,j-2)-P(0,j)$ in the triangle elements of Fig. 8.10a and the compatibility conditions corresponding to characteristic line $P_1(i-1,j-1)-P_1(i,j)$, $P_1(i+1,j-1)-P_1(i,j)$ and $P_1(i,j-2)-P_1(i,j)$ in the diamond elements of Fig. 8.10a, the stress wave propagation at the left boundary points $P(0,j)$ and the interior points $P_1(i,j)$ can be obtained, respectively. The two unknowns, $\varepsilon(0,j)$ and $v(0,j)$, at the left boundary points $P(0,j)$ can be obtained as

$$\begin{aligned} v(0,j) = v(1,j-1) &- \frac{1}{\rho_0 C_V}\left[\sigma(0,j)-\sigma(1,j-1)\right] \\ &+ \frac{\left[\sigma(1,j-1)-E_a\varepsilon(1,j-1)\right]E_M}{(E_a+E_M)\eta_M}\left[x(0,j)-x(1,j-1)\right] \end{aligned} \tag{8.56}$$

and

$$\begin{aligned} \varepsilon(0,j) = \varepsilon(0,j-2) &+ \frac{1}{E_a+E_M}\left[\sigma(0,j)-\sigma(0,j-2)\right] \\ &+ \frac{\left[\sigma(0,j-2)-E_a\varepsilon(0,j-2)\right]E_M}{(E_a+E_M)\eta_M}\left[t(0,j)-t(0,j-2)\right] \end{aligned} \tag{8.57}$$

The three unknowns, $\sigma(i,j)$, $\varepsilon(i,j)$ and $v(i,j)$, at the interior points $P_1(i,j)$ can be obtained as

$$\begin{aligned} v(i,j)-v(i-1,j-1) = &\frac{1}{\rho_0 C_V}\left[\sigma(i,j)-\sigma(i-1,j-1)\right] \\ &+ \frac{\left[\sigma(i-1,j-1)-E_a\varepsilon(i-1,j-1)\right]E_M}{(E_a+E_M)\eta_M}\left[x(i,j)-x(i-1,j-1)\right] \end{aligned} \tag{8.58}$$

$$v(i,j)-v(i+1,j-1) = -\frac{1}{\rho_0 C_V}\left[\sigma(i,j)-\sigma(i+1,j-1)\right] \tag{8.59}$$

and

$$\begin{aligned} \varepsilon(i,j)-\varepsilon(i,j-2) = &\frac{1}{E_a+E_M}\left[\sigma(i,j)-\sigma(i,j-2)\right] \\ &+ \frac{\left[\sigma(i,j-2)-E_a\varepsilon(i,j-2)\right]E_M}{(E_a+E_M)\eta_M}\left[t(i,j)-t(i,j-2)\right] \end{aligned} \tag{8.60}$$

The interior points $P_2(i, j)$ in Fig. 8.10a include points P_2^- (i, j) and points P_2^+ (i, j) at the two sides of the nonlinearly deformational macro-joint. Therefore, there are six unknowns, $\sigma^-(i,j)$, $\sigma^+(i,j)$, $v^-(i,j)$, $v^+(i,j)$, $\varepsilon^-(i,j)$ and $\varepsilon^+(i,j)$, at the interior points $P_2(i,j)$. According to Eqs. (8.61) to (8.64), which are compatibility conditions along characteristic line $P_2(i-1, j-1) - P_2^-(i, j)$, $P_2(i+1, j-1)$- $P_2^+(i, j)$, $P_2^+(i, j) - P_2^+(i, j-2)$ and $P_2^-(i, j) - P_2^-(i, j-2)$ in the separated diamond elements of Fig. 8.10a and Eqs. (8.45) and (8.47), which are the displacement discontinuity boundary conditions of the nonlinearly deformational macro-joint, the six unknowns at the interior points $P_2(i,j)$ can be obtained as

$$v^-(i,j) - v(i-1,j-1) = \frac{1}{\rho_0 C_V}\left[\sigma(i,j) - \sigma(i-1,j-1)\right] + \frac{\left[\sigma(i-1,j-1) - E_a\varepsilon(i-1,j-1)\right]E_M}{(E_a + E_M)\eta_M}dx \quad (8.61)$$

$$v^+(i,j) - v(i+1,j-1) = -\frac{1}{\rho_0 C_V}\left[\sigma(i,j) - \sigma(i+1,j-1)\right] - \frac{\left[\sigma(i+1,j-1) - E_a\varepsilon(i+1,j-1)\right]E_M}{(E_a + E_M)\eta_M}dx \quad (8.62)$$

$$\varepsilon^-(i,j) - \varepsilon^-(i,j-2) = \frac{1}{E_a + E_M}\left[\sigma(i,j) - \sigma(i,j-2)\right] + \frac{\left[\sigma(i,j-2) - E_a\varepsilon^-(i,j-2)\right]E_M}{(E_a + E_M)\eta_M}(2dt) \quad (8.63)$$

and

$$\varepsilon^+(i,j) - \varepsilon^+(i,j-2) = \frac{1}{E_a + E_M}\left[\sigma(i,j) - \sigma(i,j-2)\right] + \frac{\left[\sigma(i,j-2) - E_a\varepsilon^+(i,j-2)\right]E_M}{(E_a + E_M)\eta_M}(2dt) \quad (8.64)$$

By using Eqs. (8.45), (8.47), (8.61) and (8.62), $\sigma^-(i,j)$ and $\sigma^+(i,j)$ can be obtained as

$$\sigma(i,j) = \frac{2\rho_0 C_V k(i,j)dt}{\rho_0 C_V + 4k(i,j)dt}\left[\begin{array}{c} \frac{\sigma(i+1,j-1)+\sigma(i-1,j-1)}{\rho_0 C_V} - v(i-1,j-1) \\ +v(i+1,j-1) + \frac{1}{2dtk(i,j)}\sigma(i,j-2) \\ -\frac{\left[\sigma(i+1,j-1)-E_a\varepsilon(i+1,j-1)\right]E_M}{(E_a+E_M)\eta_M}dx \\ -\frac{\left[\sigma(i-1,j-1)-E_a\varepsilon(i-1,j-1)\right]E_M}{(E_a+E_M)\eta_M}dx \end{array}\right] \quad (8.65)$$

where $i = x_1/\Delta x$ and $j = 2, 4, 6\ldots\ldots$

Equation (8.65) shows that $\sigma(i,j)$ can be calculated from $\sigma(i, j-2)$. Therefore, the initial macro-joint stiffness of $k(i,j) = k_n$ and the initial stress of $\sigma(i, 0) = 0$ are specified, and the stress $\sigma(i, j)$ at $x = x_1$ in Fig. 8.10a can be determined in an iterative computation using Eqs. (8.65) and (8.43). Then, together with Eqs. (8.61)–(8.64), the remaining four unknowns, $v^-(i, j)$, $v^+(i, j)$, $\varepsilon^-(i, j)$ and $\varepsilon^+(i, j)$, can be obtained, respectively.

8.2.2 Waveforms of Transmitted Stress Waves Through Nonlinearly Deformational Macro-Joint and Linearly Deformational Macro-Joint

The effect of the nonlinearly deformational macro-joint on stress wave propagation is revealed by comparing it with that of a linearly deformational macro-joint for different amplitudes of the incident stress wave. The effect of the micro-defect on stress wave propagation is studied by the investigation of stress wave propagation in a double-scale discontinuous rock mass with different lengths of representative elementary volume (REV).

In the present study, the density of intact rock is $\rho_0 = 2500$ kg/m^3, and the stress wave velocity in the rock is $C_V = 4000$ m/s. The elastic moduli of the three-element viscoelastic model shown in Fig. 8.9d respectively are $E_a = 25.0$ GPa and $E_M = 15.0$ GPa and the viscous coefficient is $\eta_M = 20.0$ MPa · s. The initial macro-joint stiffness is $k_n = 3.5$ GPa/m, and the initial closure and the maximum allowance closure of the macro-joint are $d_0 = 0$ mm and $d_{max} = 0.53$ mm, respectively. In addition, the incident stress wave is assumed to be in the form of Eq. (8.49) with a frequency of $f_0 = 100.0$ Hz.

(1) Stress wave transmission with different incident wave amplitudes

Figure 8.11 compares the transmitted stress waves considering the nonlinearly and linearly deformational macro-joint in a double-scale discontinuous rock mass for different amplitudes of $A_0 = 0.5$, 1.0 and 2.5 MPa, respectively. The length of representative elementary volume (REV) is $S = 0.10\lambda_0$, where $\lambda_0 = C_V/f_0$. In Fig. 8.11, the rock mass considering the nonlinearly deformational macro-joint and micro-defect is denoted as "NLDMJ-MD". The rock mass considering the linearly deformational macro-joint and micro-defect is denoted as "LDMJ-MD". The transmission coefficient T is defined as the ratio of the amplitude of the transmitted stress wave to that of the incident stress wave, which is used to evaluate the attenuation of the transmitted stress wave. It can be observed from Fig. 8.11 that the amplitude of the transmitted stress wave is attenuated with or without considering the nonlinearly deformational behavior of the macro-joint. However, the amplitude of the transmitted stress wave considering the nonlinearly deformational macro-joint is larger than that considering the linearly deformational macro-joint.

The transmission coefficient considering the linearly deformational macro-joint remains constant, while that considering the nonlinearly deformational macro-joint increases as the amplitude of the incident stress wave increases from 0.5 to 2.5 MPa

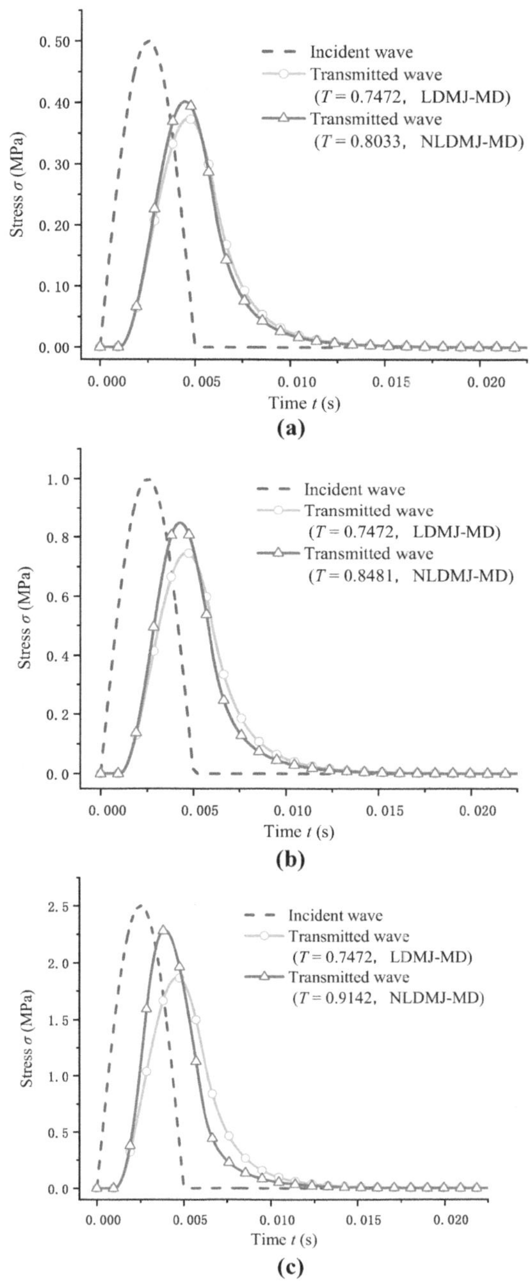

Fig. 8.11 Comparison of waves transmitted across nonlinearly and linearly deformational macro-joint in double-scale discontinuous rock mass at different amplitudes of incident stress waves. **a** $A_0 = 0.5$ MPa. **b** $A_0 = 1.0$ MPa. **c** $A_0 = 2.5$ MPa

(Fig. 8.11a–c). The transmission coefficient considering the linearly deformational macro-joint is independent of the incident stress wave amplitude, while that considering the nonlinearly deformational macro-joint is amplitude-dependent. In addition, it can be observed from Fig. 8.11a–c that the difference between transmitted stress waves considering the nonlinearly and linearly deformational macro-joint increases with the increase of the incident stress wave amplitude. For a small amplitude stress wave, such as $A_0 = 0.5$ MPa in the present study (Fig. 8.11a), the transmitted stress waves considering the nonlinearly and linearly deformational macro-joint are similar. For a large amplitude stress wave, such as $A_0 = 2.5$ MPa in the present study (Fig. 8.11c), the amplitude of the transmitted stress wave considering the nonlinearly deformational macro-joint is significantly larger than that considering the linearly deformational macro-joint.

(2) Stress wave transmission with different lengths of representative elementary volume (REV)

Figure 8.12a shows the transmitted stress wave considering the linearly deformational macro-joint in the double-scale discontinuous rock mass for different lengths of representative elementary volume (REV), i.e., $S = 0.10\lambda_0$, $0.50\lambda_0$ and $1.00\lambda_0$, respectively. Figure 8.12b shows the transmitted stress wave considering the nonlinearly deformational macro-joint in the double-scale discontinuous rock mass for different lengths of representative elementary volume (REV), i.e., $S = 0.10\lambda_0, 0.50\lambda_0$ and $1.00\lambda_0$, respectively. The incident stress wave amplitude is $A_0 = 2.5$ MPa. It is observed from Fig. 8.12a and b that the amplitude of the transmitted stress wave decreases as the length of representative elementary volume (REV) increases with or without considering the nonlinearly deformational behavior of macro-joint, which is because of the effect of micro-defect in the double-scale discontinuous rock mass on stress wave propagation. In addition, Fig. 8.12a and b shows that the effect of micro-defect on the transmitted stress wave becomes more and more significant as the length of representative elementary volume (REV) increases.

Figure 8.13 illustrates the relationship between the transmission coefficient with or without considering the micro-defect effect and the length of representative elementary volume (REV) with respect to nonlinearly and linearly deformational macro-joint in the double-scale discontinuous rock mass. The incident stress wave amplitude is $A_0 = 2.5$ MPa. It is seen from Fig. 8.13 that the transmission coefficient without considering the micro-defect effect is not affected by the length of representative elementary volume (REV), which remains constant with the increase of the length of representative elementary volume (REV), while the transmission coefficient considering the micro-defect effect decreases with the increase of the length of representative elementary volume (REV). This phenomenon can also be explained in Fig. 8.12. It can also be observed from Fig. 8.13 that the transmission coefficient without considering the micro-defect effect is larger than that considering the micro-defect effect. The difference increases as the length of representative elementary volume (REV) increases, which shows that the effect of micro-defect on the stress wave attenuation cannot be ignored when the length of representative elementary volume (REV) is large. On the other hand, it is also observed from Fig. 8.13 that

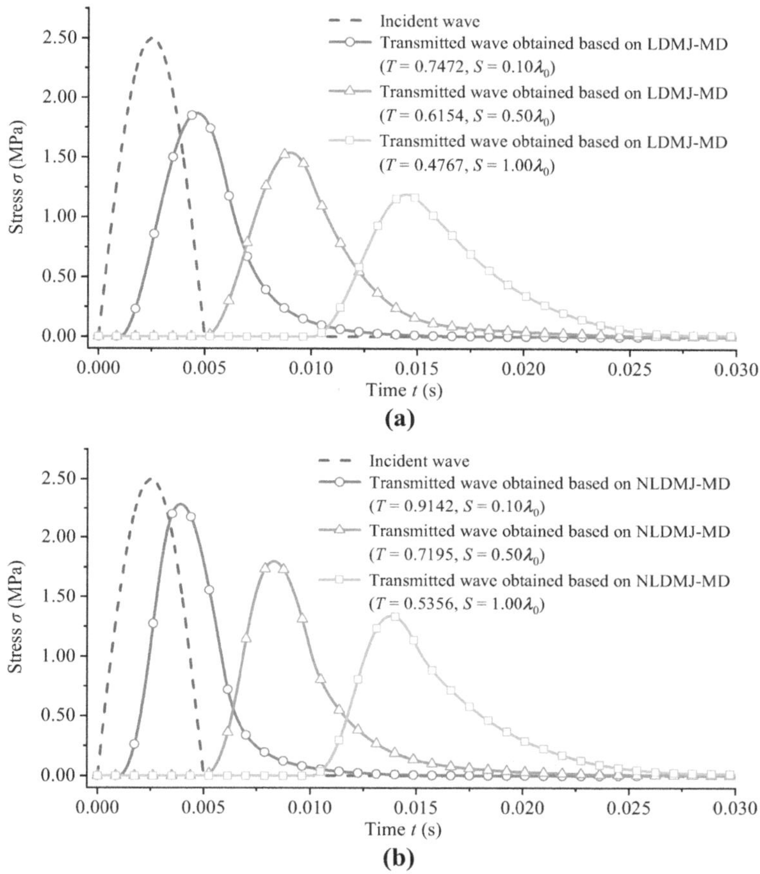

Fig. 8.12 The transmitted stress waves with different lengths of (REV). **a** The transmitted stress waves considering the linearly deformational macro-joint. **b** The transmitted stress waves considering the nonlinearly deformational macro-joint

the transmission coefficient considering the nonlinearly deformational macro-joint is always larger than that considering the linearly deformational macro-joint at the same length of representative elementary volume (REV), which can also be observed from Fig. 8.11.

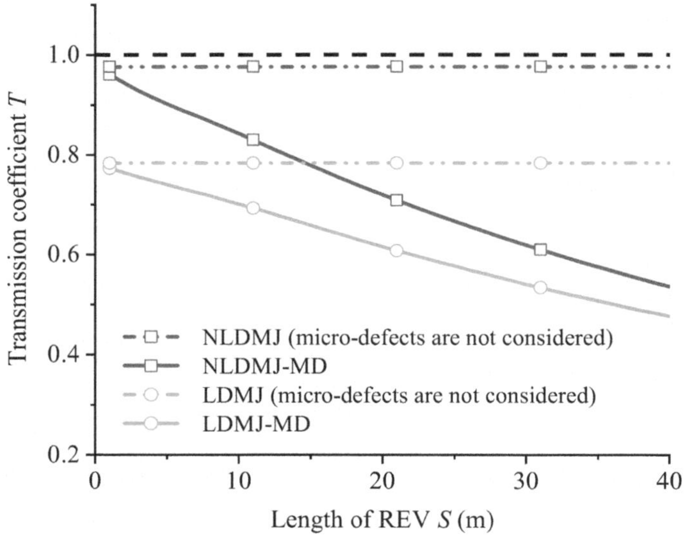

Fig. 8.13 The relationship between the transmission coefficient of stress wave and the length of representative elementary volume (REV)

8.2.3 Transmission Coefficient of Stress Wave Propagates Through a Double-Scale Discontinuous Rock Mass with a Nonlinearly Deformational Macro-Joint

(1) Effect of incident stress wave amplitude on transmission coefficient

Figure 8.14a shows the effect of the incident stress wave amplitude on the transmission coefficient obtained based on NLDMJ-MD and LDMJ-MD for different incident stress wave frequencies of $f_0 = 100$, 200 and 350 Hz, respectively. Figure 8.14b shows the effect of the incident stress wave amplitude on the transmission coefficient obtained based on NLDMJ-MD and LDMJ-MD for different initial macro-joint stiffnesses of $k_n = 2.0$, 3.5 and 7.5 GPa/m, respectively. The length of representative elementary volume (REV) is $S = 4.0$ m.

It is observed from Fig. 8.14a and b that when the stress wave propagates through the double-scale discontinuous rock mass with a linearly deformational macro-joint, the transmission coefficient remains constant with the increase of the incident stress wave amplitude. However, when the stress wave propagates through the double-scale discontinuous rock mass with a nonlinearly deformational macro-joint, the transmission coefficient increases with the increase of the incident stress wave amplitude and approaches a value smaller than 1.0, which is due to the consideration of the effect of micro-defect on stress wave propagation. When the incident stress wave amplitude is small enough, the transmission coefficients with and without considering the nonlinearly deformational macro-joint are equal, as shown in Fig. 8.14a and

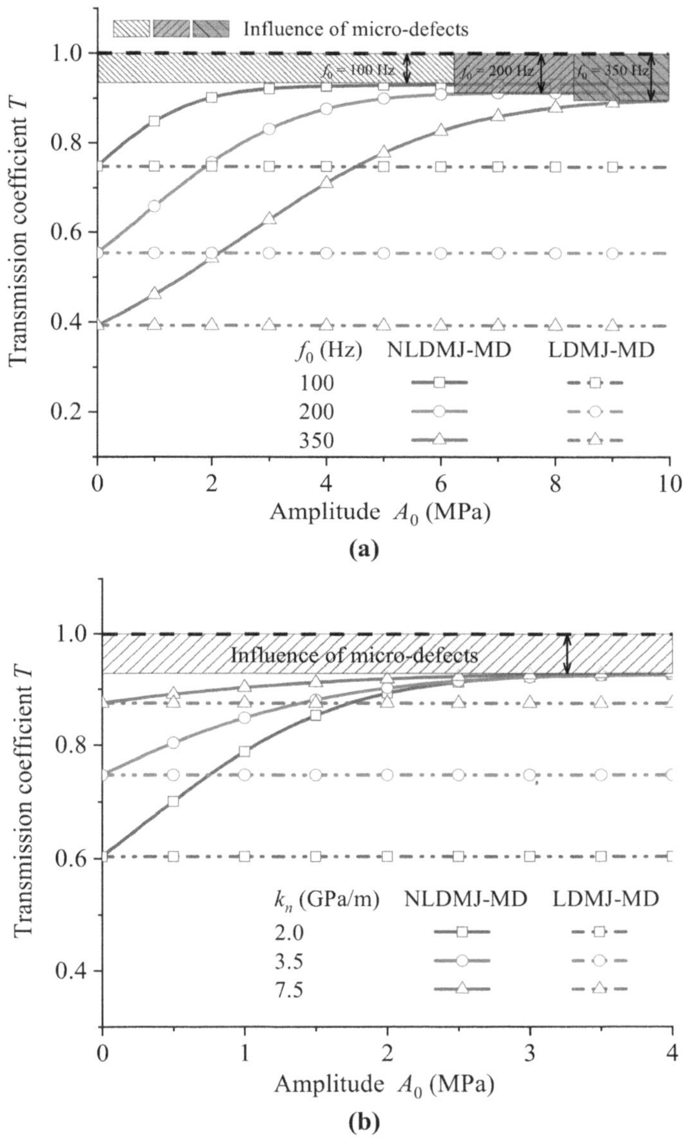

Fig. 8.14 Effect of incident stress wave amplitude on transmission coefficient. **a** $k_n = 3.5$ GPa/m. **b** $f_0 = 100$ Hz

b. However, with the increase of the incident stress wave amplitude, the transmission coefficient considering the nonlinearly deformational macro-joint is larger than that considering the linearly deformational macro-joint, and the difference between the transmission coefficients considering the nonlinearly and linearly deformational macro-joint increases, which can also be observed from Fig. 8.11.

For a given incident stress wave amplitude, the transmission coefficient decreases with the increase of incident stress wave frequency with or without considering the nonlinearly deformational behavior of macro-joint (Fig. 8.14a), and the transmission coefficient increases with the increase of initial macro-joint stiffness (Fig. 8.14b). In addition, the influence degree of micro-defect on the stress wave propagation through the double-scale discontinuous rock mass can be determined from Fig. 8.14a and b when the nonlinearly deformational behavior of macro-joint is considered and the incident stress wave amplitude is large enough. It is observed from Fig. 8.14a that the influence degree of micro-defect increases with the increase of incident stress wave frequency. It is observed from Fig. 8.14b that the influence degree of micro-defect remains unchanged with the increase of initial macro-joint stiffness.

(2) Effect of incident stress wave frequency on transmission coefficient

Figure 8.15a shows the effect of the incident stress wave frequency on the transmission coefficient obtained based on NLDMJ-MD and LDMJ-MD for different incident stress wave amplitudes of $A_0 = 1.0$, 2.0 and 3.5 MPa, respectively. Figure 8.15b shows the effect of the incident stress wave frequency on the transmission coefficient obtained based on NLDMJ-MD and LDMJ-MD for different initial macro-joint stiffnesses of $k_n = 3.5$, 7.5 and 13.5 GPa/m, respectively. The length of representative elementary volume (REV) is $S = 4.0$ m.

It is observed from Fig. 8.15a and b that the change rule of the transmission coefficient of stress wave propagation in the double-scale discontinuous rock mass with the incident stress wave frequency is the same with or without considering the nonlinearly deformational behavior of macro-joint. The transmission coefficient remains constant at 1.0 with the increases of incident stress wave frequency in the small-frequency region of $f_0 < 10^1$ Hz. Then, the coefficient decreases sharply as the incident stress wave frequency further increases in the medium-frequency region of $10^1 < f_0 < 10^4$ Hz and gradually becomes zero in the large-frequency region of $f_0 > 10^4$ Hz. The result explains the phenomenon that a small-frequency stress wave propagates farther than large large-frequency stress wave.

In addition, it can be observed from Fig. 8.15a that the transmission coefficient considering the nonlinearly deformational macro-joint decreases with the decrease of incident stress wave amplitude and approaches transmission coefficient obtained considering the linearly deformational macro-joint in the medium-frequency region. The transmission coefficient considering the linearly deformational macro-joint remains constant for different incident stress wave amplitudes, which is the same as the previous section. It is found from Fig. 8.15b that, in the medium-frequency region, the transmission coefficient increases with the increase of initial macro-joint stiffness with or without considering the nonlinearly deformational behavior of macro-joint,

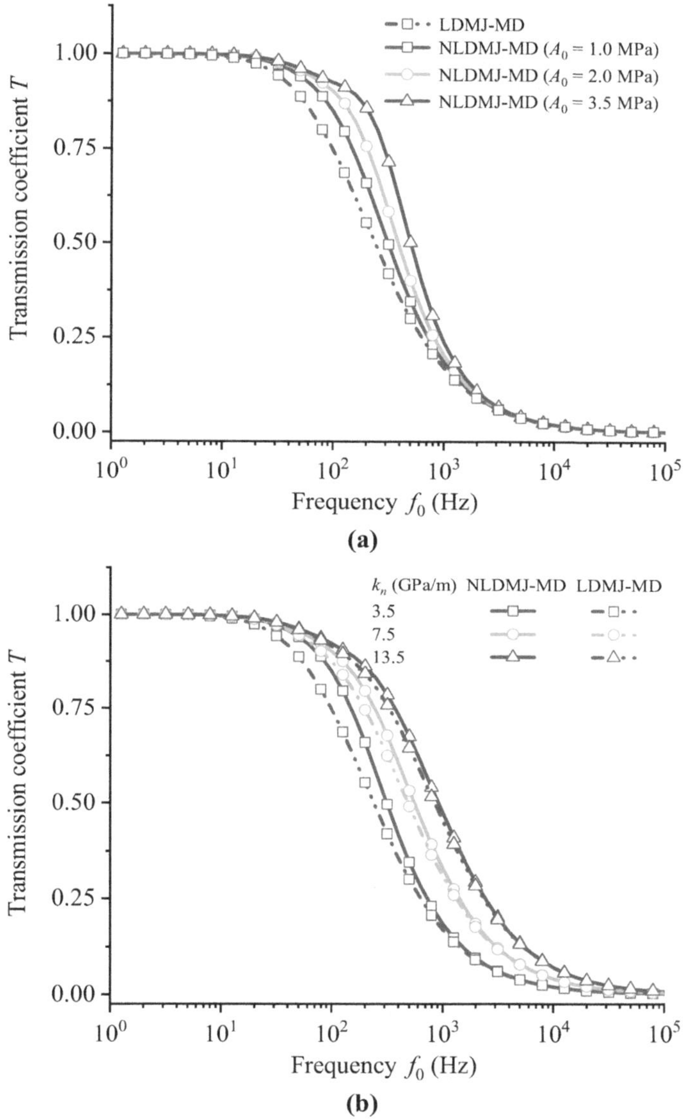

Fig. 8.15 Effect of incident stress wave frequency on transmission coefficient. **a** $k_n = 3.5$ GPa/m. **b** $A_0 = 1.0$ MPa

and the transmission coefficient considering the nonlinearly deformational macro-joint is larger than that considering the linearly deformational macro-joint at the same initial macro-joint stiffness.

(3) Effect of initial macro-joint stiffness on transmission coefficient

Figure 8.16a shows the effect of the initial macro-joint stiffness on the transmission coefficients obtained based on NLDMJ-MD and LDMJ-MD for different incident stress wave amplitudes of $A_0 = 0.5$, 1.0 and 1.5 MPa, respectively. Figure 8.16b shows the effect of the initial macro-joint stiffness on the transmission coefficient obtained based on NLDMJ-MD and LDMJ-MD for different incident stress wave frequencies of $f_0 = 50$, 150 and 300 Hz, respectively. The length of representative elementary volume (REV) is the same as previous research.

It is observed from Fig. 8.16a and b that the transmission coefficient increases rapidly with the increase of initial macro-joint stiffness at the initial stage with and without considering the nonlinearly deformational macro-joint, while it approaches a value smaller than 1.0 with the further increase of the initial macro-joint stiffness. This phenomenon can be explained by the fact that when the initial macro-joint stiffness is sufficiently large, the macro-jointed rock mass behaves as an intact rock mass and the double-scale discontinuous rock mass can be equivalented to a single-scale discontinuous rock mass with micro-defect.

It can be observed from Fig. 8.16a that, for a given initial macro-joint stiffness, the transmission coefficient considering the nonlinearly deformational macro-joint decreases and approaches that considering the linearly deformational macro-joint as the incident stress wave amplitude decreases. The transmission coefficient considering the linearly deformational macro-joint remains constant, which can also be observed in Fig. 8.14. It can be observed from Fig. 8.16b that, for a given initial macro-joint stiffness, the transmission coefficient increases as the incident stress wave frequency decreases with or without considering the nonlinearly deformational behavior of the macro-joint. The transmission coefficient considering the nonlinearly deformational macro-joint is always larger than that considering the linearly deformational macro-joint at the same frequency. When the initial macro-joint stiffness is large enough, the influence degree of micro-defect on the stress wave propagation through the double-scale discontinuous rock mass can be determined from Fig. 8.16a and b. It is observed from Fig. 8.16a that the influence degree of micro-defect remains unchanged for different incident stress wave amplitudes. It is observed from Fig. 8.16b that the influence degree of micro-defect increases with the increase of incident stress wave frequency, which can also be observed in Fig. 8.14a.

(4) Effect of the incident stress wave frequency, amplitude and initial macro-joint stiffness on ΔT

ΔT is proposed to evaluate the difference between the transmission coefficients considering the nonlinearly and linearly deformational macro-joint in the double-scale discontinuous rock mass, which is defined as

$$\Delta T = \frac{|T_{non} - T_{lin}|}{T_{non}} \times 100\% \tag{8.66}$$

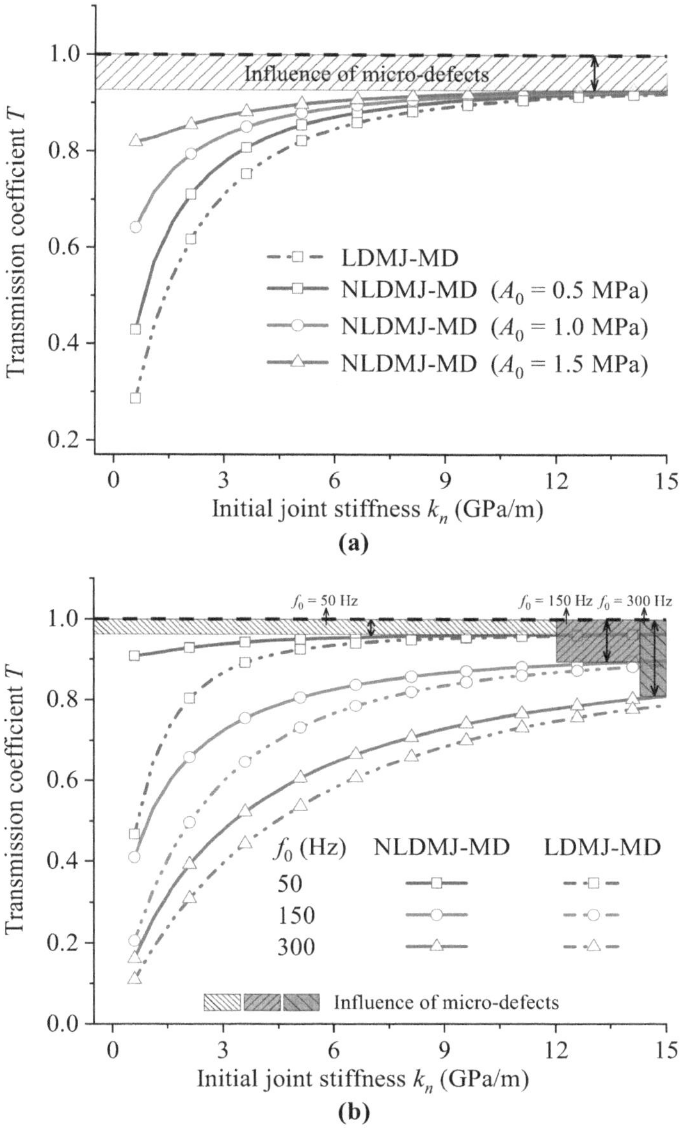

Fig. 8.16 Effect of initial macro-joint stiffness on transmission coefficient. **a** $f_0 = 100$ Hz. **b** $A_0 = 1.0$ MPa

where T_{non} and T_{lin} are denoted as transmission coefficients considering the nonlinearly and linearly deformational macro-joint in the double-scale discontinuous rock mass, respectively.

Figure 8.17a–c shows the effect of incident stress wave frequency, incident stress wave amplitude and initial macro-joint stiffness on ΔT by fixing the length of representative elementary volume (REV) at $S = 4.0$ m.

Figure 8.17a shows the effect of the incident stress wave frequency and amplitude on ΔT. The initial macro-joint stiffness is $k_n = 3.5$ GPa/m. It is observed from Fig. 8.17a that, in the medium-frequency region, ΔT decreases with the decrease of incident stress wave amplitude and gradually becomes zero. However, in the small-frequency or large-frequency region, the effect of the incident stress wave amplitude

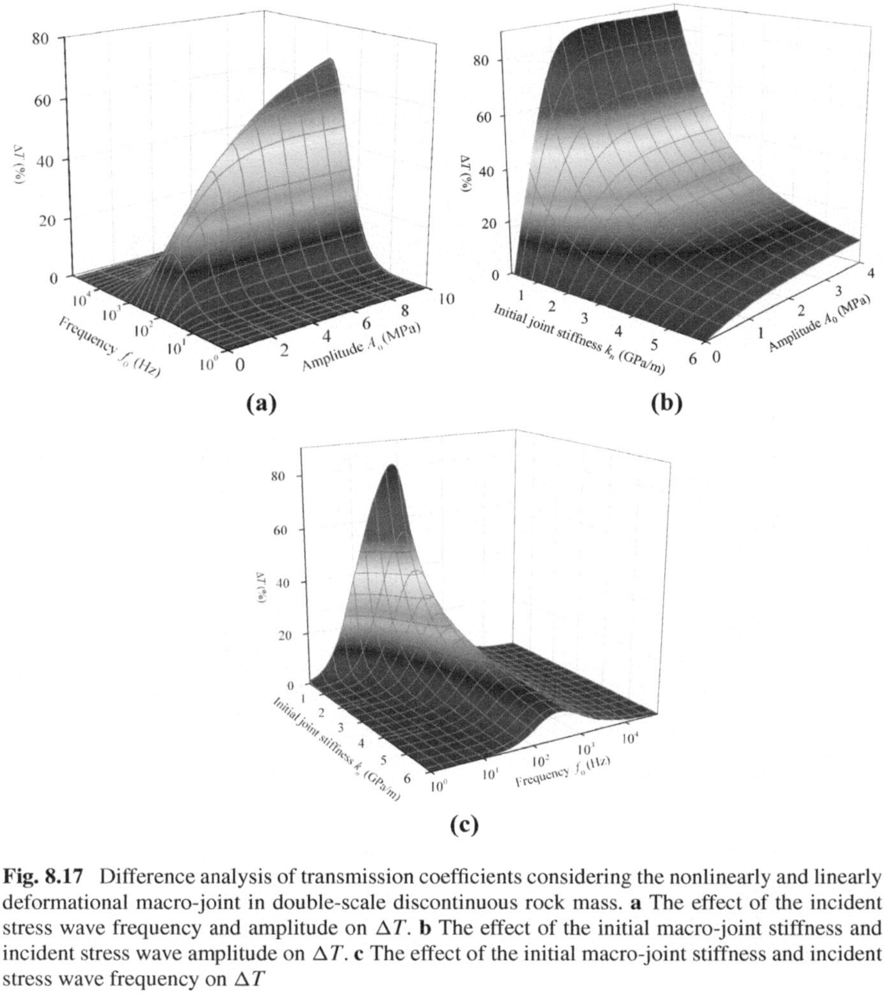

Fig. 8.17 Difference analysis of transmission coefficients considering the nonlinearly and linearly deformational macro-joint in double-scale discontinuous rock mass. **a** The effect of the incident stress wave frequency and amplitude on ΔT. **b** The effect of the initial macro-joint stiffness and incident stress wave amplitude on ΔT. **c** The effect of the initial macro-joint stiffness and incident stress wave frequency on ΔT

on ΔT can be ignored and ΔT tends to zero. In addition, it can be observed from Fig. 8.17a that, for a fixed incident stress wave amplitude, ΔT increases as the incident stress wave frequency increases to a peak value and then it decreases as the frequency further increases. This phenomenon shows that, for a small amplitude stress wave, the linearly deformational macro-joint can be used to replace the nonlinearly deformational macro-joint to study the stress wave propagation in all frequency regions. For a large amplitude stress wave, the linearly deformational macro-joint can be used to replace the nonlinearly deformational macro-joint in the small-frequency or large-frequency region.

Figure 8.17b shows the effect of the initial macro-joint stiffness and incident stress wave amplitude on ΔT. The incident stress wave frequency is $f_0 = 100$ Hz. It can be observed from Fig. 8.17b that ΔT decreases with the decrease of incident stress wave amplitude, which can also be observed from Fig. 8.17a. Moreover, it can be observed from Fig. 8.17b that, for a fixed incident stress wave amplitude, ΔT decreases with the increase of initial macro-joint stiffness. Therefore, the effect of the nonlinearly deformational macro-joint in the double-scale discontinuous rock mass can be neglected when the incident stress wave amplitude is sufficiently small. In addition, with the increase of initial macro-joint stiffness, the effect of nonlinearly deformational macro-joint and linearly deformational macro-joint on the transmission coefficient becomes more and more similar.

Figure 8.17c shows the effect of the incident stress wave frequency and initial macro-joint stiffness on ΔT. The incident stress wave amplitude is $A_0 = 1.0$ MPa. It is observed from Fig. 8.17c that, in the small-frequency or large-frequency region, the effect of the initial macro-joint stiffness on ΔT is not significant, and ΔT is close to zero. In the medium-frequency region, ΔT decreases with the increase of initial macro-joint stiffness, which can also be observed from Fig. 8.17b. In addition, it can be observed from Fig. 8.17c that, for a fixed initial macro-joint stiffness, ΔT increases as the incident stress wave frequency increases to a peak value and then it decreases as the frequency further increases, which can also be observed from Fig. 8.17a. Therefore, a linearly deformational macro-joint can be used to replace a nonlinearly deformational macro-joint in the small-frequency or large-frequency region or when the initial macro-joint stiffness is large enough.

8.3 Multiple Pulse Transmission Through Rock Mass with Double-Scale Discontinuities

With the development of deep mining engineering and tunnel engineering, the dynamic response of rock mass subjected to cyclic blast loading becomes more and more concerned [21, 22]. The cyclic load generated by the cyclic blast propagates in the rock mass in the form of stress with multiple pulses. The stress wave with multiple pulses is affected significantly by the discontinuities of micro-defects and macro-joints that exist in the rock mass [23]. On the other hand, numerous laboratory

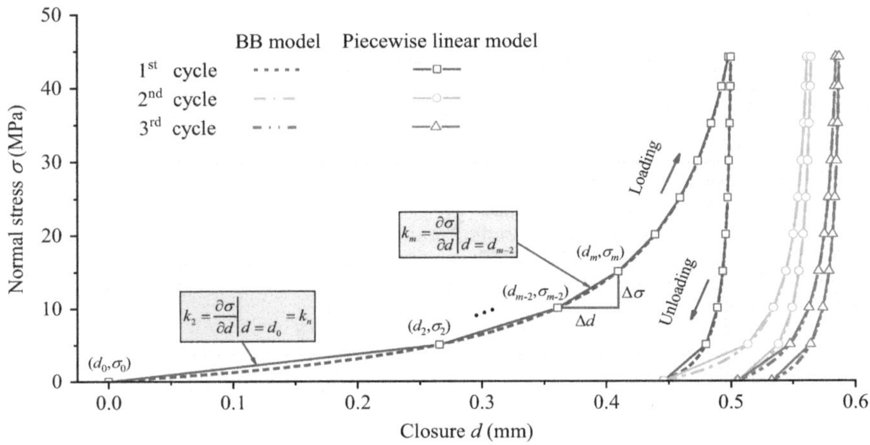

Fig. 8.19 The BB model and piecewise linear model of macro-joint under three cyclic loading and unloading

characteristic line *OA* can be obtained according to the Hugoniot relation [27, 28] and compatibility conditions along characteristic line *OA* in Fig. 8.18c. In addition, similar to the traditional three-characteristic method, the three unknowns, σ, ε and v, at left boundary points $P(0, j)$ and ordinary interior points $P_1(i, j)$ can be obtained according to the compatibility conditions corresponding to the triangle and diamond elements in Fig. 8.18c. The detailed calculation process is shown in the references [29, 30]. In the present study, in order to calculate the unknowns of interior points $P_2(i, j)$ at macro-joint, the DDM under cyclic loading and unloading is introduced into the separated diamond elements. Then, according to the compatibility conditions corresponding to the separated diamond elements, the three unknowns, σ^-, ε^- and v^-, at pre-interior points $P_2^-(i, j)$ of the macro-joint and the three unknowns, σ^+, ε^+ and v^+, at post-interior points $P_2^+(i, j)$ of the macro-joint can be obtained.

The right-running characteristic lines, left-running characteristic lines and up-running characteristic lines in the *x-t* plane can be observed from Fig. 8.18c. The right-running and left-running characteristic lines represent the trajectory of the wavefront propagating in the *x-t* plane. The up-running characteristic lines represent the movement trajectory of the medium particle. The compatibility relationship corresponding to the right-running characteristic lines, left-running characteristic lines and up-running characteristic lines can be expressed as

$$dv = \frac{1}{\rho_0 C_V} d\sigma + \left[\frac{(\sigma - E_a \varepsilon) E_M}{(E_a + E_M)\eta_M} \right] dx \tag{8.67}$$

$$dv = -\frac{1}{\rho_0 C_V} d\sigma + \left[\frac{(\sigma - E_a \varepsilon) E_M}{(E_a + E_M)\eta_M} \right] dx \tag{8.68}$$

and

$$d\varepsilon = \frac{d\sigma}{E_a + E_M} + \frac{(\sigma - E_a\varepsilon)E_M}{(E_a + E_M)\eta_M}dt \tag{8.69}$$

where σ, ε and v are the particle stress, particle strain and particle velocity, respectively. x and t are Cartesian coordinate and time, respectively. ρ_0 and C_V are the density of rock mass and the wave velocity in the rock mass, respectively. E_M and E_a respectively denote the elastic modulus of springs in Fig. 8.18b. η_M is the viscous coefficient of the dashpot in Fig. 8.18b.

The DDM is described by the stress continuity condition and the stress-closure relationship of the macro-joint. The stress-closure relationship of macro-joint under multiple pulses can be expressed as

$$d^{l,ul}_{(num)} = d^{l,ul}_{0(num)} + \frac{\sigma^{l,ul}_{(num)}}{k^{l,ul}_{n(num)} + \sigma^{l,ul}_{(num)}/d^{l,ul}_{\max(num)}}, num = 1, 2, 3 \tag{8.70}$$

The macro-joint stiffness of any point in the BB model can be expressed as

$$k^{l,ul}_{(num)} = \frac{\partial\sigma^{l,ul}_{(num)}}{\partial d^{l,ul}_{(num)}} = k^{l,ul}_{n(num)}\left[1 - \left(\frac{\sigma^{l,ul}_{(num)}}{k^{l,ul}_{n(num)}d^{l,ul}_{\max(num)} + \sigma^{l,ul}_{(num)}}\right)\right]^{-2}, num = 1, 2, 3 \tag{8.71}$$

where $num = 1, 2, 3$ respectively denote the first pulse, the second pulse and the third pulse. l and ul respectively denote the loading and unloading conditions of the (num)th pulse. d is the macro-joint closure. d_0 and d_{max} respectively denote the initial closure and the maximum allowance closure of the macro-joint. σ is the normal stress. k_n is the initial macro-joint stiffness.

According to the piecewise linear model as shown in Fig. 8.19, the macro-joint stiffness and the stress-closure relationship of the point (d_m, σ_m) in segment (d_{m-2}, σ_{m-2}) – (d_m, σ_m), respectively, can be expressed as

$$k^{l,ul}_{m(num)} = k^{l,ul}_{n(num)}\left[1 - \left(\frac{\sigma^{l,ul}_{m-2(num)}}{k^{l,ul}_{n(num)}d^{l,ul}_{\max(num)} + \sigma^{l,ul}_{m-2(num)}}\right)\right]^{-2}, num = 1, 2, 3 \tag{8.72}$$

and

$$d^{l,ul}_{m(num)} = d^{l,ul}_{m-2(num)} + \frac{\sigma^{l,ul}_{m(num)} - \sigma^{l,ul}_{m-2(num)}}{k^{l,ul}_{m(num)}}, num = 1, 2, 3 \tag{8.73}$$

where m equals 2, 4, 6⋯⋯, which denotes the point to the right of the segment (d_{m-2}, σ_{m-2}) – (d_m, σ_m) in Fig. 8.19, d_m is the macro-joint closure of the segment (d_{m-2},

σ_{m-2})-(d_m, σ_m), and k_m is the initial macro-joint stiffness of the segment (d_{m-2}, σ_{m-2}) – (d_m, σ_m).

The macro-joint stiffness and the stress-closure relation in Fig. 8.19 can be introduced into the separated diamond element in Fig. 8.18c. Assuming that the macro-joint closure $d(i, j)$ and normal stress $\sigma(i, j)$ of the point $P_2(i, j)$ in the separated diamond element correspond to the point (d_m, σ_m) in Fig. 8.19, therefore, according to the macro-joint stiffness and the stress-closure relationship of the point (d_m, σ_m), the macro-joint stiffness $k(i, j)$ and the stress-closure relationship of the point $P_2(i, j)$ in the separated diamond element can be obtained as

$$k(i,j) = k^{l,ul}_{n(num)}\left[1 - \left(\frac{\sigma(i,j-2)}{k^{l,ul}_{n(num)}d^{l,ul}_{\max(num)} + \sigma(i,j-2)}\right)\right]^{-2}, num = 1, 2, 3 \quad (8.74)$$

and

$$d(i,j) = d(i,j-2) + \frac{\sigma(i,j) - \sigma(i,j-2)}{k(i,j)} \quad (8.75)$$

In addition, the stress continuity condition at the macro-joint can be expressed as

$$\sigma^-(i,j) = \sigma^+(i,j) \quad (8.76)$$

The derivative of Eq. (8.75) with respect to t is

$$v^+(i,j) - v^-(i,j) = \frac{1}{k(i,j)}\frac{\sigma(i,j) - \sigma(i,j-2)}{2dt} \quad (8.77)$$

where $d(i,j) = u^+(i,j) - u^-(i,j)$, and $u^-(i,j)$ and $u^+(i,j)$ respectively denote the normal displacement in front of and behind the macro-joint. $\sigma^-(i,j)$ and $\sigma^+(i,j)$ respectively denote the normal stress in front of and behind the macro-joint. i equals $x_1/\Delta x$ and j equals 2, 4, 6……, which denote any point at macro-joint in Fig. 8.18c.

Together with Eqs. (8.76) and (8.77) and compatibility conditions corresponding to the separated diamond elements of Eqs. (8.78)–(8.81), the six unknowns, $\sigma^-(i, j)$, $\sigma^+(i, j)$, $v^-(i, j)$, $v^+(i, j)$, $\varepsilon^-(i, j)$ and $\varepsilon^+(i, j)$, at the interior points $P_2(i, j)$ can be obtained as.

$$\begin{aligned} v^-(i,j) - v(i-1,j-1) = {} & \frac{1}{\rho_0 C_V}\left[\sigma^-(i,j) - \sigma(i-1,j-1)\right] \\ & + \frac{\left[\sigma(i-1,j-1) - E_a\varepsilon(i-1,j-1)\right]E_M}{(E_a + E_M)\eta_M}dx \end{aligned} \quad (8.78)$$

$$v^+(i,j) - v(i+1,j-1) = -\frac{1}{\rho_0 C_V}\left[\sigma^+(i,j) - \sigma(i+1,j-1)\right]$$

$$-\frac{\left[\sigma(i+1,j-1)-E_a\varepsilon(i+1,j-1)\right]E_M}{(E_a+E_M)\eta_M}dx \quad (8.79)$$

$$\varepsilon^-(i,j)-\varepsilon^-(i,j-2)=\frac{1}{E_a+E_M}\left[\sigma^-(i,j)-\sigma(i,j-2)\right] + \frac{\left[\sigma(i,j-2)-E_a\varepsilon^-(i,j-2)\right]E_M}{(E_a+E_M)\eta_M}(2dt) \quad (8.80)$$

$$\varepsilon^+(i,j)-\varepsilon^+(i,j-2)=\frac{1}{E_a+E_M}\left[\sigma^+(i,j)-\sigma(i,j-2)\right] + \frac{\left[\sigma(i,j-2)-E_a\varepsilon^+(i,j-2)\right]E_M}{(E_a+E_M)\eta_M}(2dt) \quad (8.81)$$

At first, substituting Eqs. (8.76) and (8.77) into Eqs. (8.78) and (8.79), the unknowns $\sigma^-(i,j)$ and $\sigma^+(i,j)$ can be obtained as

$$\sigma(i,j)=\frac{2dt\rho_0 C_V k(i,j)}{\rho_0 C_V+4dtk(i,j)}\left[\frac{\sigma(i+1,j-1)+\sigma(i-1,j-1)}{\rho_0 C_V} - v(i-1,j-1)+v(i+1,j-1)+\frac{1}{2dtk(i,j)}\sigma(i,j-2) - \frac{\left[\sigma(i+1,j-1)-E_a\varepsilon(i+1,j-1)+\sigma(i-1,j-1)-E_a\varepsilon(i-1,j-1)\right]E_M}{(E_a+E_M)\eta_M}dx\right] \quad (8.82)$$

and then substituting Eqs. (8.82) into (8.78)–(8.81), the remaining four unknowns, $v^-(i,j)$, $v^+(i,j)$, $\varepsilon^-(i,j)$ and $\varepsilon^+(i,j)$, can be obtained, respectively.

8.3.2 Stress Wave Transmission with Multiple Pulses

In this section, the influence of rock mass with micro-defect and macro-joint on multiple pulse transmission with different time intervals is studied. The incident stress wave is assumed to be composed of three half-sinusoidal pulses in the form:

$$\sigma(0,t)=\begin{cases} A_0 sin(2\pi f_0 t) & 0\le t\le 1/(2f_0) \\ A_0 sin\left[2\pi f_0(t-t_{int}-1/(2f_0))\right] \quad \text{when} & 1/(2f_0)+t_{int}\le t\le 2/(2f_0)+t_{int} \\ A_0 sin\left[2\pi f_0(t-2t_{int}-2/(2f_0))\right] & 2/(2f_0)+2t_{int}\le t\le 3/(2f_0)+2t_{int} \\ 0 & \text{others} \end{cases} \quad (8.83)$$

where A_0 is the amplitude of incident pulse, f_0 is the frequency of incident pulse, and t_{int} is the time interval between two pulses.

In the calculation, according to the experimental data of stress-closure relationship of the macro-joint under normal cyclic loading and unloading [14], six BB models are fitted to describe the influence of three cyclic loading and unloading on the deformational behavior of the macro-joint as shown in Fig. 8.19. The parameter values in the fitted six BB models are shown in Table 8.3. Based on a similar approach, the cycling loading BB model can be obtained for four or more cyclic loading and unloading. In addition, the length of representative elementary volume (REV) is selected as $S = 1.0$ m and the density of rock mass is $\rho_0 = 2500.0\ \text{kg/m}^3$ in the present study. In the viscoelastic medium model shown in Fig. 8.18b, the elastic moduli are $E_a = 27.0$ GPa and $E_M = 14.0$ GPa, respectively, and the viscous coefficient is $\eta_M = 15.0\ \text{MPa} \cdot \text{s}$.

Figure 8.20a–c compares the three transmitted pulses for different time intervals of $t_{int} = 0.0 \times (1/f_0)$, $t_{int} = 0.5 \times (1/f_0)$ and $t_{int} = 1.0 \times (1/f_0)$, respectively. The frequency and amplitude of incident pulses are $f_0 = 2000$ Hz and $A_0 = 1.0$ MPa, respectively. It can be observed from Fig. 8.20a, Fig. 8.20b and c that the amplitudes of the first, second and third transmitted pulses are attenuated, while the attenuation degree of the first transmitted pulse is larger than that of the second transmitted pulse and the third transmitted pulse. The attenuation degree of the third transmitted pulse is the smallest. This phenomenon can be explained as the macro-joint is compressed by the former transmitted pulses, resulting in the macro-joint stiffness increases. Therefore, compared with the former transmitted pulses, the transmission rate of the later pulses is increased.

In addition, it is observed from Fig. 8.20a that the later transmitted pulses can be superposed with the former transmitted pulses when the time interval is sufficiently small, such as $t_{int} = 0.0 \times (1/f_0)$ in the present study. As the time interval increases, such as $t_{int} = 0.5 \times (1/f_0)$ in the present study, it is observed from Fig. 8.20b that the degree of superposition of the later transmitted pulses and the former transmitted pulses is reduced. Subsequently, when the time interval is sufficiently large, such as $t_{int} = 1.0 \times (1/f_0)$ in the present study, it is observed from Fig. 8.20c that the subsequently transmitted pulses cannot be superposed with the former transmitted pulses.

Table 8.3 The BB model parameters of macro-joint under three cyclic loading and unloading

Number of cycles	Cyclic loading/unloading path	k_n (GPa/m)	d_{max} (mm)
1st cycle	Loading	9.901	0.561
	Unloading	62.500	0.059
2nd cycle	Loading	34.483	0.127
	Unloading	76.923	0.066
3rd cycle	Loading	62.500	0.089
	Unloading	76.923	0.060

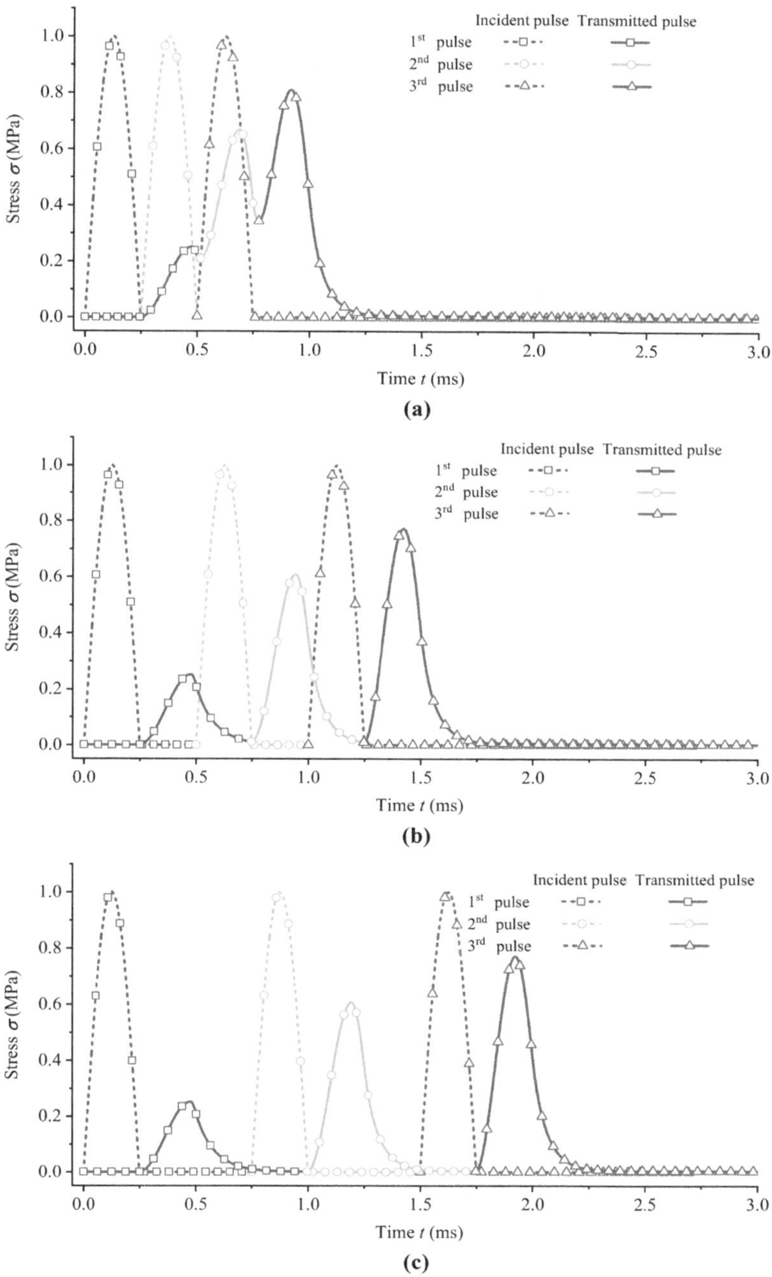

Fig. 8.20 The comparison of multiple pulse transmission with different time intervals. **a** $t_{int} = 0.0 \times (1/f_0)$. **b** $t_{int} = 0.5 \times (1/f_0)$. **c** $t_{int} = 1.0 \times (1/f_0)$

8.3.3 The Transmission Coefficient, the Effective Velocity and the Energy Transmission Coefficient of Multiple Pulses

The transmission coefficient, the effective velocity and the energy transmission coefficient of multiple pulses are discussed in detail. The transmission coefficient is the ratio of the amplitude of the transmitted pulse to the amplitude of the corresponding incident pulse. The effective velocity is the ratio of pulse propagation distance in the rock mass to the time difference between the peak value of the transmitted pulse and the corresponding incident pulse. The energy transmission coefficient is the ratio of the energy of the transmitted pulse to the energy of the corresponding incident pulse. The transmission coefficient T, the effective velocity C_{eff} and the energy transmission coefficient T_e are respectively expressed as

$$T_{(num)} = \frac{\max(\sigma_{T(num)})}{\max(\sigma_{I(num)})} \quad num = 1, 2, 3 \tag{8.84}$$

$$C_{eff\,(num)} = \frac{S}{t_{T(num)} - t_{I(num)}} \quad num = 1, 2, 3 \tag{8.85}$$

and

$$T_{e(num)} = \frac{energy_{T(num)}}{energy_{I(num)}} = \frac{\sum_{t^0_{T(num)}}^{t^1_{T(num)}} \rho_0 C_V \sigma^2_{T(num)} \Delta t}{\sum_{t^0_{I(num)}}^{t^1_{I(num)}} \rho_0 C_V \sigma^2_{I(num)} \Delta t} \quad num = 1, 2, 3 \tag{8.86}$$

where σ_T and σ_I denote the transmitted pulse and the corresponding incident pulse, respectively. S is the length of representative elementary volume (REV). t_T and t_I are the times for peak stress of the transmitted pulse and the corresponding incident pulse, respectively. ${t_T}^0$ and ${t_T}^1$ are the initial and final times of each transmitted pulse, respectively. ${t_I}^0$ and ${t_I}^1$ are the initial and final times of each incident pulse, respectively.

(1) Effect of time interval on transmission of multiple pulses

Figures 8.21, 8.22 and 8.23 show the effect of time interval on the transmission coefficient, the effective velocity and the energy transmission coefficient of three pulses, respectively. The frequency and amplitude of the three incident pulses are $f_0 = 2000$ Hz and $A_0 = 1.0$ MPa, respectively.

It is seen from Figs. 8.21 and 8.22 that the transmission coefficient and effective velocity obtained based on the first pulse remain constant as the increase of time interval. However, when the time interval is small (e.g., $t_{\text{int}} \le 0.5 \times (1/f_0)$), the transmission coefficient and effective velocity obtained based on the second pulse and the third pulse decrease as the time interval increases. When the time interval is large (e.g., $t_{\text{int}} \ge 0.5 \times (1/f_0)$), the transmission coefficient and effective velocity obtained

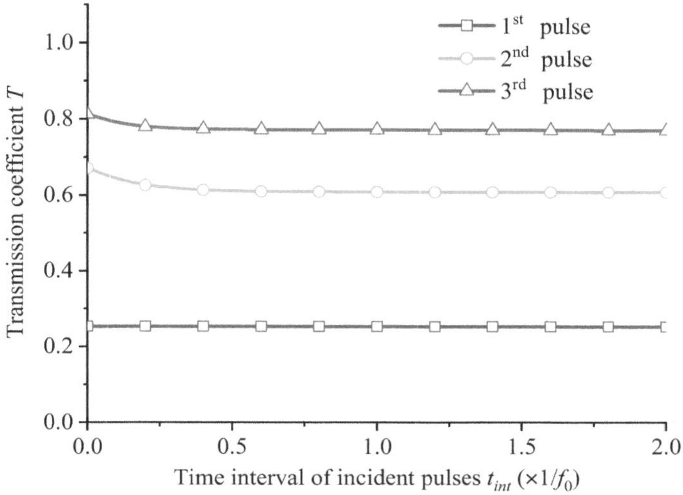

Fig. 8.21 Effect of the time interval on transmission coefficient

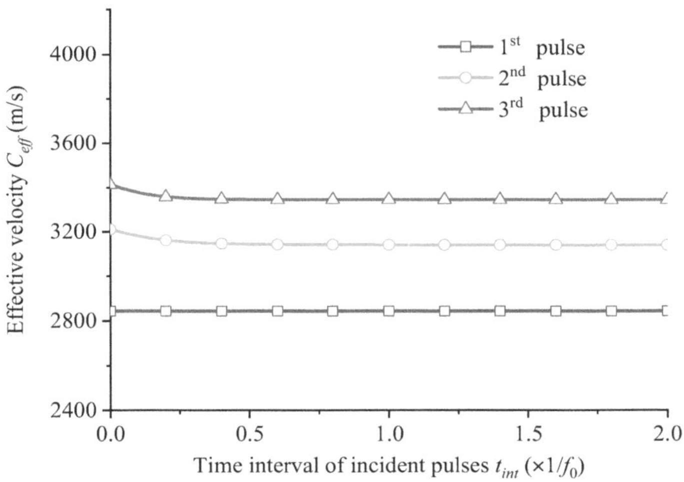

Fig. 8.22 Effect of the time interval on effective velocity

based on the second pulse and the third pulse remain constant as the increase of the time interval. For a given time interval, the transmission coefficient and effective velocity increase in turn with the transmission of the first pulse, the second pulse and the third pulse.

It is seen from Fig. 8.23 that when the time interval is small (e.g., $t_{\text{int}} \leq 0.5 \times (1/f_0)$), the energy transmission coefficient obtained based on the first pulse increases

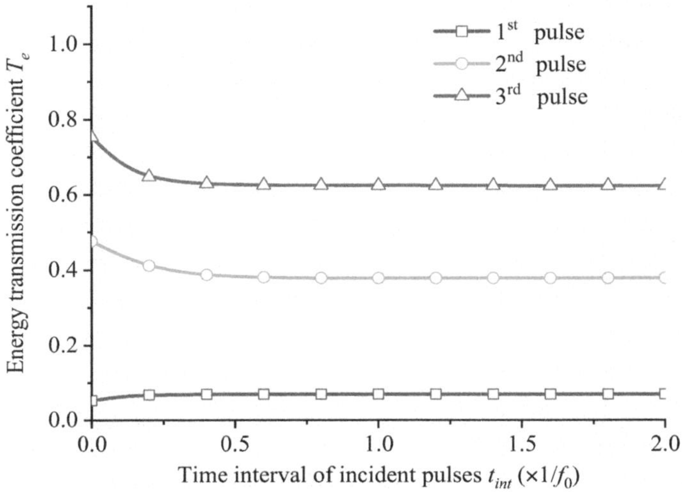

Fig. 8.23 Effect of the time interval on energy transmission coefficient

slightly as the increase of the time interval. However, the energy transmission coefficient obtained based on the second pulse and third pulse decreases as the time interval increases. When the time interval is large (e.g., $t_{\text{int}} \geq 0.5 \times (1/f_0)$), the energy transmission coefficient obtained based on the first pulse, the second pulse and third pulse remains constant as the increase of time interval. In addition, for a given time interval, the energy transmission coefficient increases in turn with the transmission of the first pulse, the second pulse and the third pulse.

(2) Effect of frequency on transmission of multiple pulses

Figures 8.24, 8.25 and 8.26 show the effect of the incident pulse frequency on the transmission coefficient, the effective velocity and the energy transmission coefficient of three pulses, respectively. The amplitude of three incident pulses is $A_0 = 1.0$ MPa. The time interval is $t_{int} = 1.0 \times (1/f_0)$. The frequency-dependence of the transmission coefficient, the effective velocity and the energy transmission coefficient of three pulses is investigated in this section.

It is observed from Figs. 8.24 and 8.26 that the effect of the incident pulse frequency on the transmission coefficient is similar to that on the energy transmission coefficient, and both the transmission coefficient and the energy transmission coefficient of three pulses are highly frequency-dependent. Figures 8.24 and 8.26 show that both the transmission coefficient and the energy transmission coefficient of three pulses are almost a constant of 1.0 in the small-frequency region, then that of three pulses decreases rapidly with the further increase of frequency in the medium-frequency region and gradually becomes zero in the large-frequency region. In the small-frequency region or large-frequency region, the transmission coefficient and the energy transmission coefficient of the first pulse, the second pulse and the third

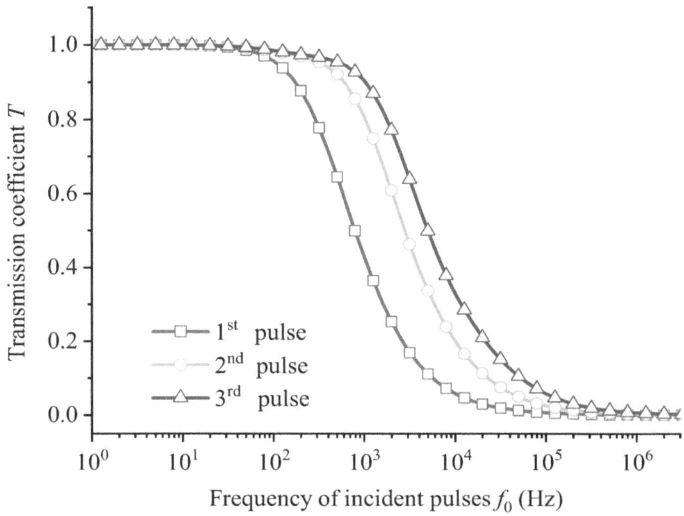

Fig. 8.24 Effect of incident pulse frequency on transmission coefficient

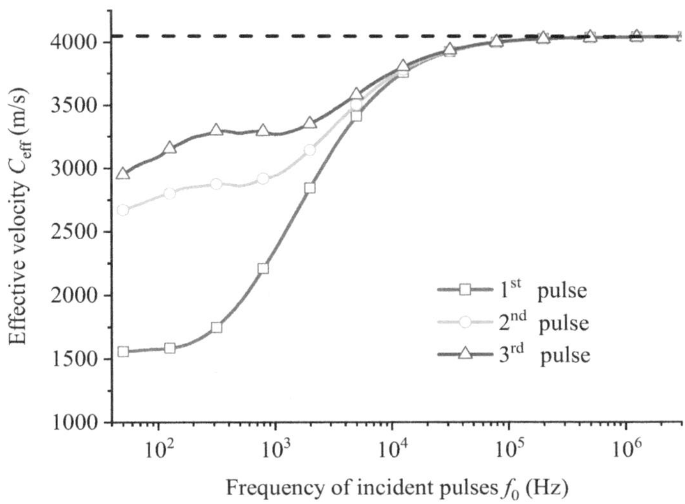

Fig. 8.25 Effect of incident pulse frequency on effective velocity

pulse are almost the same. However, in the medium-frequency region, the transmission coefficient and the energy transmission coefficient of the first pulse are the smallest, followed by the second pulse, while the third pulse is the largest.

It is observed from Fig. 8.25 that the effective velocity of three pulses is the smallest in the small-frequency region, and increases slightly as the incident pulse

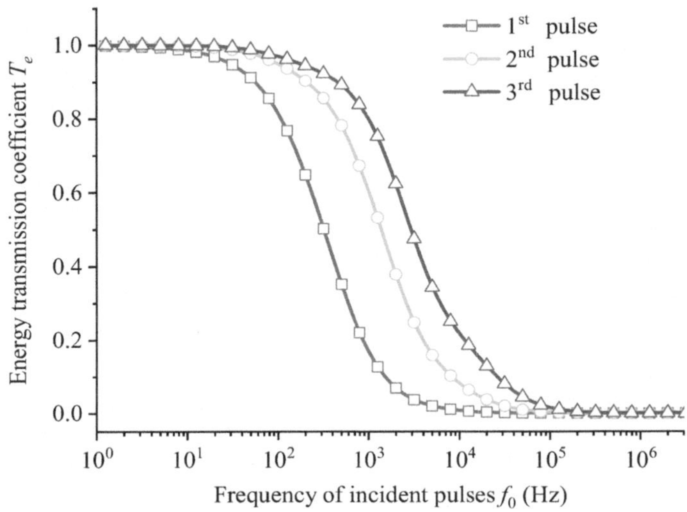

Fig. 8.26 Effect of incident pulse frequency on energy transmission coefficient

frequency increases, and then that of three pulses increases rapidly with the further increase of frequency. When the frequency is sufficiently large, the effective velocity of the three pulses approaches the wave velocity in intact rock, which is 4049.7 m/s. Moreover, it is also observed from Fig. 8.25 that the effective velocity of the third pulse is larger than the first pulse and the second pulse in the small-frequency or medium-frequency region, and the effective velocity of the first pulse is smallest. Figure 8.25 shows the frequency-dependence of the effective velocity of the first pulse, the second pulse and the third pulse decrease in turn.

(3) Effect of amplitude on transmission of multiple pulses

Figures 8.27, 8.28 and 8.29 show the effect of the incident pulse amplitude on the transmission coefficient, the effective velocity and the energy transmission coefficient of three pulses, respectively. The frequency of three incident pulses is $f_0 = 350$ Hz. The time interval is $t_{int} = 1.0 \times (1/f_0)$. The amplitude-dependence of the transmission coefficient, the effective velocity and the energy transmission coefficient of three pulses is investigated in this section.

It is observed from Fig. 8.27 that the transmission coefficient of the first pulse increases rapidly as the incident pulse amplitude increases at the initial stage (e.g., $A_0 \leq 15.0$ MPa), and subsequently remains a constant smaller than 1.0 as the incident pulse amplitude further increases. However, the transmission coefficient of the second pulse and the third pulse increases slightly as the incident pulse amplitude increases at the initial stage (e.g., $A_0 \leq 5.0$ MPa), and then approaches a constant of smaller than 1.0. This phenomenon can be explained that the macro-joint stiffness is large when the second pulse and third pulse propagate, and the rock mass with micro-defect and macro-joint can be equivalented as the rock mass with micro-defect; it results in a

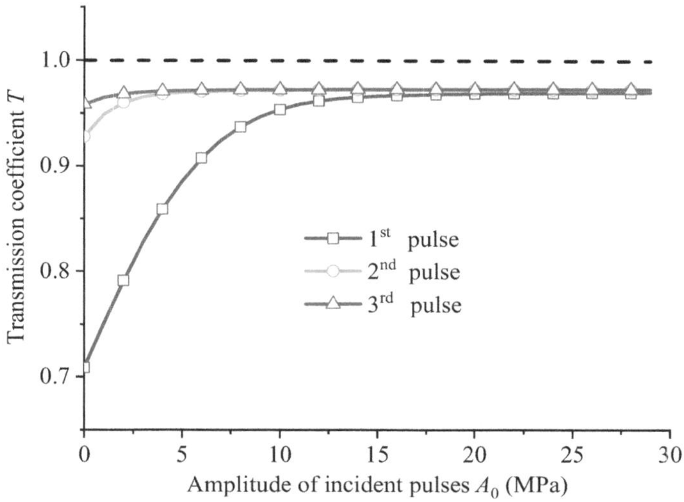

Fig. 8.27 Effect of incident pulse amplitude on transmission coefficient

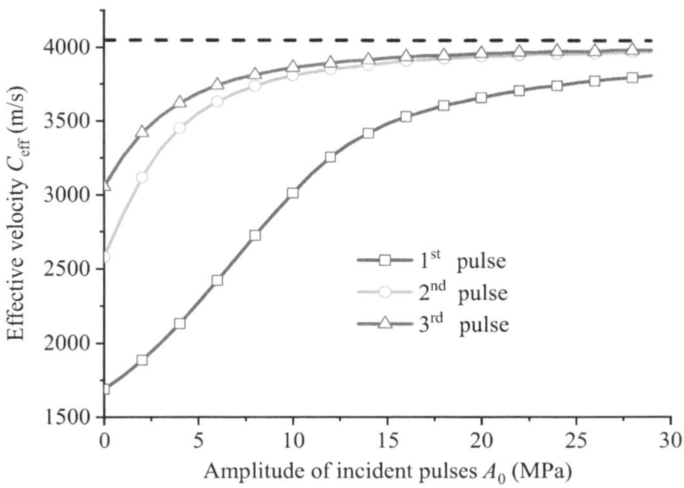

Fig. 8.28 Effect of incident pulse amplitude on effective velocity

large transmission of the second pulse and third pulse. In addition, for the macro-joint with nonlinear deformation, the macro-joint stiffness also increases as the incident pulse amplitude increases. When the incident pulse amplitude is large enough, the transmission coefficient of the three pulses increases to a constant smaller than 1.0, which is because of the effect of micro-defect in the rock mass. It is also observed from Fig. 8.27 that, for a given small incident pulse amplitude, the transmission

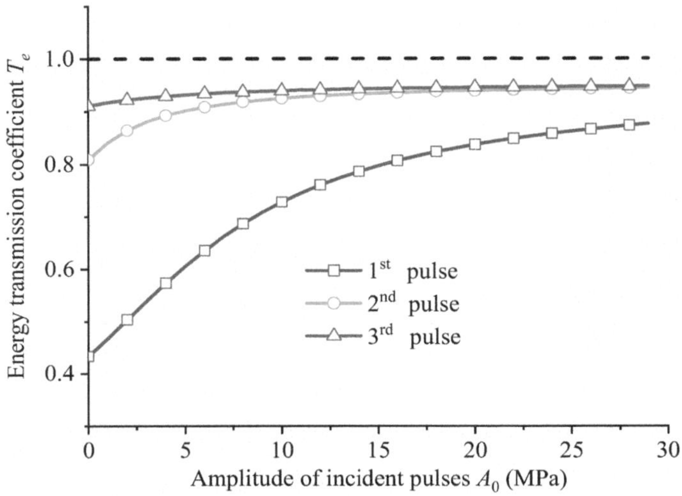

Fig. 8.29 Effect of incident pulse amplitude on energy transmission coefficient

coefficient of the first pulse is obviously smaller than that of the second pulse and the third pulse. The transmission coefficient of the third pulse is the largest.

It is observed from Fig. 8.28 that the effective velocities of the first pulse, the second pulse and the third pulse increase as the incident pulse amplitude increases. When the incident pulse amplitude is sufficiently large, the effective velocities of the second and third pulses approach the wave velocity in intact rock, which is 4049.7 m/s. However, the effective velocity of the first pulse is obviously smaller than the wave velocity in intact rock. In addition, it is also observed from Fig. 8.28 that, for a given incident pulse amplitude, the effective velocity of the first pulse is the smallest, followed by the second pulse and the third pulse being the largest.

It is observed from Fig. 8.29 that the energy transmission coefficients of the first pulse, the second pulse and the third pulse increase as the incident pulse amplitude increases. Compared with the energy transmission coefficient of the second pulse and third pulse, the energy transmission coefficient of the first pulse increases significantly as the incident pulse amplitude increases. The energy transmission coefficient of the third pulse does not increase significantly. When the incident pulse amplitude is small, the energy transmission coefficient of the first pulse is the smallest, followed by the second pulse and the third pulse being the largest. When the incident pulse amplitude is large enough, the energy transmission coefficient of the second pulse and third pulse remains a constant smaller than 1.0, which is obviously larger than the energy transmission coefficient of the first pulse.

Figures 8.27, 8.28 and 8.29 show that the transmission coefficient, the effective velocity and the energy transmission coefficient of three pulses are amplitude-dependent. However, the amplitude-dependence decreases as the number of pulses increases. The amplitude-dependence of the first pulse is larger than the second pulse and the third pulse, and the amplitude-dependence of the third pulse is the smallest.

8.4 Effect of Unloading Behavior on Stress Wave Transmission in a Double-Scale Discontinuous Rock Mass

The investigation of stress wave transmission through rock mass is an important topic in the field of rock dynamics. As a heterogeneous and discontinuous material, rock mass contains various scale discontinuities such as micro-defects and macro-joints, which affect the stress wave propagation [13, 31–33]. The micro-defect leads to stress wave amplitude attenuation and waveform dispersion. In addition, the macro-joint exhibits different loading and unloading deformations under dynamic loads [34–37]. The unloading deformation behavior of the macro-joint significantly affects the energy of stress wave transmission [25]. It is crucial to study stress wave transmission properties in double-scale discontinuous rock mass considering the unloading effect for dynamic load energy evaluation in rock engineering.

The present section investigates the effect of unloading behavior on stress wave transmission in a double-scale discontinuous rock mass. The piecewise linear loading and unloading model of macro-joint is introduced. The DDM based on the piecewise linear loading and unloading models is introduced into the split three-characteristic method to study stress wave propagation. The validity of the present study is carried out by comparison with two traditional studies considering the macro-joint only. The stress wave with different waveform transmissions through rock mass is analyzed. The effect of incident wave waveform, incident wave frequency, incident wave amplitude and propagation distance on the energy transmission coefficient and amplitude transmission coefficient is discussed, respectively.

8.4.1 Method for Stress Wave Transmission Considering Unloading Behavior

A natural rock mass is shown in Fig. 8.30. The natural rock mass contains micro-defects and a single macro-joint. When a dynamic load is exerted on the natural rock mass, the dynamic load propagates through the natural rock mass in the form of an incident stress wave. The incident stress wave consists of the loading part and unloading part as shown in Fig. 8.30. The inelastic deformation of the macro-joint during loading leads to incomplete recovery of the deformation during unloading. Therefore, the deformation properties of the macro-joint corresponding to the unloading part are different from those corresponding to the loading part. In

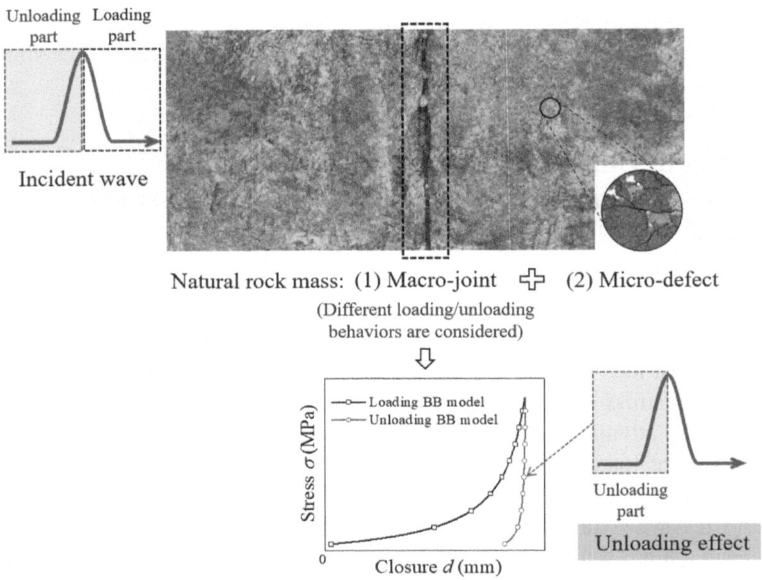

Fig. 8.30 The stress wave propagation through a natural rock mass contains macro-joint with different loading/unloading behaviors and micro-defects

general, typical loading and unloading deformation curves of the macro-joint are described by the loading BB model and unloading BB model as shown in Fig. 8.30.

In order to study the stress wave transmission considering unloading behavior, the research scheme of stress wave propagation through a double-scale discontinuous rock mass considering unloading behavior is proposed as shown in Fig. 8.31. Figure 8.31a shows a representative elementary volume (REV) of the rock mass containing micro-defects and a single macro-joint. Figure 8.31b shows the computational model of stress wave propagation in the representative elementary volume (REV). It can be seen from Fig. 8.31b that the micro-defected rock mass is equivalent to a viscoelastic model. The equivalent viscoelastic model consists of a spring and a Maxwell model in parallel. In addition, it can also be seen from Fig. 8.31b that the macro-joint is considered with different loading and unloading models for stress wave propagation. The loading model and unloading model of macro-joint can be seen in Fig. 8.31c. In order to improve computational efficiency, a piecewise linear loading model and unloading model are proposed. The piecewise linear loading model and unloading model are combined with a split three-characteristic method to study stress wave propagation. Figure 8.31d shows the split three-characteristic method, which can be obtained based on the equivalent viscoelastic model. The split three-characteristic method is used to study stress wave propagation through the double-scale discontinuous rock mass [20]. It can be seen from Fig. 8.31d that the split three-characteristic method consists of three basic elements such as the triangle element, diamond element and separated diamond element. When the initial and

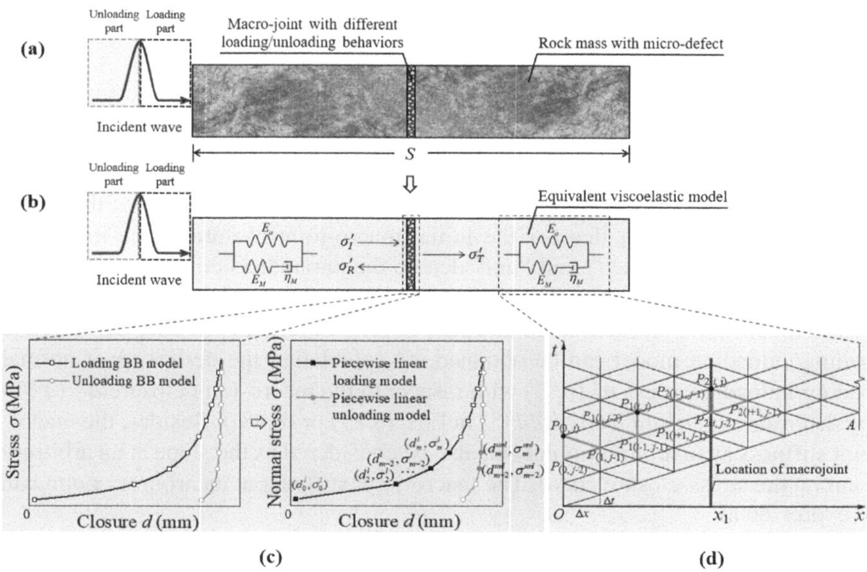

Fig. 8.31 Research scheme of stress wave propagation through a double-scale discontinuous rock mass considering unloading behavior. **a** The representative elementary volume (REV) of the rock mass contains micro-defects and a single macro-joint. **b** The computational model of stress wave propagation. **c** The loading and unloading models of macro-joint and the piecewise linear loading and unloading models of macro-joint. **d** The split three-characteristic method

boundary conditions of the rock mass are known, the triangle element and diamond element can be used to calculate the stress waves at the boundary and interior points of the rock mass, respectively [7]. However, to investigate the stress wave propagation at macro-joint with different loading and unloading behaviors, the DDM based on the piecewise linear loading/unloading model is proposed to be introduced into the separated diamond element.

(1) The DDM based on the piecewise linear loading/unloading model

The DDM is widely used to study the stress wave propagation at the macro-joint. It assumes that the stress field is continuous and the displacement field is discontinuous on two sides of the macro-joint. Therefore, the DDM consists of a displacement discontinuity condition and a stress continuity condition. The displacement discontinuity condition can be obtained according to the deformation properties of the macro-joint. In the present study, the deformation properties of the macro-joint can be represented by the loading/unloading model shown in Fig. 8.31c. The stress-closure curve of the loading model and unloading model can be respectively expressed as

$$d^l = \frac{\sigma^l}{k_n^l + \sigma^l / d_{\max}^l} \tag{8.87}$$

and

$$d^{unl} = d_1 + \frac{\sigma^{unl}}{k_n^{unl} + \sigma^{unl}/d_{\max}^{unl}} \tag{8.88}$$

where d and $d_{\max}$ denote the macro-joint closure and the macro-joint maximum allowable closure, respectively. σ denotes the normal stress. k_n denotes the initial macro-joint stiffness. d_1 denotes the initial macro-joint closure in the unloading model. The superscripts "*l*" and "*unl*" denote the parameters corresponding to the loading model and unloading model, respectively.

The macro-joint stiffness at an arbitrary point on the stress-closure curve of the loading/unloading model can be obtained, by calculating the derivative of normal loading/unloading stress σ^l (σ^{unl}) with respect to the macro-joint closure d^l (d^{unl}) (by determining $\partial\sigma^l/\partial d^l$ ($\partial\sigma^{unl}/\partial d^{unl}$) in Eqs. (8.87) or (8.88)). Besides, the macro-joint stiffness at an arbitrary point can also be considered as the slope at an arbitrary point on the stress-closure curve. The macro-joint stiffness at an arbitrary point can be expressed as

$$k^l = \frac{\partial\sigma^l}{\partial d^l} = k_n^l\left(1 + \frac{\sigma^l/d_{\max}^l}{k_n^l}\right)^2 \tag{8.89}$$

and

$$k^{unl} = \frac{\partial\sigma^{unl}}{\partial d^{unl}} = k_n^{unl}\left(1 + \frac{\sigma^{unl}/d_{\max}^{unl}}{k_n^{unl}}\right)^2 \tag{8.90}$$

In the present study, the loading/unloading model is divided by linear segments to form the piecewise linear loading/unloading model as shown in Fig. 8.31c. For example, the piecewise linear loading model includes linear segments $\left(d_0^i, \sigma_0^l\right) - \left(d_2^l, \sigma_2^l\right)$, …, $\left(d_{m-2}^l, \sigma_{m-2}^l l\right) - \left(d_m^l, \sigma_m^l\right)$, , etc. Assuming that the point $(d_{m-2}^l, \sigma_{m-2}^l)$ in the loading model is known, the slope of the linear segment $(d_{m-2}^l, \sigma_{m-2}^l) - (d_m^l, \sigma_m^l)$ can be determined from the slope of point $(d_{m-2}^l, \sigma_{m-2}^l)$. When $d_m^l - d_{m-2}^l = \Delta d$ is sufficiently small, the above calculation holds. Therefore, bringing the point (d_m^l, σ_m^l) into Eq. (8.89), the slope of the linear segment $(d_{m-2}^l, \sigma_{m-2}^l) - (d_m^l, \sigma_m^l)$ in the piecewise linear loading model can be obtained as

$$k_m^l = \frac{\partial\sigma^l}{\partial d^l} = k_n^l\left(1 + \frac{\sigma_{m-2}^l/d_{\max}^l}{k_n^l}\right)^2 \tag{8.91}$$

Similarly, assuming that the point $(d_m^{unl}, \sigma_m^{unl})$ in the unloading model is known, the slope of the linear segment $(d_{m-2}^{unl}, \sigma_{m-2}^{unl}) - (d_m^{unl}, \sigma_m^{unl})$ can be determined from the slope of point $(d_m^{unl}, \sigma_m^{unl})$. Bringing the point $(d_m^{unl}, \sigma_m^{unl})$ into Eq. (8.90), the slope of the linear segment $(d_{m-2}^{unl}, \sigma_{m-2}^{unl}) - (d_m^{unl}, \sigma_m^{unl})$ in the piecewise linear unloading model can be obtained as

$$k_m^{unl} = \frac{\partial \sigma^{unl}}{\partial d^l} = k_n^{unl}\left(1 + \frac{\sigma_m^{unl}/d_{\max}^{unl}}{k_n^{unl}}\right)^2 \tag{8.92}$$

According to Eqs. (8.91) and (8.92), the stress-closure curves of the linear segments $(d_{m-2}^l, \sigma_{m-2}^l) - (d_m^l, \sigma_m^l)$ and $(d_{m-2}^{unl}, \sigma_{m-2}^{unl}) - (d_m^{unl}, \sigma_m^{unl})$ can be expressed as

$$d_m^{l/unl} - d_{m-2}^{l/unl} = \frac{\sigma_m^{l/unl} - \sigma_{m-2}^{l/unl}}{k_m^{l/unl}} \tag{8.93}$$

For the loading part of the stress wave propagation, the value of closure and stress at the points $P_2(i, j\text{-}2)$ and $P_2(i, j)$ in the separated diamond element (in Fig. 8.31e) can be expressed as $(d_{m-2}^l, \sigma_{m-2}^l)$ and (d_m^l, σ_m^l), respectively. For the unloading part of the stress wave propagation, the value of closure and stress at the points $P_2(i, j - 2)$ and $P_2(i, j)$ in the separated diamond element can be expressed as $(d_m^{unl}, \sigma_m^{unl})$ and $(d_{m-2}^{unl}, \sigma_{m-2}^{unl})$, respectively. According to Eq. (8.93), the stress-closure curve corresponding to $P_2(i, j) - P_2(i, j - 2)$ segment can be obtained, in which the value of closure and stress at the points $P_2(i, j\ 2)$ is known. The stress and closure relationship at the point $P_2(i, j)$ can be expressed as

$$d^{l/unl}(i,j) = d^{l/unl}(i,j-2) + \frac{\sigma^{l/unl}(i,j) - \sigma^{l/unl}(i,j-2)}{k^{l/unl}(i,j)} \tag{8.94}$$

where i denotes the position of point $P_2(i, j)$ on the x-axis in Fig. 8.31e. j denotes the position of point $P_2(i, j)$ on the t-axis. The coordinates of the point $P_2(i, j)$ can be expressed as $(i \times \Delta x, j \times \Delta t)$.

The $d^{l/unl}(i, j)$ in Eq. (8.94) can be expressed as

$$d^{l/unl}(i,j) = u^{l/unl,+}(i,j) - u^{l/unl,-}(i,j) \tag{8.95}$$

where u denotes the normal displacement. The sign "−" denotes the left side of the macro-joint, and the sign "+" denotes the right side of the macro-joint.

The $k^{l/unl}(i, j)$ in Eq. (8.94) is the macro-joint stiffness, which can be expressed as

$$k^{l/unl}(i,j) = k_n^{l/unl}\left[1 + \frac{\sigma^{l/unl}(i,j-2)/d_{\max}^{l/unl}}{k_n^{l/unl}}\right]^2 \tag{8.96}$$

According to Eqs. (8.94)–(8.96), the displacement discontinuity condition at the point $P_2(i, j)$ can be expressed as

$$u^{l/unl,+}(i,j) - u^{l/unl,-}(i,j) = d^{l/unl}(i,j-2) + \frac{\sigma^{l/unl}(i,j) - \sigma^{l/unl}(i,j-2)}{k^{l/unl}(i,j)} \tag{8.97}$$

Because the stress-closure relationship at the point $P_2(i, j-2)$ is known, the derivative of Eq. (8.97) with respect to t is

$$v^{l/unl,+}(i,j) - v^{l/unl,-}(i,j) = \frac{1}{k^{l/unl}(i,j)} \frac{\partial \sigma^{l/unl}(i,j)}{\partial t} \tag{8.98}$$

where v denotes the velocity.

Equation (8.98) can be expressed in differential form as

$$v^{l/unl,+}(i,j) - v^{l/unl,-}(i,j) = \frac{1}{k^{l/unl}(i,j)} \frac{\sigma^{l/unl}(i,j) - \sigma^{l/unl}(i,j-2)}{2dt} \tag{8.99}$$

In addition, the stress continuity condition at the point $P_2(i, j)$ can be expressed as

$$\sigma^{l/unl,-}(i,j) = \sigma^{l/unl,+}(i,j) \tag{8.100}$$

The displacement discontinuity condition Eq. (8.97) and stress continuity condition Eq. (8.100) form the DDM.

(2) Split three-characteristic method combined with DDM

The DDM is introduced to the location of macro-joint in the split three-characteristic method to study stress wave propagation. During the calculation, the displacement discontinuity condition can be expressed in the form of the velocity discontinuity of Eq. (8.99). By combining Eqs. (8.99) and (8.100) with the compatibility conditions corresponding to the characteristic lines in the separated diamond element of Fig. 8.31d, the stress, velocity and strain of stress wave propagation through the point $P_2(i, j)$ at the location of the macro-joint can be obtained. Point $P_2(i, j)$ at the location of the macro-joint includes point P_2^+ (i, j) to the left of the macro-joint and point P_2^+ (i, j) to the right of the macro-joint. Therefore, the stress, velocity and strain at the point $P_2(i, j)$, including the stress $\sigma^{l/unl,-}(i, j)$, velocity $v^{l/unl,-}(i, j)$ and strain $\varepsilon^{l/unl,-}(i, j)$ at the left of the macro-joint, and the stress $\sigma^{l/unl,+}(i, j)$, velocity $v^{l/unl,+}(i, j)$, strain $\varepsilon^{l/unl,+}(i, j)$ at the right of the macro-joint.

The compatibility conditions corresponding to the characteristic lines in the separated diamond elements can be obtained according to the constitutive equation of the equivalent viscoelastic model in Fig. 8.31b, motion equation of stress wave and continuity equation of stress wave [7].

The constitutive equation of the equivalent viscoelastic model can be expressed as

$$\frac{\partial \varepsilon}{\partial t} - \frac{1}{E_a + E_M} \frac{\partial \sigma}{\partial t} + \frac{E_a E_M \varepsilon}{(E_a + E_M)\eta_M} - \frac{E_M \sigma}{(E_a + E_M)\eta_M} = 0 \tag{8.101}$$

The motion equation of stress wave can be expressed as

$$\rho_0 \frac{\partial v}{\partial t} - \frac{\partial \sigma}{\partial x} = 0 \tag{8.102}$$

The continuity equation of stress wave can be expressed as

$$\frac{\partial \varepsilon}{\partial t} - \frac{\partial v}{\partial x} = 0 \tag{8.103}$$

The compatibility condition corresponding to the characteristic line $P_2(i,j) - P_2(i - 1, j - 1)$ in the separated diamond element is

$$\begin{aligned} v^{l/unl,-}(i,j) - v(i-1,j-1) = & \frac{1}{\rho_0 C_V}\left[\sigma^{l/unl,-}(i,j) - \sigma(i-1,j-1)\right] \\ & + \frac{\left[\sigma(i-1,j-1) - E_a \varepsilon(i-1,j-1)\right]E_M}{(E_a + E_M)\eta_M} dx \end{aligned} \tag{8.104}$$

where ρ_0 is the rock density. C_V is the wave velocity of the rock mass. E_a and E_M are the elastic modulus of spring in the equivalent viscoelastic model. η_M is the viscous coefficient in the equivalent viscoelastic model.

The compatibility condition corresponding to the characteristic line $P_2^+ (i, j) - P_2(i + 1, j - 1)$ in the separated diamond element is

$$\begin{aligned} v^{l/unl,+}(i,j) - v(i+1,j-1) = & -\frac{1}{\rho_0 C_V}\left[\sigma^{l/unl,+}(i,j) - \sigma(i+1,j-1)\right] \\ & - \frac{\left[\sigma(i+1,j-1) - E_a \varepsilon(i+1,j-1)\right]E_M}{(E_a + E_M)\eta_M} dx \end{aligned} \tag{8.105}$$

The compatibility condition corresponding to the characteristic line $P_2(i,j) - P_2^+ (i, j–2)$ in the separated diamond element is

$$\begin{aligned} \varepsilon^{l/unl,-}(i,j) - \varepsilon^{l/unl,-}(i,j-2) = & \frac{1}{E_a + E_M}\left[\sigma^{l/unl,-}(i,j) - \sigma^{l/unl,-}(i,j-2)\right] \\ & + \frac{\left[\sigma^{l/unl,-}(i,j-2) - E_a \varepsilon^{l/unl,-}(i,j-2)\right]E_M}{(E_a + E_M)\eta_M} 2dt \end{aligned} \tag{8.106}$$

The compatibility condition corresponding to the characteristic line $P + 2(i,j) - P_2^+ (i, j - 2)$ in the separated diamond element is

$$\varepsilon^{l/unl,+}(i,j) - \varepsilon^{l/unl,+}(i,j-2) = \frac{1}{E_a + E_M}\left[\sigma^{l/unl,+}(i,j) - \sigma^{l/unl,+}(i,j-2)\right]$$

$$+\frac{\left[\sigma^{l/unl,+}(i,j-2)-E_a\varepsilon^{l/unl,+}(i,j-2)\right]E_M}{(E_a+E_M)\eta_M}2dt \tag{8.107}$$

The stresses $\sigma^{l/unl,-}(i, j)$ and $\sigma^{l/unl,+}(i, j)$ on two sides of the macro-joint can be obtained by introducing Eqs. (8.99) and (8.100) into Eqs. (8.104) and (8.105). The stresses can be expressed as

$$\begin{aligned}\sigma^{l/unl}(i,j)=&\frac{2dt\rho_0C_Vk^{l/unl}(i,j)}{\rho_0C_V+4dtk^{l/unl}(i,j)}\left\{\frac{\sigma(i+1,j-1)+\sigma(i-1,j-1)}{\rho_0C_V}+\frac{\sigma^{l/unl}(i,j-2)}{2dtk^{l/unl}(i,j)}\right.\\&+v(i+1,j-1)-\frac{\left[\sigma(i+1,j-1)-E_a\varepsilon(i+1,j-1)\right]E_M}{(E_a+E_M)\eta_M}dx\\&\left.-v(i-1,j-1)-\frac{\left[\sigma(i-1,j-1)-E_a\varepsilon(i-1,j-1)\right]E_M}{(E_a+E_M)\eta_M}dx\right\}\end{aligned} \tag{8.108}$$

Subsequently, the velocities $v^{l/unl,-}(i, j)$ and $v^{l/unl,+}(i, j)$ on two sides of the macro-joint can be obtained by introducing Eq. (8.108) into Eqs. (8.104) and (8.105). The strains $\varepsilon^{l/unl,-}(i, j)$ and $\varepsilon^{l/unl,+}(i, j)$ on two sides of the macro-joint can be obtained by introducing Eqs. (8.108) into (8.106) and (8.107).

In this section, the stress wave $\sigma^{l/unl}(i, j)$ in the rock mass is investigated. The energy transmission coefficient T_e and amplitude transmission coefficient T_σ are discussed. The energy transmission coefficient T_e is defined as

$$T_e=\frac{energy_{tran}}{energy_{inc}}=\frac{\sum_{t^0_{tran}}^{t^1_{tran}}\rho_0C_V(\sigma_{tran})^2\Delta t}{\sum_{t^0_{inc}}^{t^1_{inc}}\rho_0C_V(\sigma_{inc})^2\Delta t} \tag{8.109}$$

where σ_{tran} denotes the transmitted wave. σ_{inc} denotes the incident wave. t^0_{tran} and t^l_{tran} denote the initial time and final time of the transmitted wave, respectively. t^0_{inc} and t^l_{inc} denote the initial time and final time of the incident wave, respectively.

The amplitude transmission coefficient T_σ is defined as

$$T_\sigma=\frac{\max(\sigma_{tran})}{\max(\sigma_{inc})} \tag{8.110}$$

where "max" denotes the maximum value of the stress wave.

8.4.2 The Stress Wave Transmission with Different Incident Wave Waveforms

In this section, the present study is compared with traditional studies to reveal the effect of macro-joint unloading behavior and micro-defects on stress wave propagation. In the traditional study I, the macro-joint with the different loading and unloading deformations is considered, and the micro-defects are not considered [25]. In the traditional study II, the macro-joint with the same loading and unloading deformations is considered, and the micro-defects are not considered [4]. Different from traditional studies, which consider single-scale discontinuities, in the present study, the macro-joint with the different loading and unloading deformations and the micro-defect are comprehensively considered.

In the calculations, the rock mass is considered an equivalent viscoelastic medium in the present study. The parameter values for the equivalent viscoelastic model in Fig. 8.31b are $E_a = 28.0$ GPa, $E_M = 37.0$ GPa and $\eta_M = 15.0$ MPa · s, respectively. Therefore, the dynamic elastic modulus of the rock mass in the present study is $E_a + E_M$. In addition, the rock mass is considered an elastic medium in the traditional studies I and II. The dynamic elastic modulus of the rock mass in the traditional studies I and II is selected as $E_a + E_M = 65.0$ GPa. The density of intact rock is $\rho_0 =$ 2500 kg/m^3. The initial macro-joint stiffnesses of the loading model and unloading model are $k_n^l = 5.6$ and $k_n^{unl} = 28.8$ GPa/m, respectively. The maximum allowance closure of the loading model and unloading model are $d_{\max}^l = 0.48$ and $d_{max}^{unl} =$ 0.12 mm, respectively. The length of representative elementary volume (REV) is S = 15.0 m, which can also be expressed as the propagation distance.

In addition, assuming that the rock mass is initially undisturbed, therefore, the initial condition of stress in the rock mass is $\sigma(x, 0) = 0$. The initial condition of velocity in the rock mass is $v(x, 0) = 0$. The initial condition of strain in the rock mass is $\varepsilon(x, 0) = 0$. In addition, the incident wave is assumed to be a rectangular wave, half-sinusoidal wave and triangular wave. The transmission properties of the incident waves of three waveforms are investigated. The boundary condition of the rectangular incident wave can be expressed as

$$\sigma(0, t) = \begin{cases} A_0, & \text{when} \quad 0 \le t \le 1/(2f_0) \\ 0 & \quad\quad\quad \text{others} \end{cases} \tag{8.111}$$

where A_0 is the amplitude of the incident wave. f_0 is the frequency of the incident wave.

The boundary condition of the half-sinusoidal incident wave can be expressed as

$$\sigma(0, t) = \begin{cases} A_0 sin(2\pi f_0 t), & \text{when} \quad 0 \le t \le 1/(2f_0) \\ 0 & \quad\quad\quad \text{others} \end{cases} \tag{8.112}$$

The boundary condition of the triangular incident wave can be expressed as

$$\sigma(0,t) = \begin{cases} 4A_0 f_0 t & \\ 4A_0 f_0\left[1/(2f_0) - t\right], & \text{when} \\ 0 & \end{cases} \begin{matrix} 0 \le t \le 1/(4f_0) \\ 1/(4f_0) < t \le 1/(2f_0) \\ \text{others} \end{matrix} \quad (8.113)$$

The amplitude is $A_0 = 0.5$ MPa and the frequency is $f_0 = 100$ Hz for rectangular incident wave, half-sinusoidal incident wave and triangular incident wave.

Figure 8.32a shows the rectangular incident wave and rectangular transmitted wave. In Fig. 8.32, the physical meaning of the horizontal coordinate is the time of rock particle vibration, and the physical meaning of the vertical coordinate is the stress of the rock particle at a certain time. The rectangular transmitted wave is obtained from the present study, traditional study I and traditional study II, respectively. It is observed from Fig. 8.32a that, for the present study considering the effect of micro-defect, the amplitude of the rectangular transmitted wave is attenuated. However, for the traditional study I and traditional study II without considering the effect of micro-defect, the amplitudes of the rectangular transmitted waves are not significantly changed. It can be seen from Fig. 8.32a that the rectangular transmitted waves in the present study, traditional study I and traditional study II cannot maintain the initial waveform. The rectangular transmitted wave appears as a waveform spreading, which can be regarded as waveform dispersion. The more distinct the spread of the waveform indicates the larger the dispersion degree of the wave. Therefore, it can be seen from Fig. 8.32a that the dispersion degree of the rectangular transmitted wave in the present study is larger than that in the traditional study I and traditional study II. In addition, the dispersion degree of the rectangular transmitted wave in the traditional study II without considering the unloading effect is larger than that in the traditional study I considering the unloading effect.

Figure 8.32b shows the half-sinusoidal incident wave and half-sinusoidal transmitted wave. The half-sinusoidal transmitted wave is obtained from the present study, traditional study I and traditional study II, respectively. It is observed from Fig. 8.32b that, for the present study, traditional study I and traditional study II, the amplitudes of the half-sinusoidal transmitted waves are attenuated. The attenuation degree of the amplitudes of the half-sinusoidal transmitted wave in the present study is the largest, which is caused by considering the effect of the micro-defect. In addition, the attenuation degree of the amplitudes of the half-sinusoidal transmitted waves in the traditional study I and traditional study II is small and the same. It is also observed from Fig. 8.32b that the half-sinusoidal transmitted waves in the present study, traditional study I and traditional study II have dispersion. The dispersion degree of the half-sinusoidal transmitted wave in the present study is the largest. In addition, the dispersion degree of the half-sinusoidal transmitted wave in the traditional study II without considering the unloading effect is larger than that in the traditional study I considering the unloading effect. This is similar to the rectangular transmitted wave.

Figure 8.32c shows the triangular incident wave and triangular transmitted wave. The triangular transmitted wave is obtained from the present study, traditional study I and traditional study II, respectively. Similar to the half-sinusoidal transmitted wave (in Fig. 8.32b), the amplitudes of the triangular transmitted waves in the present study,

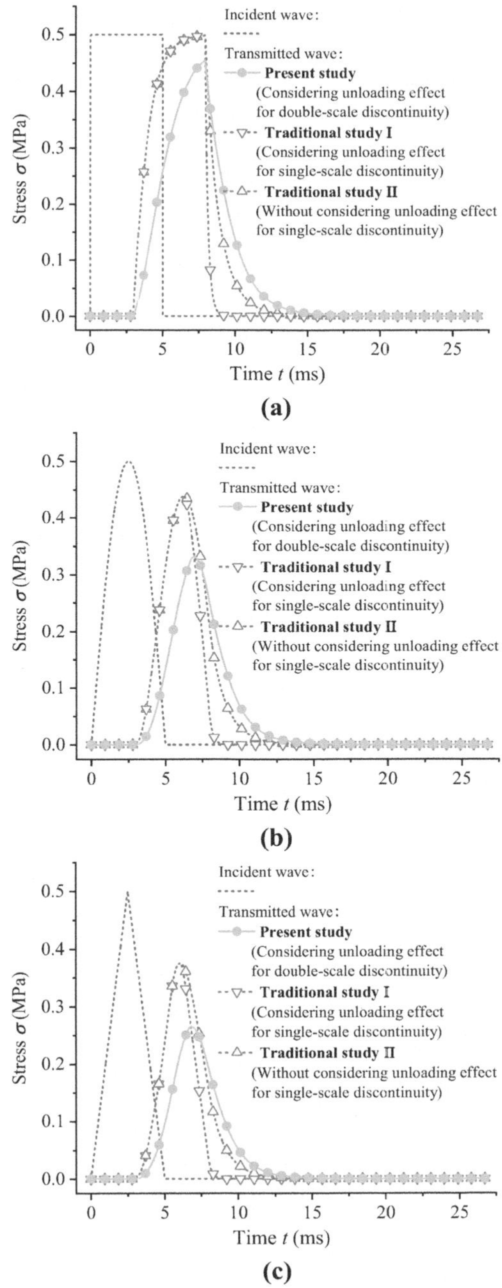

Fig. 8.32 The stress wave transmission with different incident wave waveforms. **a** Rectangular wave. **b** Half-sinusoidal wave. **c** Triangular wave

traditional study I and traditional study II are attenuated. Due to the consideration of the effect of micro-defects, the amplitude of the triangular transmitted waves in the present study is the smallest. The amplitude of the triangular transmitted waves in traditional study I and traditional study II is large. In addition, the triangular transmitted waves in the present study, traditional study I and traditional study II have dispersion. The dispersion degree of the triangular transmitted wave in the present study is the largest. In addition, the dispersion degree of the triangular transmitted wave in the traditional study II without considering the unloading effect is larger than that in the traditional study I considering the unloading effect.

8.4.3 The Energy Transmission Coefficient and Amplitude Transmission Coefficient

(1) The effect of incident wave waveform on the stress wave transmission

Figure 8.33 shows the effect of the incident wave waveform on the energy transmission coefficient. The incident wave waveforms are rectangular, half-sinusoidal and triangular, respectively. The frequency and amplitude of the three waveform incident wave are $f_0 = 100$ Hz and $A_0 = 0.5$ MPa, respectively. The propagation distance is $S = 15.0$ m. The energy transmission coefficient is obtained based on the present study, traditional study I and traditional study II, respectively. It is observed from Fig. 8.33 that in both the present study and the traditional study, the energy transmission coefficient of the rectangular wave is the largest, followed by the half-sinusoidal wave and the energy transmission coefficient of the triangular wave being the smallest. It can also be observed from Fig. 8.33 that, for an arbitrarily given incident wave waveform, the energy transmission coefficient of the present study is the smallest, followed by the traditional study I considering the unloading effect. The energy transmission coefficient of the traditional study II without considering the unloading effect is the largest. Comparing traditional study I with traditional study II shows that the energy transmission coefficient decreases when the unloading behavior of the macro-joint is considered. Therefore, the difference in the energy transmission coefficients between traditional study I and traditional study II is caused by the unloading effect. The difference in the energy transmission coefficients between the present study and traditional study I is caused by the micro-defect. In addition, the difference in the energy transmission coefficients between the present study and the traditional study II is caused by a combination of micro-defects and the unloading effect.

Figure 8.34 shows the effect of the incident wave waveform on the amplitude transmission coefficient. The incident wave waveforms are rectangular, half-sinusoidal and triangular, respectively. The values of frequency, amplitude and propagation distance are the same as in Fig. 8.33. The amplitude transmission coefficient is obtained based on the present study, traditional study I and traditional study II, respectively. It is observed from Fig. 8.34 that, in both the present study and traditional study, the amplitude transmission coefficient of the rectangular wave is the

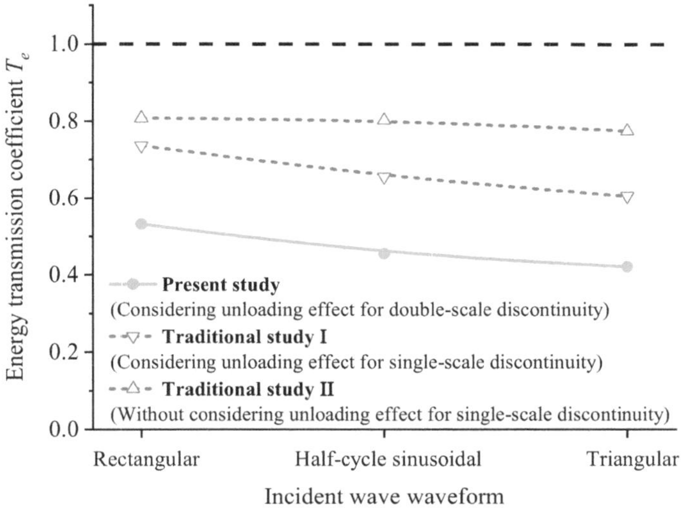

Fig. 8.33 The effect of incident wave waveform on the energy transmission coefficient

largest, followed by the half-sinusoidal wave and the amplitude transmission coefficient of the triangular wave being the smallest. Interestingly, the amplitude transmission coefficient of the rectangular wave approaches 1.0 for the two traditional studies. In addition, it is found from Fig. 8.34 that, for an arbitrarily given incident wave waveform, the amplitude transmission coefficient obtained based on the present study is the smallest. The amplitude transmission coefficient obtained based on the traditional study I considering the unloading effect is identical to that obtained based on the traditional study II without considering the unloading effect. That is, whether or not the unloading behavior is considered does not affect the amplitude transmission coefficient under arbitrary waveforms. The amplitude transmission coefficient of the present study is smaller than that of the traditional study I and traditional study II results from considering the effect of micro-defects.

(2) The effect of incident wave frequency on the stress wave transmission

Figure 8.35 shows the effect of incident wave frequency on the energy transmission coefficient. The half-sinusoidal wave is selected as the incident wave in the present study. This is because the stress wave with an arbitrary waveform can be represented as the summation of a series of half-sinusoidal waves by the Fourier transform. The half-sinusoidal wave is solved separately and then recombined to obtain the transmission properties of incident waves of arbitrary waveforms across the rock mass. The amplitude of the half-sinusoidal incident wave is $A_0 = 0.5$ MPa. The propagation distance is $S = 15.0$ m. The energy transmission coefficient is obtained based on the present study, traditional study I and traditional study II, respectively. It is observed from Fig. 8.35 that the energy transmission coefficients first drop sharply when the incident wave frequency is small, then slowly decrease and gradually tend

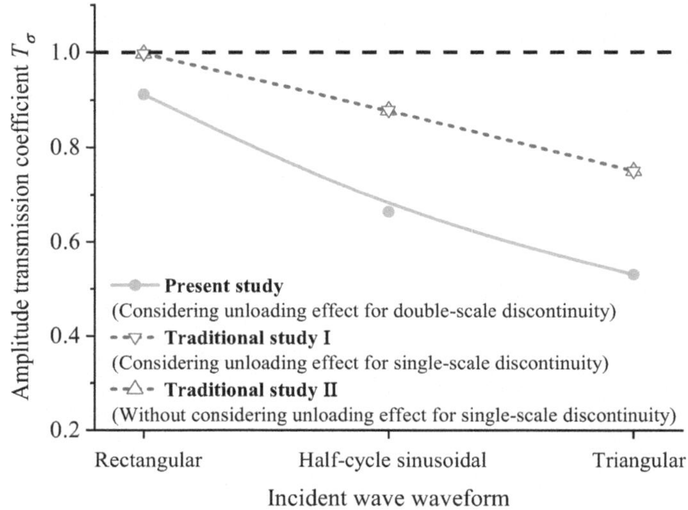

Fig. 8.34 The effect of incident wave waveform on the amplitude transmission coefficient

to zero as the incident wave frequency increases further. It is also observed from Fig. 8.35 that, for a given incident wave frequency, the energy transmission coefficient of the present study is the smallest, followed by the traditional study I. The energy transmission coefficient of the traditional study II is the largest. Similar to the previous sections, the energy transmission coefficient of the traditional study I is smaller than traditional study II due to the consideration of the unloading effect in the traditional study I. The energy transmission coefficient of the present study is smaller than the traditional study I due to the consideration of the effect of micro-defects in the present study. In addition, the energy transmission coefficient of the present study is smaller than the traditional study II due to the consideration of the micro-defect and unloading effect in the present study. Note that when the incident wave frequency is larger, the energy transmission coefficients of the present study and the traditional study I are close to each other, which indicates that the effect of micro-defect can be neglected in this case.

Figure 8.36 shows the effect of incident wave frequency on the amplitude transmission coefficient. The half-sinusoidal wave is selected as the incident wave. The values of amplitude and propagation distance are the same as in Fig. 8.35. The amplitude transmission coefficient is obtained based on the present study, traditional study I and traditional study II, respectively. It is observed from Fig. 8.36 that the amplitude transmission coefficient decreases rapidly as the incident wave frequency increases for a smaller incident wave frequency. As the incident wave frequency increases further, the amplitude transmission coefficient slowly decreases and gradually tends to stabilize. This phenomenon shows that higher frequency waves are more easily filtered out when stress waves propagate through the rock mass. In addition, it is observed from Fig. 8.36 that, for a given incident wave frequency, the amplitude

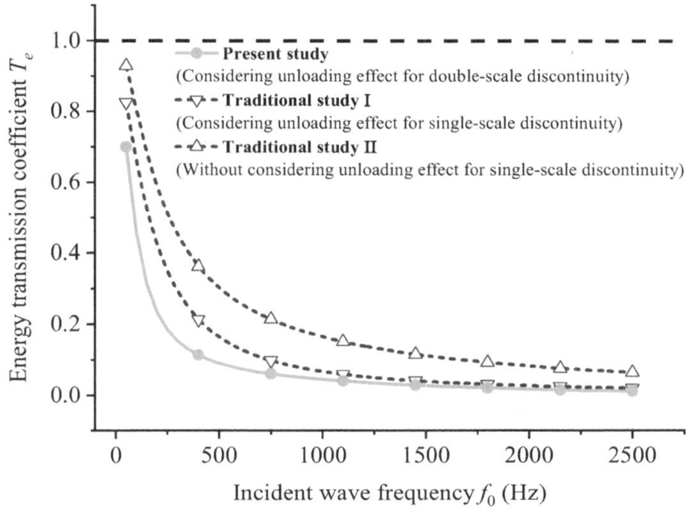

Fig. 8.35 The effect of incident wave frequency on the energy transmission coefficient

transmission coefficients are the same for the two traditional studies. For arbitrary frequencies, the amplitude transmission coefficients are not affected whether or not the unloading effect is considered. The amplitude transmission coefficient of the present study is smaller than that of the two traditional studies, which is due to the consideration of the micro-defect in the present study.

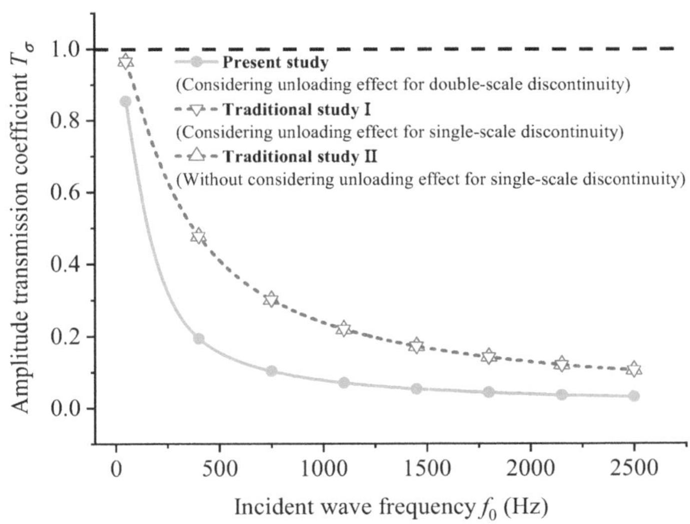

Fig. 8.36 The effect of incident wave frequency on the amplitude transmission coefficient

(3) The effect of incident wave amplitude on the stress wave transmission

Figure 8.37 shows the effect of incident wave amplitude on the energy transmission coefficient. The half-sinusoidal wave is selected as the incident wave. The frequency of the half-sinusoidal incident wave is $f_0 = 100$ Hz. The propagation distance is $S = 15.0$ m. The energy transmission coefficient is obtained based on the present study, traditional study I and traditional study II, respectively. It is observed from Fig. 8.37 that the energy transmission coefficient increases as the incident wave amplitude increases. The energy transmission coefficient of the two traditional studies is gradually close to 1.0 as the incident wave amplitude increases. However, the energy transmission coefficient of the present study is always significantly smaller than 1.0. The phenomenon can be explained that when the incident wave amplitude is large, the macro-joint stiffness of the nonlinear deformation macro-joint is larger, which makes the macro-joint have a smaller effect on stress wave propagation. In this case, the macro-joint can be neglected, the traditional study is approximated as the propagation of stress waves in the intact rock, and the present study is approximated as the propagation of stress waves in micro-defected rock mass. Therefore, the energy transmission coefficient of the present study is always much smaller than 1.0, which is due to the effect of micro-defects considered in the present study. The phenomenon illustrates the effect of micro-defects cannot be ignored. In addition, it is observed from Fig. 8.37 that, for a given incident wave amplitude, the energy transmission coefficient of the present study, traditional study I and traditional study II successively increases.

Figure 8.38 shows the effect of incident wave amplitude on the amplitude transmission coefficient. The half-sinusoidal wave is selected as the incident wave. The

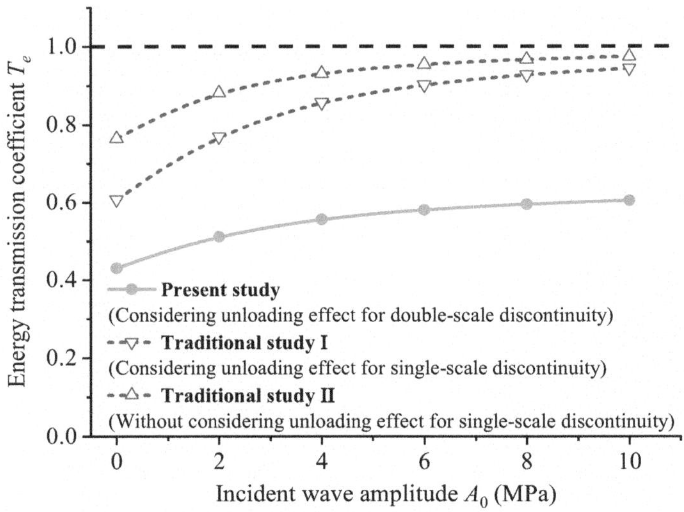

Fig. 8.37 The effect of incident wave amplitude on the energy transmission coefficient

values of frequency and propagation distance are the same as in Fig. 8.37. The amplitude transmission coefficient is obtained based on the present study, traditional study I and traditional study II, respectively. It is observed from Fig. 8.38 that the amplitude transmission coefficients increase as the incident wave amplitude increases. The amplitude transmission coefficients of the two traditional studies increase rapidly with the increase of incident wave amplitude at the initial stage, while that increases to a constant 1.0 with a further increase of the incident wave amplitude. However, the amplitude transmission coefficient of the present study increases slowly as the incident wave amplitude increases and approaches a stable value smaller than 1.0. This phenomenon can be explained by the fact that the incident wave with a large amplitude leads to the large stiffness of the nonlinearly macro-joint, which reduces the effect of the macro-joint on the stress wave propagation. Therefore, the amplitude transmission coefficients of the two traditional studies approach 1.0 for the large incident wave amplitude. However, due to the consideration of the micro-defect, the amplitude transmission coefficient of the present study is always smaller than 1.0. In addition, it is observed from Fig. 8.38 that, for a given incident wave amplitude, the amplitude transmission coefficients are the same for the traditional study I and traditional study II. The amplitude transmission coefficient of the present study is smaller than that of the two traditional studies.

(4) The effect of propagation distance on the stress wave transmission

Figure 8.39 shows the effect of propagation distance on the energy transmission coefficient. The half-sinusoidal wave is selected as the incident wave. The frequency and amplitude of the half-sinusoidal incident wave are $f_0 = 100$ Hz and $A_0 = 0.5$ MPa, respectively. The energy transmission coefficient is obtained based on the present

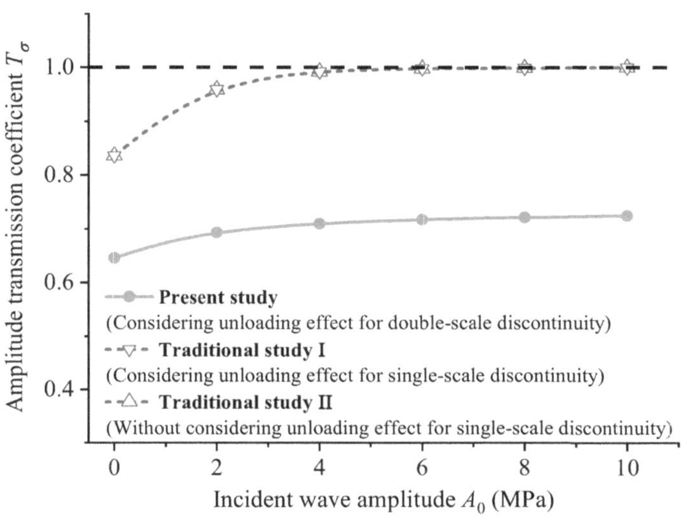

Fig. 8.38 The effect of incident wave amplitude on the amplitude transmission coefficient

study, traditional study I and traditional study II, respectively. It is observed from Fig. 8.39 that the energy transmission coefficient of the present study decreases as the propagation distance increases due to the consideration of the effect of micro-defects. However, the energy transmission coefficient of the traditional study I and traditional study II is not affected by the propagation distance, which remains constant as the propagation distance increases. Therefore, the energy transmission coefficients of the two traditional studies are independent of propagation distance. It is observed from Fig. 8.39 that, for a given propagation distance, the energy transmission coefficient for the present study, traditional study I and traditional study II successively increases. It should be noted that when the propagation distance is small enough, the energy transmission coefficient of the present study is close to that of the traditional study I. This phenomenon indicates that the effect of micro-defects on the energy transmission coefficient can be neglected when the propagation distance is small enough. Therefore, the traditional study I can be used as a special case of the present study to investigate the energy transmission coefficient for the small propagation distance.

Figure 8.40 shows the effect of propagation distance on the amplitude transmission coefficient. The half-sinusoidal wave is selected as the incident wave. The values of frequency and amplitude of the half-sinusoidal incident wave are the same as in Fig. 8.40. The amplitude transmission coefficient is obtained based on the present study, traditional study I and traditional study II, respectively. It is observed from Fig. 8.40 that the amplitude transmission coefficient of the present study decreases as the propagation distance increases. However, the amplitude transmission coefficient of the traditional study I and traditional study II remains constant as the incident wave amplitude increases. Therefore, the amplitude transmission coefficients of the two

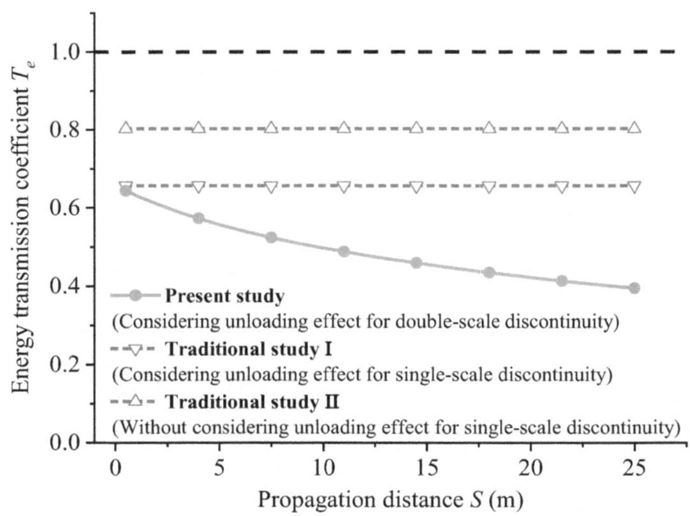

Fig. 8.39 The effect of propagation distance on the energy transmission coefficient

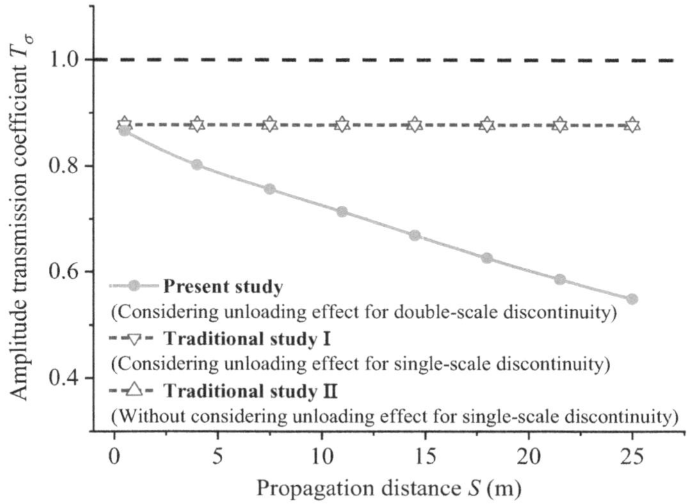

Fig. 8.40 The effect of propagation distance on the amplitude transmission coefficient

traditional studies are independent of propagation distance. It is also observed from Fig. 8.40 that, for a given propagation distance, the amplitude transmission coefficients are the same for the traditional study I and traditional study II. The amplitude transmission coefficient of the present study is smaller than that of the two traditional studies. In addition, when the propagation distance is small enough, the amplitude transmission coefficient of the present study is close to that of the two traditional studies. The effect of micro-defect on the amplitude transmission coefficient can be neglected for a small propagation distance. Therefore, the traditional study I and traditional study II can be used as a special case of the present study to investigate the amplitude transmission coefficient for the small propagation distance.

8.5 Combined Static-Dynamic Loading Effect on the Wave Transmission Properties in the Rock Mass with Macro-Joint and Micro-Defect

The study of stress wave propagation properties has received increasing attention as a basis for the stability analysis of underground rock engineering. Deep rock mass usually consists of various discontinuities all of which affect stress wave propagation, including micro-defect and macro-joint [38, 39]. In addition, the deep rock mass is subjected to both static loads, i.e., in situ stress, and dynamic loads, i.e., impact and blasting [40, 41]. It makes the deep rock mass instantaneously superimposed with dynamic loads on top of the original static loads, in which the propagation properties of stress waves are different from those when subjected to dynamic loads alone. It is vital to obtain accurate wave propagation properties in double-scale discontinuous rock mass under combined static-dynamic loads for the safety and stability of deep rock engineering.

This section presents a study of the combined static-dynamic loading effect on wave transmission properties in the double-scale discontinuous rock mass. The split three-characteristic method, which considers the effect of static prestresses, is proposed. The relationship of static prestress with amplitude and energy transmission coefficients is investigated. Subsequently, the effect of incident wave amplitude, incident wave frequency and propagation distance on amplitude and energy transmission coefficients is revealed under different static prestresses. In addition, the differences in the amplitude and energy transmission coefficients with and without considering micro-defects are analyzed. The combined effect of incident wave amplitude, incident wave frequency and propagation distance with static prestress on the difference in amplitude and energy transmission coefficient is further discussed.

8.5.1 Split Three-Characteristic Method Considering Static Prestress

Figure 8.41 schematically shows a double-scale discontinuous rock mass under combined static-dynamic loads. The rock mass with double-scale discontinuities consists of a micro-defect and a macro-joint. Static prestress and dynamic loads are applied simultaneously on the rock mass with double-scale discontinuities. Static prestress is applied to two sides of the rock mass, and the dynamic loads propagate through the rock mass in the form of stress waves. Typically, stress wave propagation properties are analyzed by treating the micro-defected rock mass as an equivalent viscoelastic medium and by adopting the DDM at the macro-joint. The equivalent viscoelastic medium is represented by a viscoelastic model containing the modulus of elasticity and the coefficient of viscosity. The DDM assumes that the stresses on two sides of the macro-joint are equal, but the displacements are not equal due to the closure of the macro-joint. However, as the static prestress increases, the micro-defect is closed, and the macro-joint is compressed, as shown in Fig. 8.41. Therefore, under different static prestresses, the mechanical properties of the micro-defected rock mass and macro-joint differ, which leads to different equivalent viscoelastic mediums and DDMs for studying stress wave propagation under different static prestresses. Figure 8.42 shows the mechanical behaviors of the micro-defected rock mass and macro-joint under static prestress. The wave velocity of the micro-defected rock mass under static prestress is shown in Fig. 8.42a. The preexisting micro-defect weakens the wave velocity of the rock mass. The micro-defect is closed under static prestress, which causes the wave velocity of the micro-defected rock mass to increase as the static prestress increases. It is effective for the micro-defected rock mass in the elastic deformation stage. The deformation properties of the macro-joint under static prestress are shown in Fig. 8.42b. The macro-joint is compressed under static

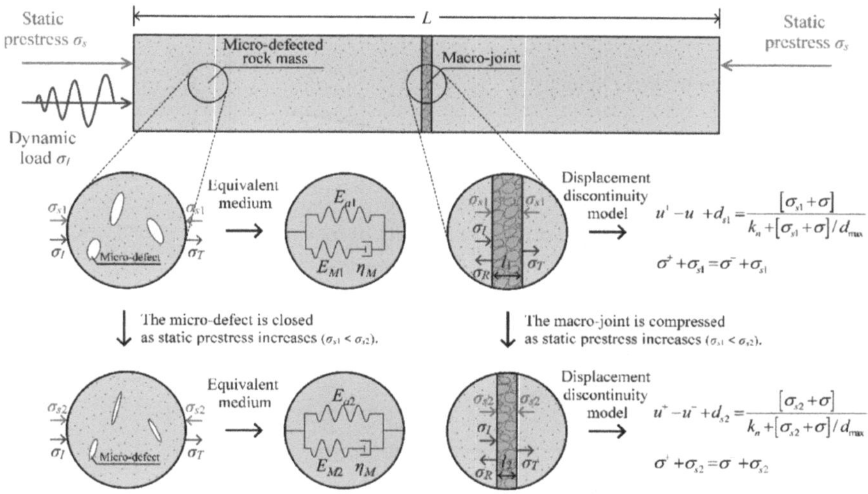

Fig. 8.41 Schematic of a double-scale discontinuous rock mass under combined static-dynamic loads

prestress, which causes a change in the initial mechanical properties of the macro-joint under static prestress. Therefore, the traditional equivalent viscoelastic model and DDM are inappropriate for investigating wave propagation under the effect of static prestress. The equivalent medium model of a double-scale discontinuous rock mass under combined static-dynamic loads is proposed, as shown in Fig. 8.43, in which the equivalent viscoelastic model and DDM that considers the effect of static prestress are introduced.

The equivalent viscoelastic model that considers the effect of static prestress in Fig. 8.43 is composed of a Maxwell model and spring in parallel. It can effectively study the propagation of stress waves through a micro-defected rock mass when the wavelength of stress waves is much larger than the maximum geometry of the micro-defect. The equivalent viscoelastic model can describe not only the dynamic mechanical properties of the rock mass at large strain rates, but also the static mechanical properties of the rock mass at small strain rates. In addition, in order to describe the mechanical properties of rock mass under static prestress effectively, the relationship between the elastic modulus and the static prestress is introduced. The constitutive equation of the equivalent viscoelastic model that considers the effect of static prestress in Fig. 8.43 can be expressed as

$$\frac{\partial \varepsilon}{\partial t} - \frac{1}{E_a(\sigma_s) + E_M(\sigma_s)}\frac{\partial \sigma}{\partial t} + \frac{E_a(\sigma_s)E_M(\sigma_s)\varepsilon}{[E_a(\sigma_s) + E_M(\sigma_s)]\eta_M} - \frac{E_M(\sigma_s)\sigma}{[E_a(\sigma_s) + E_M(\sigma_s)]\eta_M} = 0 \tag{8.114}$$

where ε denotes the particle strain of the rock mass, σ denotes the particle stress of the rock mass, σ_s denotes the static prestress, t denotes the time, η_M denotes the

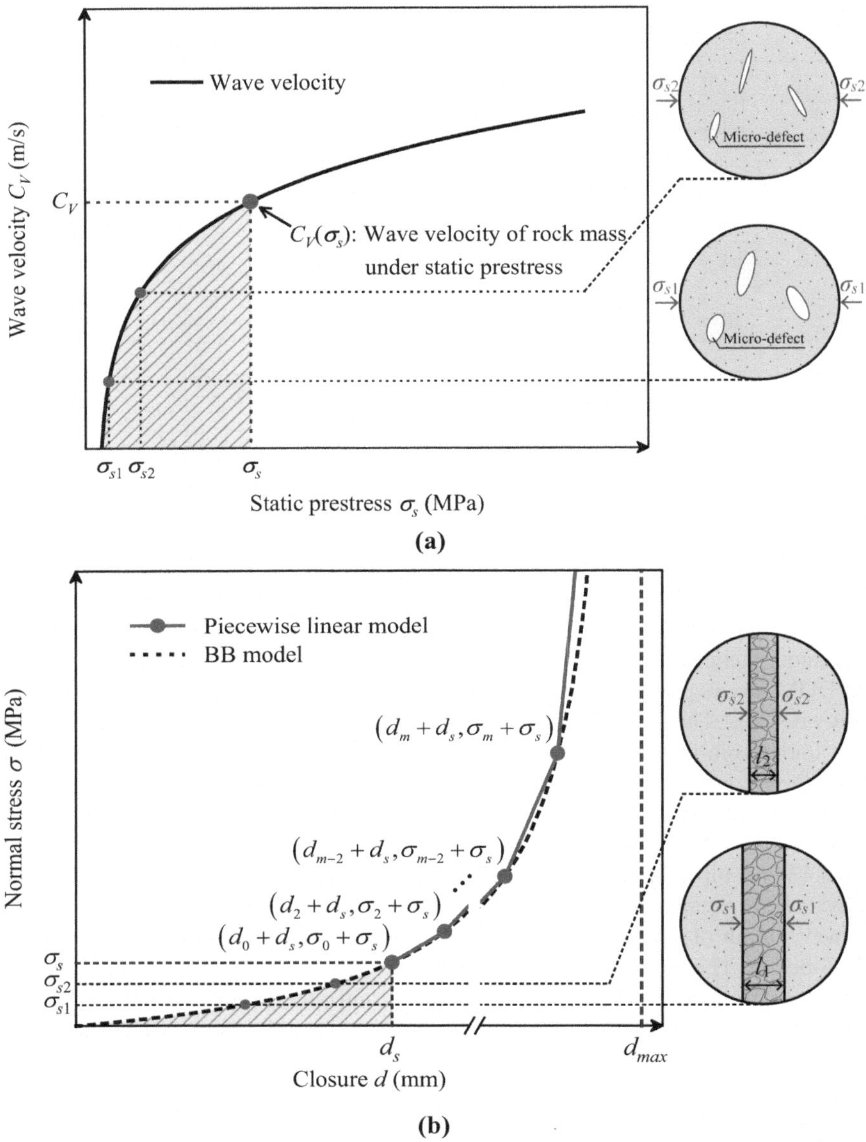

Fig. 8.42 Mechanical behaviors of the micro-defected rock mass and macro-joint under static prestress. **a** Wave velocity of micro-defected rock mass under static prestress. **b** Deformation properties of macro-joint under static prestress

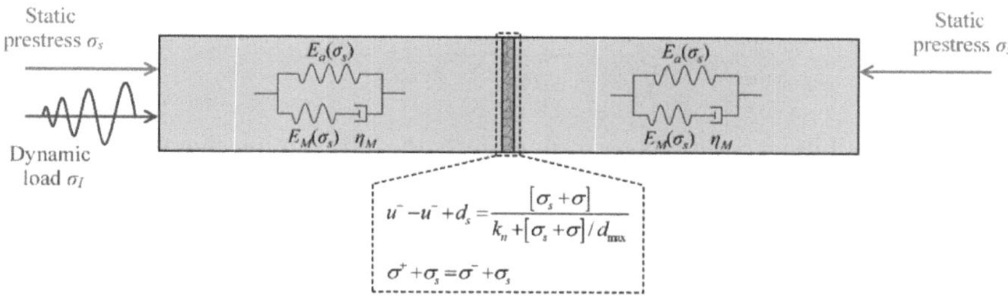

Fig. 8.43 Equivalent medium model of double-scale discontinuous rock mass under combined static-dynamic loads

viscous coefficient in the equivalent viscoelastic model in Fig. 8.43, and $E_M(\sigma_s)$ and $E_a(\sigma_s)$ denote the elastic modulus of the spring under static prestress in Fig. 8.43.

The motion equation of stress wave propagation can be expressed as

$$\rho_0 \frac{\partial v}{\partial t} - \frac{\partial \sigma}{\partial x} = 0 \tag{8.115}$$

where ρ_0 denotes the density of the rock mass, v denotes the particle velocity of the rock mass, and x denotes the coordinate of the particle of the rock mass.

The continuity equation of stress wave propagation can be expressed as

$$\frac{\partial \varepsilon}{\partial t} - \frac{\partial v}{\partial x} = 0 \tag{8.116}$$

The governing equations consist of Eqs. (8.114)–(8.116). Based on the characteristic method, Wang et al. [7], the right-running characteristic lines, left-running characteristic lines and up-running characteristic lines can be obtained according to the governing equations. The compatibility conditions that correspond to the three sets of characteristic lines can also be obtained.

The right-running characteristic lines can be expressed as

$$\frac{dx}{dt} = \sqrt{\frac{E_a(\sigma_s) + E_M(\sigma_s)}{\rho_0}} = C_V(\sigma_s) \tag{8.117}$$

where $C_V(\sigma_s)$ denotes the wave velocity under static prestress.

The compatibility conditions that correspond to the right-running characteristic lines can be expressed as

$$dv = \frac{d\sigma}{\rho_0 C_V(\sigma_s)} + \frac{[\sigma - E_a(\sigma_s)\varepsilon]E_M(\sigma_s)}{[E_a(\sigma_s) + E_M(\sigma_s)]\eta_M} dx \tag{8.118}$$

The left-running characteristic lines can be expressed as

$$\frac{dx}{dt} = -\sqrt{\frac{E_a(\sigma_s) + E_M(\sigma_s)}{\rho_0}} = -C_V(\sigma_s) \tag{8.119}$$

The compatibility conditions corresponding to the left-running characteristic lines can be expressed as

$$dv = -\frac{d\sigma}{\rho_0 C_V(\sigma_s)} + \frac{[\sigma - E_a(\sigma_s)\varepsilon]E_M(\sigma_s)}{[E_a(\sigma_s) + E_M(\sigma_s)]\eta_M}dx \tag{8.120}$$

The up-running characteristic lines can be expressed as

$$dx = 0 \tag{8.121}$$

The compatibility conditions corresponding to the up-running characteristic lines can be expressed as

$$d\varepsilon = \frac{d\sigma}{E_a(\sigma_s) + E_M(\sigma_s)} + \frac{[\sigma - E_a(\sigma_s)\varepsilon]E_M(\sigma_s)}{[E_a(\sigma_s) + E_M(\sigma_s)]\eta_M}dt \tag{8.122}$$

The three-characteristic method that considers static prestress is composed of Eqs. (8.117)–(8.122). The right-running characteristic lines Eq. (8.117) and the left-running characteristic lines Eq. (8.119) represent the trajectory of the wavefront propagating with velocity $C_V(\sigma_s)$. The up-running characteristic lines Eq. (8.121) represent the trajectory of the particle. In addition, the compatibility conditions Eqs. (8.118) and (8.120) represent the relationship between particle stress and particle velocity along the right-running characteristic lines as in Eq. (8.117) and the left-running characteristic lines as in Eq. (8.119). The compatibility conditions as in Eq. (8.122) represent the relationship between particle stress and particle strain along the up-running characteristic lines as in Eq. (8.121).

Assume the existence of a macro-joint at $x = x_J$, which forms the split three-characteristic method that considers static prestress, as shown in Fig. 8.44. The DDM that considers the effect of static prestress is introduced to the position of $x = x_J$ for calculating the stress wave propagation across macro-joint. The left and right sides of $x = x_J$ are the three characteristic lines for calculating the stress wave propagation in the micro-defected rock mass. It can be seen from Eqs. (8.117) and (8.119) that the slope of the right-running and left-running characteristic lines is determined by the wave velocity. The wave velocity increases as the static prestress increases. When the static prestress σ_{s1} increases to σ_{s2}, the right-running characteristic lines change from $\mathrm{d}x/\mathrm{d}t = C_V(\sigma_{s1})$ to $\mathrm{d}x/\mathrm{d}t = C_V(\sigma_{s2})$. The left-running characteristic lines change from $\mathrm{d}x/\mathrm{d}t = -C_V(\sigma_{s1})$ to $\mathrm{d}x/\mathrm{d}t = -C_V(\sigma_{s2})$. These changes differ from the traditional three-characteristic method, in which the right-running and left-running characteristic lines are fixed.

In addition, the split three-characteristic method consists of three basic elements, namely the triangle, diamond and separated diamond elements. Figure 8.45 shows in detail the three basic elements under different static prestresses. Figure 8.45a and

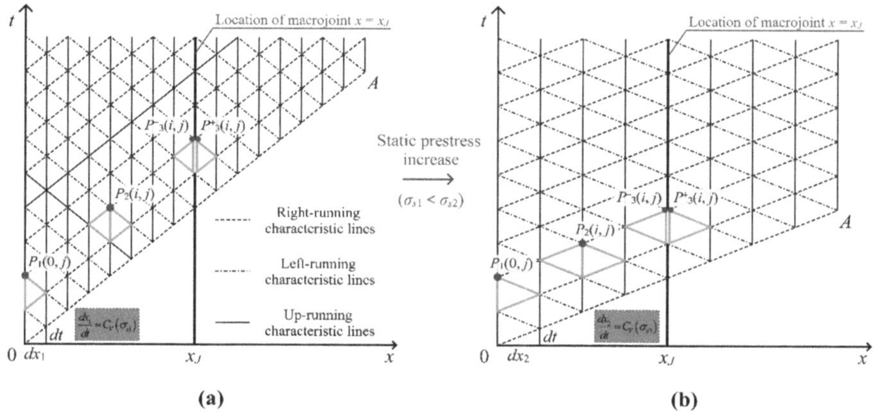

Fig. 8.44 Split three-characteristic method considering static prestress. **a** Static prestress is σ_{s1}. **b** Static prestress is σ_{s2}

b shows the triangle element and diamond element under static prestresses σ_{s1} and σ_{s2}, which are used to calculate the stress, strain and velocity at the left boundary point $P_1(0, j)$ and the interior point $P_2(i, j)$ of the micro-defected rock mass, similar to the traditional method [7]. Figure 8.45c shows the separated diamond element under static prestresses σ_{s1} and σ_{s2}, which is utilized in combination with the DDM to calculate the stress, strain and velocity at the interior point $P_3(i, j)$ of the macro-joint. When the static prestress is σ_{s1}, the time step and grid distances are $\mathrm{d}t$ and $\mathrm{d}x_1$, respectively. However, when the static prestress increases to σ_{s2}, the grid distance increases to $\mathrm{d}x_2$ with the constant time step $\mathrm{d}t$.

(1) **Stress wave propagation through a micro-defected rock mass under static prestress**

When the boundary conditions and initial conditions are known, the stress, strain and velocity on the boundary point $P_1(0, j)$ and internal point $P_2(i, j)$ of the micro-defected rock mass can be obtained based on the triangle elements and diamond elements.

Under initial conditions, the static prestress at the left end of the rock mass is expressed as

$$\sigma_s(0, 0) = \sigma_s \tag{8.123}$$

The static prestress at the right end of the rock mass is expressed as

$$\sigma_s(L, 0) = \sigma_s \tag{8.124}$$

In addition, the stress initial condition is

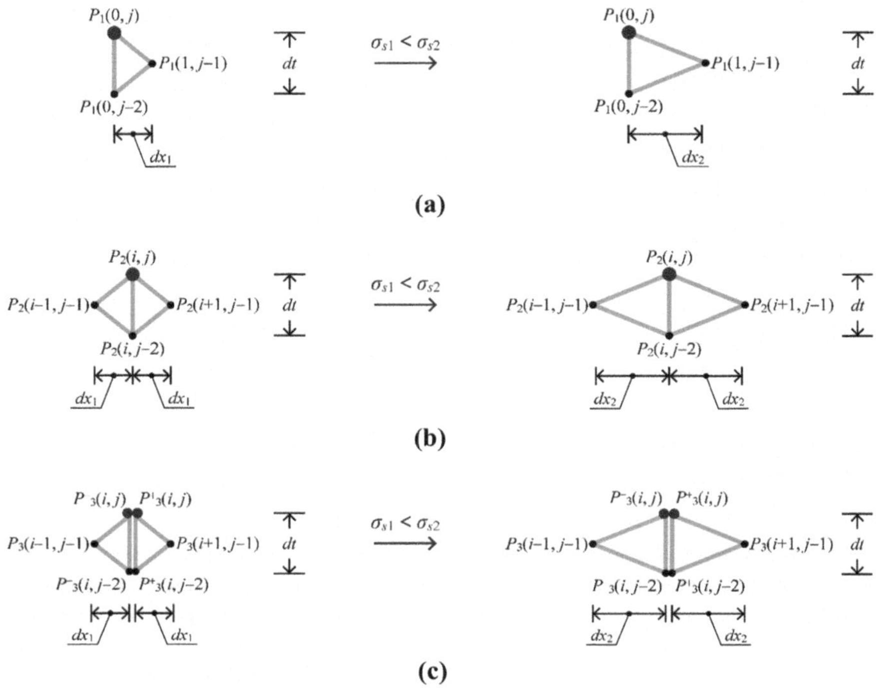

Fig. 8.45 Three basic elements under different static prestresses. **a** Triangular elements. **b** Diamond elements. **c** Separated diamond elements

$$\sigma(x, 0) = 0,\ 0 < x \le L \tag{8.125}$$

where L denotes the propagation distance of the stress wave in the rock mass.

The strain initial condition is

$$\varepsilon(x, 0) = 0,\ 0 < x \le L \tag{8.126}$$

The velocity initial condition is

$$v(x, 0) = 0,\ 0 < x \le L \tag{8.127}$$

In addition, a half-sine stress wave is applied to the left end face of the rock mass as boundary conditions

$$\sigma(0, t) = \begin{cases} A_0 sin(2\pi f_0 t) \\ 0 \end{cases},\ \text{when}\ \begin{matrix} 0 \le t \le 1/(2f_0) \\ t > 1/(2f_0) \end{matrix} \tag{8.128}$$

where A_0 denotes the amplitude of the incident wave and f_0 denotes the frequency of the incident wave.

The relationship of velocity with stress and strain along characteristic line $0A$ can be expressed as

$$v = -\frac{\sigma}{\rho_0 C_V(\sigma_s)} \tag{8.129}$$

and

$$v = -C_V(\sigma_s)\varepsilon \tag{8.130}$$

According to Eq. (8.128), stress $\sigma = 0$ when time $t = 0$. Therefore, stress $\sigma = 0$, strain $\varepsilon = 0$ and velocity $v = 0$ are satisfied along the characteristic line $0A$.

The stress, strain and velocity on the boundary point $P_1(0, j)$ of the micro-defected rock mass can be obtained according to the stress initial conditions, and the compatibility conditions that correspond to characteristic lines $P_1(0, j) - P_1(1, j - 1)$ and $P_1(0, j) - P_1(0, j - 2)$. The differential expressions for the two compatibility conditions under static prestress can be described as

$$\begin{aligned} v(0,j) = v(1,j-1) - \frac{\sigma(0,j) - \sigma(1,j-1)}{\rho_0 C_V(\sigma_s)} \\ + \frac{\left[\sigma(1,j-1) - E_a(\sigma_s)\varepsilon(1,j-1)\right]E_M(\sigma_s)}{[E_a(\sigma_s) + E_M(\sigma_s)]\eta_M}\left[x(0,j) - x(1,j-1)\right] \end{aligned} \tag{8.131}$$

and

$$\begin{aligned} \varepsilon(0,j) = \varepsilon(0,j-2) + \frac{\sigma(0,j) - \sigma(0,j-2)}{E_a(\sigma_s) + E_M(\sigma_s)} \\ + \frac{\left[\sigma(0,j-2) - E_a(\sigma_s)\varepsilon(0,j-2)\right]E_M(\sigma_s)}{[E_a(\sigma_s) + E_M(\sigma_s)]\eta_M}\left[t(0,j) - t(0,j-2)\right] \end{aligned} \tag{8.132}$$

The stress, strain and velocity at point $P_1(0, j - 2)$ and point $P_1(1, j - 1)$ in Eqs. (8.131) and (8.132) are known. Therefore, the velocity $v(0, j)$ and strain $\varepsilon(0, j)$ at point $P_1(0, j)$ can be obtained according to Eqs. (8.131) and (8.132). In addition, the stress $\sigma(0, j)$ can be obtained according to the stress initial conditions.

The stress, strain and velocity on the internal point $P_2(i, j)$ inside the micro-defected rock mass can be obtained according to the compatibility conditions that correspond to characteristic lines $P_2(i, j) - P_2(i - 1, j - 1)$, $P_2(i, j) - P_2(i + 1, j - 1)$ and $P_2(i, j) - P_2(i, j - 2)$. The differential expressions for the three compatibility conditions under static prestress can be expressed as

$$v(i,j) = v(i-1,j-1) + \frac{\sigma(i,j) - \sigma(i-1,j-1)}{\rho_0 C_V(\sigma_s)} + \frac{[\sigma(i-1,j-1) - E_a(\sigma_s)\varepsilon(i-1,j-1)]E_M(\sigma_s)}{[E_a(\sigma_s) + E_M(\sigma_s)]\eta_M}[x(i,j) - x(i-1,j-1)] \quad (8.133)$$

$$v(i,j) = v(i+1,j-1) - \frac{\sigma(i,j) - \sigma(i+1,j-1)}{\rho_0 C_V(\sigma_s)} + \frac{[\sigma(i+1,j-1) - E_a(\sigma_s)\varepsilon(i+1,j-1)]E_M(\sigma_s)}{[E_a(\sigma_s) + E_M(\sigma_s)]\eta_M}[x(i,j) - x(i+1,j-1)] \quad (8.134)$$

and

$$\varepsilon(i,j) = \varepsilon(i,j-2) + \frac{\sigma(i,j) - \sigma(i,j-2)}{E_a(\sigma_s) + E_M(\sigma_s)} + \frac{[\sigma(i,j-2) - E_a(\sigma_s)\varepsilon(i,j-2)]E_M(\sigma_s)}{[E_a(\sigma_s) + E_M(\sigma_s)]\eta_M}[t(i,j) - t(i,j-2)] \quad (8.135)$$

The stress, strain and velocity at point $P_2(i, j-2)$, point $P_2(i-1, j-1)$ and point $P_2(i+1, j-1)$ are known. The stress $\sigma(i, j)$, velocity $v(i, j)$ and strain $\varepsilon(i, j)$ at the internal point $P_2(i, j)$ can be obtained by solving Eqs. (8.133)–(8.135).

(2) **Stress wave propagation through the macro-joint under static prestress**

The stress wave propagation at point $P_3(i, j)$ at the location of the macro-joint can be investigated by introducing the DDM that considers static prestress into the separated diamond element. The DDM that considers static prestress assumes that the particle stress on the left side point $P_3^-(i, j)$ and right side point $P_3^+(i, j)$ of the macro-joint are equal, while the particle displacements on the left side point $P_3^-(i, j)$ and right side point $P_3^+(i, j)$ of the macro-joint are unequal due to the closure of the macro-joint. The closure of the macro-joint with respect to stress is represented by the BB model, as shown in Fig. 8.42b. The relationship between the closure of the macro-joint and stress under the effect of static prestress can be expressed as

$$d_s + d = \frac{\sigma_s + \sigma}{k_{n0} + (\sigma_s + \sigma)/d_{\max}} \quad (8.136)$$

where d_s is the closure of the macro-joint under static prestress, d is the closure of the macro-joint under a stress wave, $d_{\max}$ is the maximum allowable closure of the macro-joint, and k_{n0} is the initial stiffness of the macro-joint under zero static prestress.

The derivative of normal stress σ with respect to the closure of macro-joint d in Eq. (8.136) is considered the macro-joint stiffness, which can be expressed as

$$k = \frac{d\sigma}{dd} = k_{n0}\left(1 + \frac{\sigma_s + \sigma}{k_{n0}d_{max}}\right)^2 \quad (8.137)$$

Substituting $\sigma = 0$ into Eq. (8.137), the initial macro-joint stiffness $k_n(\sigma_s)$ under an arbitrary static prestress can be obtained as

$$k_n(\sigma_s) = k_{n0}\left(1 + \frac{\sigma_s}{k_{n0}d_{max}}\right)^2 \tag{8.138}$$

The relationship between the closure of the macro-joint and stress is treated in segments to form a piecewise linear model, as shown in Fig. 8.42b, in which the macro-joint stiffness of segment m can be expressed as

$$k_m = \frac{d\sigma}{dd} = k_{n0}\left(1 + \frac{\sigma_s + \sigma_{m-2}}{k_{n0}d_{max}}\right)^2 \tag{8.139}$$

The relationship between the closure of the macro-joint and stress for segment m can be expressed as

$$(d_s + d_m) - (d_s + d_{m-2}) = \frac{[\sigma_s + \sigma_m(x,t)] - [\sigma_s + \sigma_{m-2}(x,t)]}{k_m} \tag{8.140}$$

Assuming that the closure of the macro-joint and stress for point $P_3(i,j)$ in Fig. 8.45 correspond to $(d_m + d_s, \sigma_m + \sigma_s)$ in Fig. 8.42b, the macro-joint stiffness for point $P_3(i,j)$ can be expressed as

$$k(i,j) = \frac{d\sigma}{dd} = k_{n0}\left[1 + \frac{\sigma_s + \sigma(i,j-2)}{k_{n0}d_{max}}\right]^2 \tag{8.141}$$

In addition, the relationship between the closure of the macro-joint and stress for point $P_3(i,j)$ in Fig. 8.45 can be expressed as

$$[d_s + d(i,j)] - [d_s + d(i,j-2)] = \frac{[\sigma_s + \sigma(i,j)] - [\sigma_s + \sigma(i,j-2)]}{k(i,j)} \tag{8.142}$$

According to Eq. (8.142), the difference between the particle displacements on the left side point P_3^- (i,j) and the particle displacements on the right side point P_3^+ (i,j) of the macro-joint can be obtained as

$$u^-(i,j) - u^+(i,j) = d(i,j-2) + \frac{\sigma(i,j) - \sigma(i,j-2)}{k(i,j)} \tag{8.143}$$

where superscript "−" denotes the left side of the macro-joint, and the superscript "+" denotes the right side of the macro-joint.

The derivative of u with respect to t in Eq. (8.143) is

$$v^+(i,j) - v^-(i,j) = \frac{1}{k(i,j)}\frac{\sigma(i,j) - \sigma(i,j-2)}{2dt} \tag{8.144}$$

The particle stress on the left side point P_3^- (i, j) and right side point P_3^+ (i, j) of the macro-joint can be expressed as

$$\sigma_s + \sigma^-(i,j) = \sigma_s + \sigma^+(i,j) \tag{8.145}$$

The stress, strain and velocity on the left side point P_3^- (i, j) and right side point $P + 3(i, j)$ of the macro-joint can be obtained according to Eqs. (8.144) and (8.145), and the compatibility conditions that correspond to characteristic lines P_3^- $(i, j) - P_3(i-1, j-1)$, $P + 3(i, j) - P_3(i+1, j-1)$ and $P_3^{-,+}$ $(i, j) - P_3^{-,+}$ $(i, j-2)$. The differential expressions for the four compatibility conditions under static prestress can be expressed as

$$v^-(i,j) = v(i-1,j-1) + \frac{\sigma(i,j) - \sigma(i-1,j-1)}{\rho_0 C_V(\sigma_s)} + \frac{[\sigma(i-1,j-1) - E_a(\sigma_s)\varepsilon(i-1,j-1)]E_M(\sigma_s)}{[E_a(\sigma_s) + E_M(\sigma_s)]\eta_M}[x(i,j) - x(i-1,j-1)] \tag{8.146}$$

$$v^+(i,j) = v(i+1,j-1) - \frac{\sigma(i,j) - \sigma(i+1,j-1)}{\rho_0 C_V(\sigma_s)} - \frac{[\sigma(i+1,j-1) - E_a(\sigma_s)\varepsilon(i+1,j-1)]E_M(\sigma_s)}{[E_a(\sigma_s) + E_M(\sigma_s)]\eta_M}[x(i,j) - x(i+1,j-1)] \tag{8.147}$$

$$\varepsilon^-(i,j)=\varepsilon^-(i,j-2) + \frac{\sigma(i,j) - \sigma(i,j-2)}{\mathrm{E_a}(\sigma_s)+E_M(\sigma_s)} + \frac{[\sigma(i,j-2) - E_a(\sigma_s)\varepsilon^-(i,j-2)]E_M(\sigma_s)}{[\mathrm{E_a}(\sigma_s)+E_M(\sigma_s)]\eta_M}[t(i,j) - t(i,j-2)] \tag{8.148}$$

and

$$\varepsilon^+(i,j) = \varepsilon^+(i,j-2) + \frac{\sigma(i,j) - \sigma(i,j-2)}{\mathrm{E_a}(\sigma_s)+E_M(\sigma_s)} + \frac{[\sigma(i,j-2) - E_a(\sigma_s)\varepsilon^+(i,j-2)]E_M(\sigma_s)}{[\mathrm{E_a}(\sigma_s)+E_M(\sigma_s)]\eta_M}[t(i,j) - t(i,j-2)] \tag{8.149}$$

The stress, strain and velocity of point $P_3^{-,+}$ $(i, j-2)$, point $P_3(i-1, j-1)$ and point $P_3(i+1, j-1)$ are known. In addition, substituting Eqs. (8.144) and (8.145) into (8.146) and (8.147), the stress $\sigma(i, j)$ on the left side point P_3^- (i, j) and right side point P_3^+ (i, j) can be obtained as

$$\sigma(i,j) = \frac{2dt\rho_0 C_V(\sigma_s)k(i,j)}{\rho_0 C_V(\sigma_s) + 4dtk(i,j)}\left\{\frac{\sigma(i+1,j-1) + \sigma(i-1,j-1)}{\rho_0 C_V(\sigma_s)} + \frac{1}{2dtk(i,j)}\sigma(i,j-2) + v(i+1,j-1) - \frac{[\sigma(i+1,j-1) - E_a(\sigma_s)\varepsilon(i+1,j-1)]E_M(\sigma_s)}{[E_a(\sigma_s) + E_M(\sigma_s)]\eta_M}dx\right.$$

$$- v(i-1,j-1) - \frac{[\sigma(i-1,j-1) - E_a(\sigma_s)\varepsilon(i-1,j-1)]E_M(\sigma_s)}{[E_a(\sigma_s) + E_M(\sigma_s)]\eta_M}dx\Bigg\} \tag{8.150}$$

Equation (8.150) is introduced into Eqs. (8.146) and (8.147), and the velocity $v^-(i, j)$ on the left side point P_3^- (i, j) and the velocity $v^+(i, j)$ on the right side point P_3^+ (i, j) can be obtained. Equation (8.150) is introduced into Eqs. (8.148) and (8.149), and the strain $\varepsilon^-(i, j)$ on the left side point P_3^- (i, j) and the strain $\varepsilon^+(i, j)$ on the right side point P_3^+ (i, j) can be obtained. Therefore, the stress, strain and velocity in the double-scale discontinuous rock mass under the combined static-dynamic loading conditions can be obtained.

8.5.2 Static Prestress Effect on the Wave Transmission

In this section, the relationship of amplitude and energy transmission coefficients with static prestress is investigated based on the proposed method. The amplitude transmission coefficient is defined as

$$T_\sigma = \frac{\max(\sigma_T)}{\max(\sigma_I)} \tag{8.151}$$

where σ_I is the incident stress wave and σ_T is the transmitted stress wave.

The energy transmission coefficient is defined as

$$T_e = \frac{\sum_{t_T^{ini}}^{t_T^{fin}} \rho_0 C_V \sigma_T^2 \Delta t}{\sum_{t_I^{ini}}^{t_I^{fin}} \rho_0 C_V \sigma_I^2 \Delta t} \tag{8.152}$$

where t_I is the time corresponding to the incident stress wave, t_T is the time corresponding to the transmitted stress wave, "*ini*" denotes the initial time of the stress wave, and "*fin*" denotes the final time of the stress wave.

The validity of the present study is evaluated by comparing the results from the traditional study. Figure 8.46 shows the representative elementary volume (REV) of the rock mass under static prestress in the present study and traditional study. In the present study, the propagation properties of stress waves through double-scale discontinuous rock mass considering the micro-defect are studied based on the proposed method, as shown in Fig. 8.46a. In the traditional study, the propagation properties of stress waves through single-scale discontinuous rock mass without considering the micro-defect are studied based on the DDM, as shown in Fig. 8.46b. Granite is selected as the rock material to investigate stress wave propagation. Under the initial conditions, the parameters of double-scale discontinuous rock mass considering the micro-defect are as shown in Table 8.4. The parameters of single-scale discontinuous rock mass without considering the micro-defect are as shown in Table 8.5. Table 8.4 and Table 8.5 show the parameters of rock mass under zero static prestress. When the static prestress is considered, the wave velocity $C_V(\sigma_s)$ of the micro-defected rock mass is affected due to the closure of the micro-defect. The wave velocity $C_V(\sigma_s)$ under static prestress can be obtained as

$$C_V(\sigma_s) = p_1 \sigma_s^{p_2} \tag{8.153}$$

where the static prestress is selected as $\sigma_s \geq 0.1$ MPa in the present study, p_1 and p_2 are the fitting coefficients, p_1 is selected as 4068.7, and p_2 is selected as 0.0624 [42].

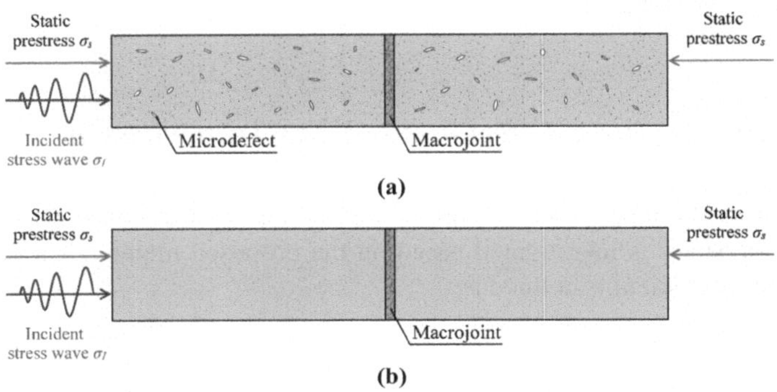

Fig. 8.46 The representative elementary volume (REV) of the rock mass under static prestress. **a** Double-scale discontinuous rock mass considering micro-defect in the present study. **b** Single-scale discontinuous rock mass without considering micro-defect in the traditional study

Table 8.4 Parameters of double-scale discontinuous rock mass considering micro-defect

Parameters of macro-joint	Value	Parameters of micro-defected rock mass	Value
Initial macro-joint stiffness k_{n0}	3.0GPa/m	Initial elastic modulus E_a	10.2 GPa
		Initial elastic modulus E_M	20.4 GPa
Maximum allowance closure d_{max}	0.53 mm	Initial wave velocity C_{V0}	3500 m/s
		Viscous coefficient η_M	16.0 MPa · s
		Density ρ_0	2500 kg/m^3

Table 8.5 Parameters of single-scale discontinuous rock mass without considering micro-defect

Parameters of macro-joint	Value	Parameters of rock mass	Value
Initial macro-joint stiffness k_{n0}	3.0 GPa/m	Elastic modulus E	30.6 GPa
		Wave velocity C_V	3500 m/s
Maximum allowance closure d_{max}	0.53 mm		
		Density ρ_0	2500 kg/m^3

In addition, the elastic moduli $E_a(\sigma_s)$ and elastic moduli $E_M(\sigma_s)$ affected by the wave velocity are $C_V(\sigma_s)^2\rho_0/3$ and $2C_V(\sigma_s)^2\rho_0/3$, respectively. The initial macro-joint stiffness $k_n(\sigma_s)$ affected by the wave velocity can be obtained according to Eq. (8.138).

Figure 8.47a shows the variation in the amplitude transmission coefficient T_σ with static prestresses σ_s for different incident wave amplitudes. Figure 8.47b shows the variation in the energy transmission coefficient T_e as the static prestress σ_s for different incident wave amplitudes. The incident wave amplitudes A_0 are selected as 0.5, 4.0 and 7.0 MPa. The propagation distance L is fixed at 3.0 m. The incident wave frequency f_0 is fixed at 1000 Hz. Figure 8.47a and b shows that the static prestress significantly affects the amplitude and energy transmission coefficients, regardless of whether the micro-defect in the rock mass is considered. When the static prestress is small, the amplitude and energy transmission coefficient increase rapidly with an increase in static prestress. When the static prestress is large, the amplitude and energy transmission coefficient remain constant with an increase in static prestress. However, when the static prestress is large, the amplitude and energy transmission coefficient considering the micro-defects reach a value smaller than 1.0. The amplitude and energy transmission coefficient without considering the micro-defects tend to be 1.0. This phenomenon illustrates that when the static prestress is large, the macro-jointed rock mass without considering the micro-defect exhibits a weak and negligible effect on wave propagation as intact rock mass. However, the effect of double-scale discontinuous rock mass considering the micro-defect on wave propagation is not negligible. It is also observed from Fig. 8.47a and b that regardless of whether the micro-defect in the rock mass is considered, the amplitude and energy transmission coefficient that correspond to incident wave amplitudes of 0.5, 4.0 and 7.0 MPa are sequentially increasing for a given smaller static prestress. The amplitude and energy transmission coefficient that correspond to incident wave amplitudes of 0.5, 4.0 and 7.0 MPa are kept constant for a given larger static prestress. Moreover, this finding contradicts the amplitude and energy transmission coefficients without considering the micro-defect. The amplitude and energy transmission coefficients considering the micro-defect are invariably small. However, when the static prestress is small, the difference between the results obtained considering the micro-defects and those without considering the micro-defects is smaller. The difference increases as the static prestress increases, and the difference tends to stabilize when the static prestress is sufficiently large. Therefore, under different incident wave amplitudes, the effect of the micro-defect on wave propagation is small for small static prestresses. The effect of the micro-defect on wave propagation is larger for large static prestresses.

Figure 8.48a shows the variation in the amplitude transmission coefficient T_σ with static prestresses σ_s for different incident wave frequencies. Figure 8.48b shows the variation in the energy transmission coefficient $\boldsymbol{T_e}$ as static prestress σ_s for different incident wave frequencies. The incident wave frequencies f_0 are selected as 300, 700 and 1500 Hz. The propagation distance L is fixed at 3.0 m. The incident wave amplitude A_0 is fixed at 0.5 MPa. The results show that regardless of whether the micro-defect in the rock mass is considered, the amplitude and energy transmission coefficients increase with an increase in static prestress in the initial stage and

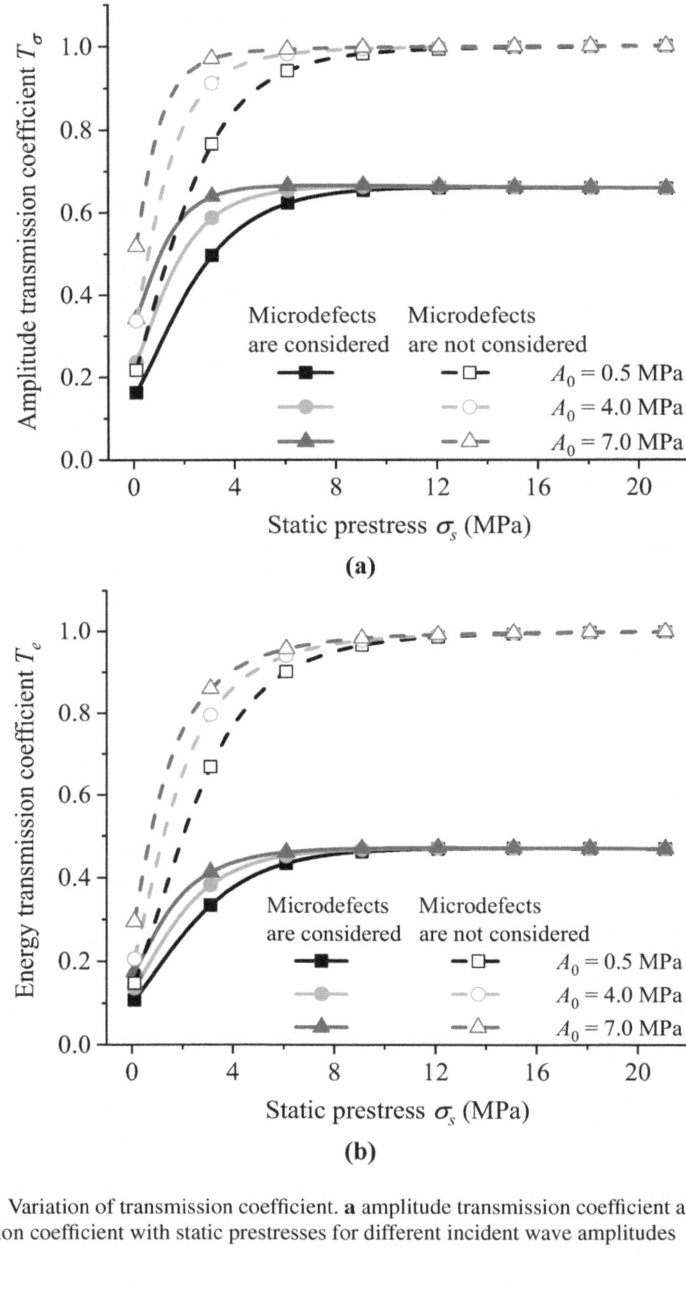

Fig. 8.47 Variation of transmission coefficient. **a** amplitude transmission coefficient and **b** energy transmission coefficient with static prestresses for different incident wave amplitudes

then remain constant with a further increase in static prestress. This phenomenon is also shown in Fig. 8.47. As shown in Fig. 8.48a and b, for small static prestresses, the amplitude and energy transmission coefficients with and without considering the micro-defects, which correspond to incident wave frequencies of 300, 700 and 1500 Hz, are in decreasing order. However, for large static prestresses, the amplitude and energy transmission coefficients without considering the micro-defects, which correspond to incident wave frequencies of 300, 700 and 1500 Hz, are in decreasing order. The amplitude and energy transmission coefficients considering the micro-defects, which correspond to incident wave frequencies of 300, 700 and 1500 Hz, are kept constant. Similar to Figs. 8.47, 8.48 shows that the difference between the results obtained with and without considering the micro-defects increases as the static prestress increases. Therefore, under different incident wave frequencies, the effect of micro-defect is small when the static prestress is small and is large when the static prestress is large.

Figure 8.49a shows the variation in the amplitude transmission coefficient T_σ as static prestress σ_s for different propagation distances. Figure 8.49b shows the variation in the energy transmission coefficient T_e as static prestress σ_s for different propagation distances. The propagation distances L are selected as 1.0, 4.0 and 8.0 m. The incident wave frequency f_0 is fixed at 1000 Hz. The incident wave amplitude A_0 is fixed at 0.5 MPa. The static prestress effect on the amplitude and energy transmission coefficient can also be observed in Figs. 8.47 and 8.48. Moreover, Fig. 8.49a and b shows that when the micro-defects are considered, the amplitude and energy transmission coefficients corresponding to propagation distances of 1.0, 4.0 and 8.0 m are in decreasing order for a given arbitrary static prestress. However, when the micro-defects are not considered, the amplitude and energy transmission coefficients are constant at different propagation distances for a given arbitrary static prestress. Therefore, the wave propagation properties are independent of the propagation distance when the micro-defects are not considered. In addition, when the static prestress is small, the results obtained with and without considering the micro-defects are similar. When the static prestress is large, the results obtained with and without considering the micro-defects vary significantly. Therefore, for different propagation distances, Figs. 8.47 and 8.48 show that the effect of micro-defect is small for small static prestresses and is large for large static prestresses.

8.5.3 The Energy Transmission Coefficient and Amplitude Transmission Coefficient

(1) Effect of incident wave amplitude on wave transmission

Figure 8.50a shows the effect of the incident wave amplitude A_0 on the amplitude transmission coefficients T_σ. Figure 8.50b shows the effect of the incident wave amplitude A_0 on the energy transmission coefficients T_e. The static prestress σ_s is selected as 0, 1.0 and 5.0 MPa. The propagation distance L is fixed at 3.0 m. The

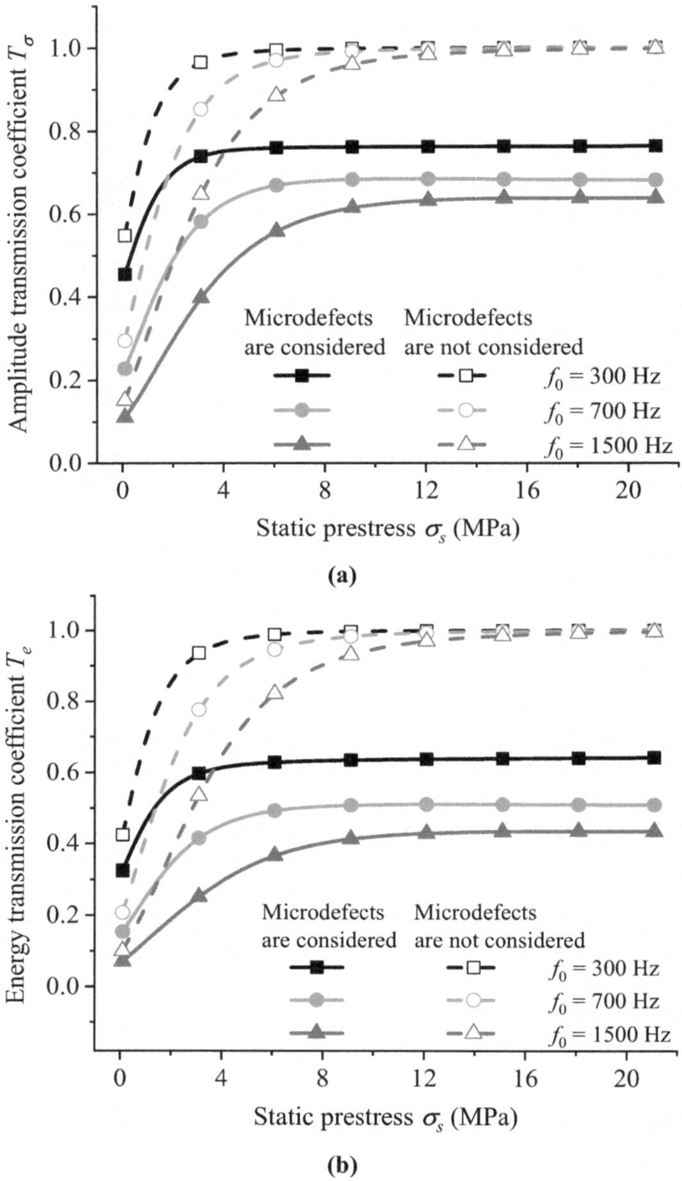

Fig. 8.48 Variation of transmission coefficients. **a** Amplitude transmission coefficient. **b** Energy transmission coefficient with static prestresses for different incident wave frequencies

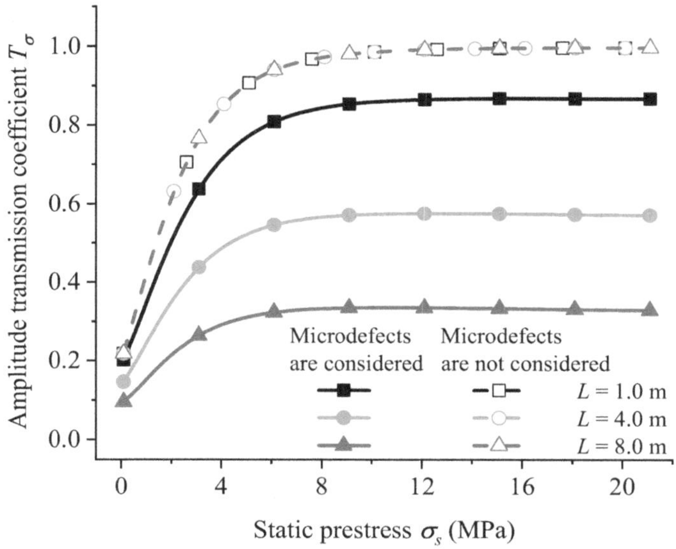

(a)

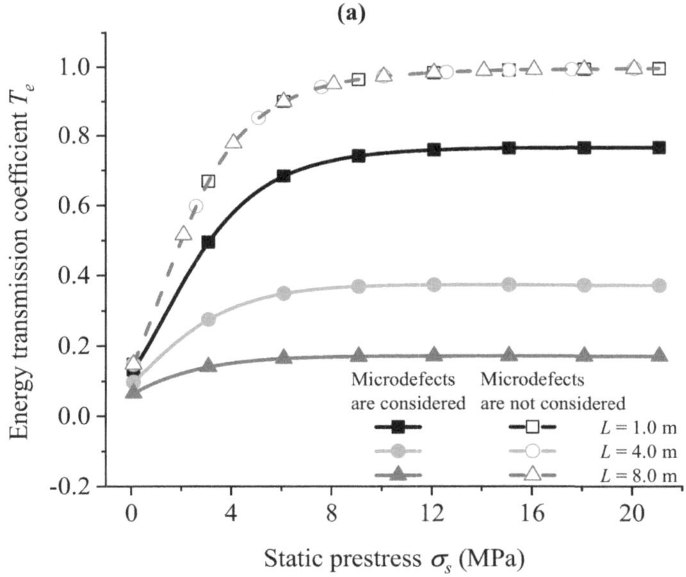

(b)

Fig. 8.49 Variation of transmission coefficient. **a** Amplitude transmission coefficient. **b** Energy transmission coefficient with static prestresses for different propagation distances incident wave frequency f_0 is fixed at 200 Hz. Figure 8.50a and b shows that when the static prestress is zero or small, such as $\sigma_s = 0$ and $\sigma_s = 1.0$ MPa, the amplitude and energy transmission coefficients with and without considering micro-defects increase rapidly and then stabilize with an increase in the incident wave amplitude. When the static prestress is large, such as $\sigma_s = 5.0$ MPa, the amplitude and energy transmission coefficients with and without considering the micro-defects remain constant with an increase in the incident wave amplitude. The amplitude and energy transmission coefficients considering the micro-defects remain constant and are <1.0. However, the amplitude and energy transmission coefficients without considering the micro-defects converge to 1.0. This phenomenon is explained by the notion that the dependence of the amplitude and energy transmission coefficients on the incident wave amplitude is not significant when the static prestress is large. As shown in Fig. 8.50a, for a given smaller incident wave amplitude, the amplitude transmission coefficient with and without considering micro-defects, which correspond to static prestresses of 0, 1.0 and 5.0 MPa, increases sequentially. However, the amplitude transmission coefficients are constant under different static prestresses for a given larger incident wave amplitude. Therefore, when the incident wave amplitude is large, the effect of the static prestress on the amplitude transmission coefficient is negligible, while its effect on the energy transmission coefficient is not negligible. According to Fig. 8.50b, for a given arbitrary incident wave amplitude, the energy transmission coefficient with and without considering micro-defects, which correspond to static prestresses of 0, 1.0 and 5.0 MPa, increases sequentially. Figure 8.50a and b indicates that the amplitude and energy transmission coefficients without considering micro-defects are always larger than those considering the micro-defects. The difference between the results obtained with and without considering the micro-defects increases as the incident wave amplitude increases for small static prestresses. However, the difference remains constant as the incident wave amplitude increases for large static prestresses.

(2) Effect of incident wave frequency on wave transmission

Figure 8.51a shows the effect of the incident wave frequency f_0 on the amplitude transmission coefficients T_σ. Figure 8.51b shows the effect of the incident wave frequency f_0 on the energy transmission coefficients T_e. The static prestress σ_s is selected as 0 and 2.0 MPa. The propagation distance L is fixed at 3.0 m. The incident wave amplitude A_0 is fixed at 0.5 MPa. Figure 8.51a and b shows that the amplitude and energy transmission coefficients remain constant at 1.0 when the incident wave frequency is small (e.g., $f_0 < 10^1$ Hz). The amplitude and energy transmission coefficients decrease sharply as the incident wave frequency further increases (e.g., $10^1 < f_0 < 10^5$ Hz). The amplitude and energy transmission coefficients gradually become zero when the incident wave frequency is large (e.g., $f_0 > 10^5$ Hz). Figure 8.51a and b shows that, for a given smaller and larger incident wave frequency, the amplitude and energy transmission coefficients with and without considering the micro-defects are constant under different static prestresses. For a given intermediate frequency, the amplitude and energy transmission coefficients with and without considering the micro-defects, which correspond to static prestresses of 0 and 2.0 MPa, increase

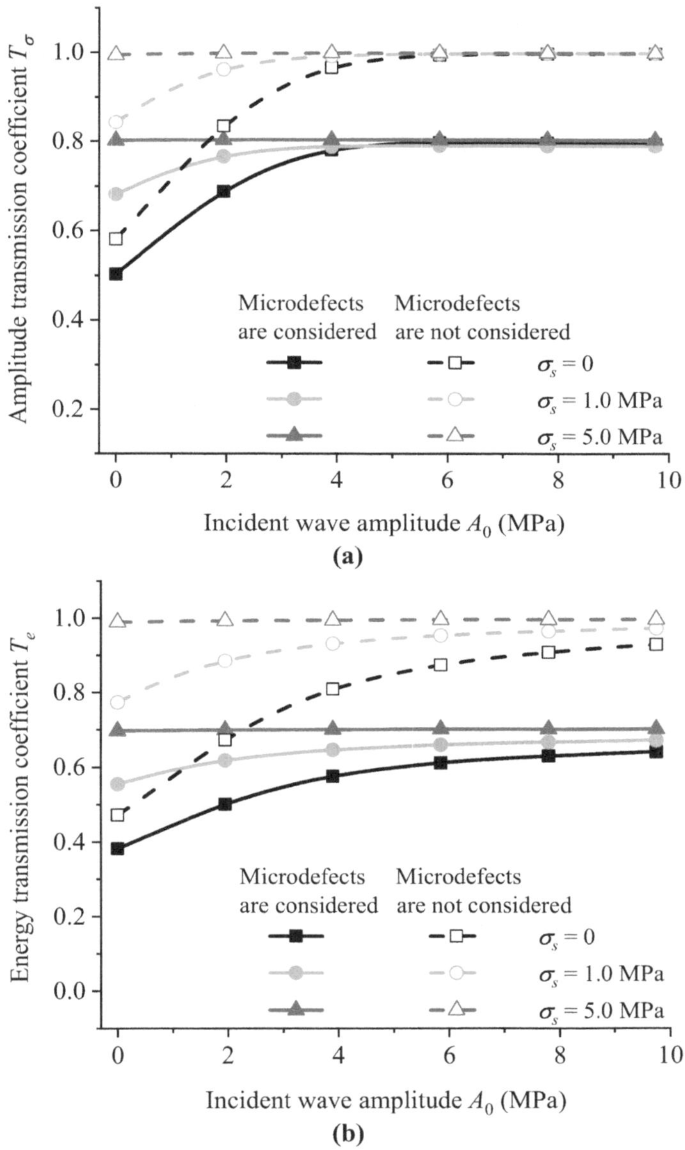

Fig. 8.50 Effect of incident wave amplitude on transmission coefficient. **a** Amplitude transmission coefficient. **b** Energy transmission coefficient

sequentially. Figure 8.51a and b shows that, for a given smaller or larger incident wave frequency, the difference between the results obtained with and without considering the micro-defects can be overlooked. For a given intermediate frequency, the amplitude and energy transmission coefficients considering the micro-defects are less than those without considering the micro-defects.

(3) Effect of propagation distance on wave transmission

Figure 8.52a shows the effect of the propagation distance L on the amplitude transmission coefficients T_σ. Figure 8.52b shows the effect of the propagation distance L on the energy transmission coefficients T_e. The static prestress σ_s is selected as 0, 1.0 and 5.0 MPa. The incident wave frequency f_0 is fixed at 200 Hz. The incident wave amplitude A_0 is fixed at 0.5 MPa. Figure 8.52a and b shows that the amplitude and energy transmission coefficient considering micro-defects decrease with an increase in the propagation distance. The amplitude and energy transmission coefficient without considering micro-defects remain constant with an increase in the propagation distance, which is not affected by the propagation distance. As shown in Fig. 8.52a and b, for a given propagation distance, the amplitude and energy transmission coefficients with and without considering micro-defects, which correspond to static prestresses of 0, 1.0 and 5.0 MPa, increase sequentially. In addition, the amplitude and energy transmission coefficients considering micro-defects are less than those without considering micro-defects. The difference between the results obtained with and without considering micro-defects decreases as the propagation distance decreases. It is noteworthy that the results with and without considering micro-defects are approximately equal for a small propagation distance. Therefore, when the propagation distance is sufficiently small, the results obtained by considering micro-defects in the present study degrade to those obtained by the traditional study without considering micro-defects.

(4) Combined effect of incident wave amplitude and static prestress on the difference in transmission coefficient

ΔT_σ is proposed to indicate the difference between the amplitude transmission coefficients T_σ with and without considering micro-defects. ΔT_σ is defined as

$$\Delta T_\sigma = \frac{\left| T_\sigma^{single} - T_\sigma^{double} \right|}{T_\sigma^{single}} \times 100\% \tag{8.154}$$

ΔT_e is proposed to indicate the difference between the energy transmission coefficients T_e with and without considering micro-defects. ΔT_e is defined as

$$\Delta T_e = \frac{\left| T_e^{single} - T_e^{double} \right|}{T_e^{single}} \times 100\% \tag{8.155}$$

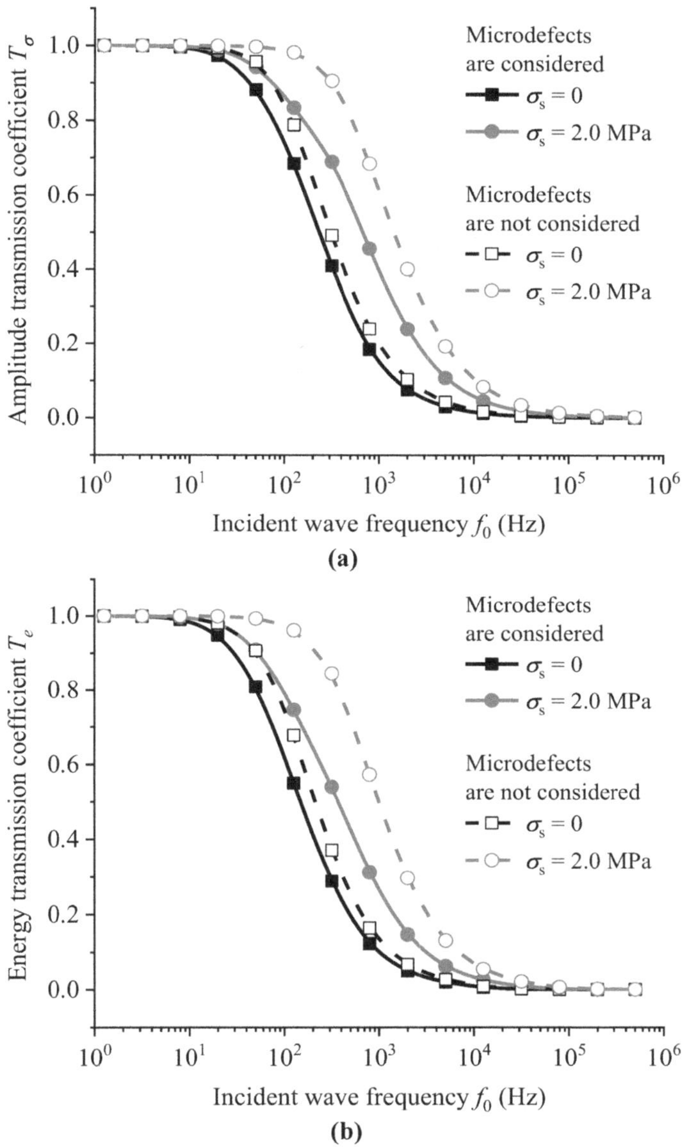

Fig. 8.51 Effect of incident wave frequency on transmission coefficient. **a** Amplitude transmission coefficient. **b** Energy transmission coefficient

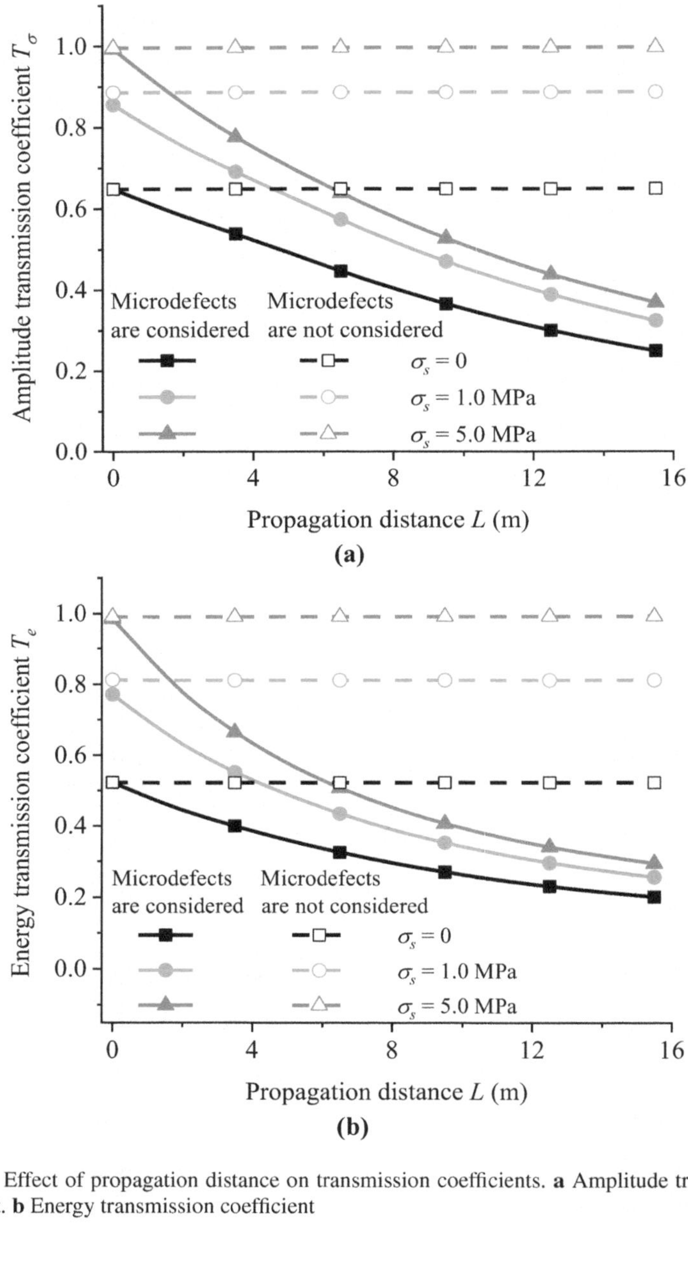

Fig. 8.52 Effect of propagation distance on transmission coefficients. **a** Amplitude transmission coefficient. **b** Energy transmission coefficient

where $T_{\sigma}^{\text{double}}$ and T_{e}^{double} denote the amplitude transmission coefficient and energy transmission coefficient, respectively, of stress wave propagation through a double-scale discontinuous rock mass, namely macro-joint and micro-defect in the rock mass are considered. $T_{\sigma}^{\text{single}}$ and T_{e}^{single} denote the amplitude transmission coefficient and energy transmission coefficient, respectively, of stress wave propagation through a single-scale discontinuous rock mass, namely macro-joints in the rock mass are considered solely.

Figure 8.53a shows the combined effect of the incident wave amplitude A_0 and static prestress σ_s on the difference in the amplitude transmission coefficient ΔT_{σ}. Figure 8.53b shows the combined effect of the incident wave amplitude A_0 and static prestress σ_s on the difference in the energy transmission coefficient ΔT_e. The propagation distance L is fixed at 3.0 m. The incident wave frequency is fixed at 1000 Hz. Figure 8.53a shows that, for a given incident wave amplitude, the difference in the amplitude transmission coefficient ΔT_{σ} increases to a peak value as the static prestress increases and then decreases and tends to a constant value as the static prestress further increases. For a given smaller static prestress, the difference in the amplitude transmission coefficient ΔT_{σ} increases as the incident wave amplitude increases and then shows a slightly decreasing trend as the incident wave amplitude further increases. However, for a given larger static prestress, the difference in the amplitude transmission coefficient ΔT_{σ} remains constant. As shown in Fig. 8.53a, when the static prestress and incident wave amplitude are small, the difference in the amplitude transmission coefficient ΔT_{σ} is small. This finding shows the relatively small effect of micro-defects on wave propagation.

Figure 8.53b shows that, for a given incident wave amplitude, the difference in the energy transmission coefficient ΔT_e rapidly increases as the static prestress increases and tends to a constant value when the static prestress is sufficiently large. For a given small static prestress, the difference in the energy transmission coefficient ΔT_e increases as the incident wave amplitude increases. For a given large static prestress, the difference in the energy transmission coefficient ΔT_e remains a constant value as the incident wave amplitude increases. Figure 8.53a shows that when the static prestress and incident wave amplitude are small, the difference in the energy transmission coefficient ΔT_e is small.

(5) Combined effect of incident wave frequency and static prestress on the difference in transmission coefficient

Figure 8.54a shows the combined effect of the incident wave frequency f_0 and static prestress σ_s on the difference in amplitude transmission coefficient ΔT_{σ}. Figure 8.54b shows the combined effect of the incident wave frequency f_0 and static prestress σ_s on the difference in the energy transmission coefficient ΔT_e. The incident wave amplitude is fixed at 0.5 MPa. The propagation distance is fixed at 6.0 m. Figure 8.54a and b shows that when the incident wave frequency is small (e.g., $f_0 < 10^1$ Hz), the difference in the amplitude transmission coefficient ΔT_{σ} and the difference in the energy transmission coefficient ΔT_e are close to zero as the static prestress increases. This phenomenon indicates that micro-defects in the rock mass can be disregarded for small-frequency stress waves The traditional study

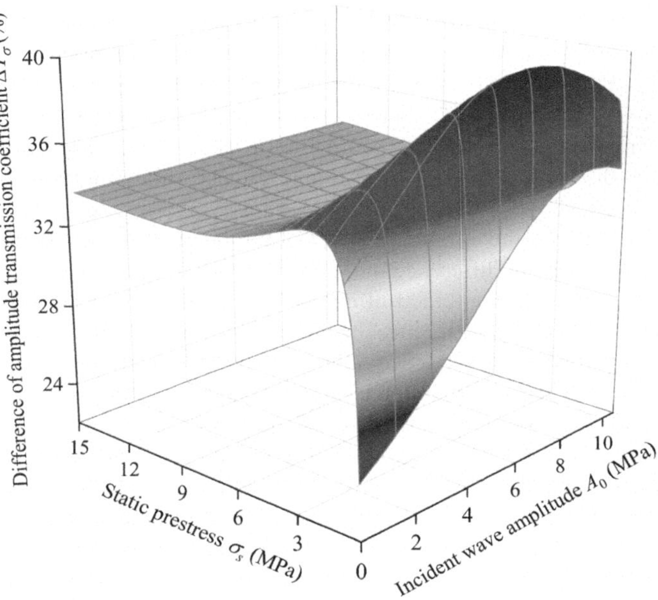

(a)

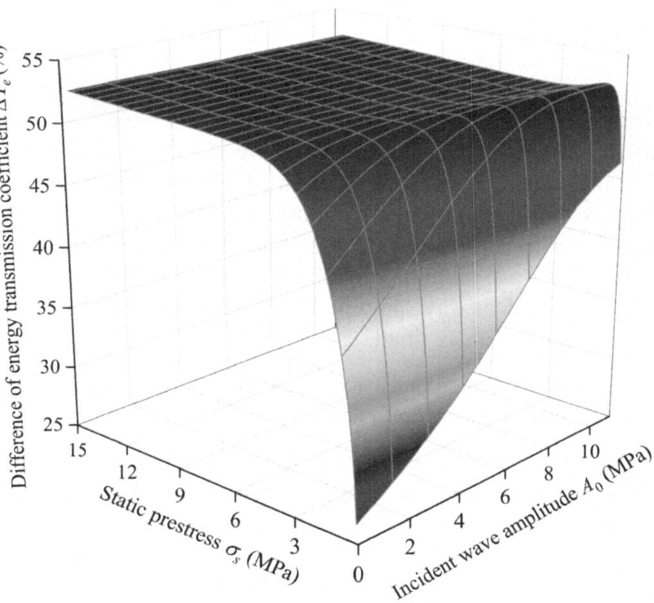

(b)

Fig. 8.53 Combined effect of incident wave amplitude and static prestress on transmission coefficients. **a** Difference in the amplitude transmission coefficient. **b** Difference in the energy transmission coefficient without considering micro-defects can be applied instead of the present study to investigate the stress wave propagation. However, when the incident wave frequency is large (e.g., $f_0 > 10^1$ Hz), the difference in amplitude transmission coefficient ΔT_σ and the difference in energy transmission coefficient ΔT_e increase as the static prestress increases at the initial stage, which approaches a constant value as the static prestress further increases. As shown in Fig. 8.54a and b, for a given static prestress, the difference in the amplitude transmission coefficient ΔT_σ and the difference in the energy transmission coefficient ΔT_e remain constant at zero when the incident wave frequency is small (e.g.,$f_0 < 10^1$ Hz). The difference in the amplitude transmission coefficient ΔT_σ and the difference in the energy transmission coefficient ΔT_e increase as the incident wave frequency increases when the incident wave frequency is large (e.g., $f_0 > 10^1$ Hz). Figure 8.54a and b shows that when the incident wave frequency is small, the difference in the amplitude transmission coefficient ΔT_σ and the difference in the energy transmission coefficient ΔT_e are small for arbitrary static prestress. When the incident wave frequency and static prestress are large, the difference in the amplitude transmission coefficient ΔT_σ and the difference in the energy transmission coefficient ΔT_e are large.

(6) Combined effect of propagation distances and static prestress on the difference in transmission coefficient

Figure 8.55a shows the combined effect of the propagation distances L and static prestress σ_s on the difference in the amplitude transmission coefficient ΔT_σ. Figure 8.55b shows the combined effect of the propagation distances L and static prestress σ_s on the difference in the energy transmission coefficient ΔT_e. The incident wave amplitude is fixed at 0.5 MPa. The incident wave frequency is fixed at 1000 Hz. Figure 8.55a and b shows that, for a given small propagation distance, the difference in the amplitude transmission coefficient ΔT_σ and the difference in the energy transmission coefficient ΔT_e increase slightly and then decrease with an increase in static prestress and tend to a constant value. For a given large propagation distance, the difference in the amplitude transmission coefficient ΔT_σ and the difference in the energy transmission coefficient ΔT_e increase with an increase in static prestress and then tend to a constant value. In addition, Fig. 8.55a and b shows that, for a given arbitrary static prestress, the difference in the amplitude transmission coefficient ΔT_σ and the difference in the energy transmission coefficient ΔT_e increase as the propagation distance increases. When the static prestress and propagation distance are small, the difference in the amplitude transmission coefficient ΔT_σ and the difference in the energy transmission coefficient ΔT_e are small and close to zero. The effect of the micro-defect in the rock mass on wave propagation can be disregarded. When the static prestress and propagation distance are large, the difference in the amplitude transmission coefficient ΔT_σ and the difference in the energy transmission coefficient ΔT_e are large.

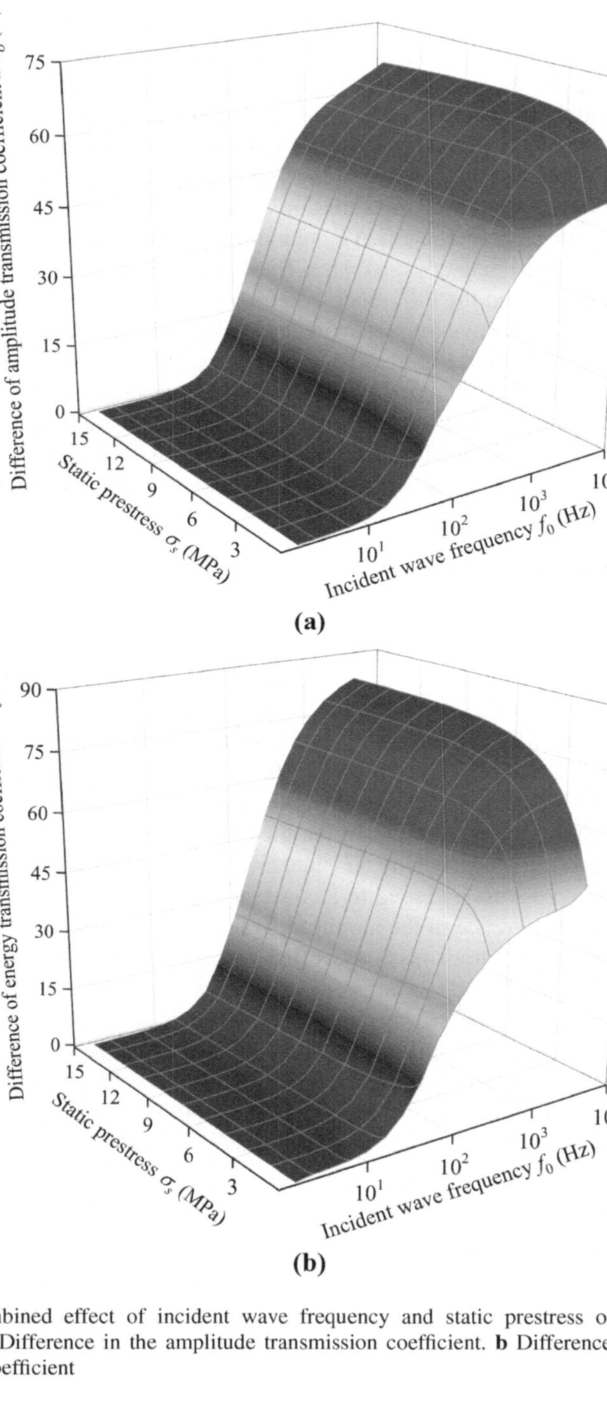

Fig. 8.54 Combined effect of incident wave frequency and static prestress on transmission coefficients. **a** Difference in the amplitude transmission coefficient. **b** Difference in the energy transmission coefficient

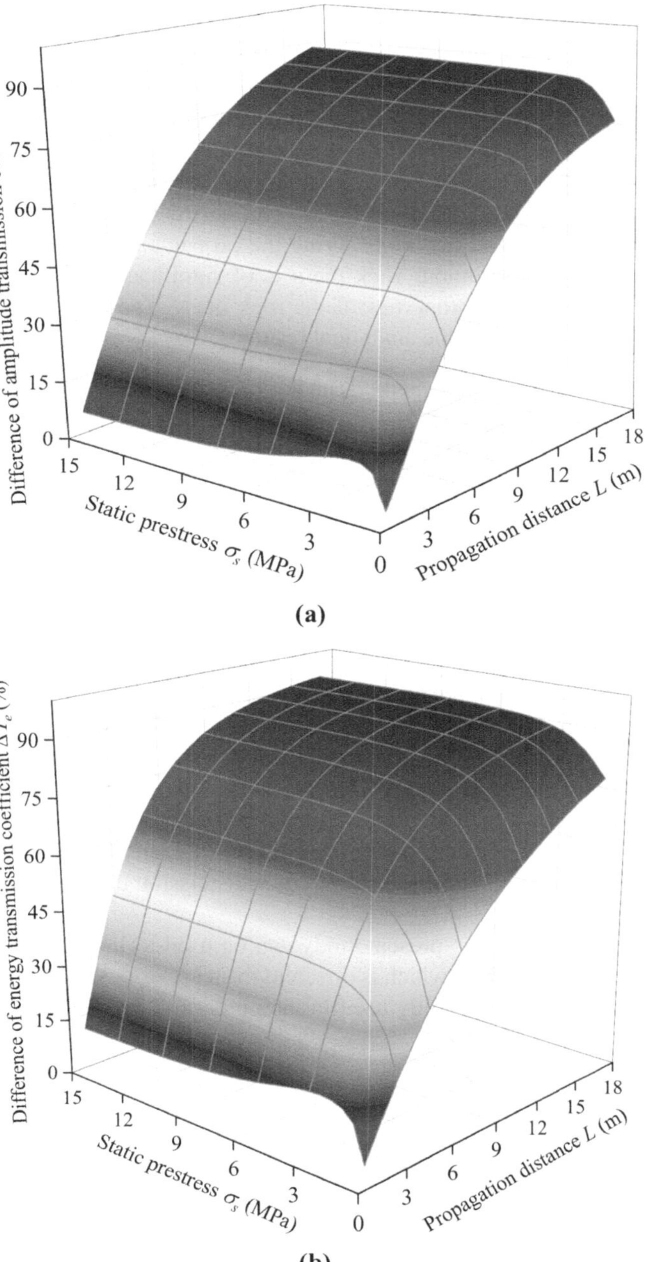

Fig. 8.55 Combined effect of propagation distance and static prestress on **a** difference in the amplitude transmission coefficient and **b** difference in the energy transmission coefficient

8.6 Conclusions

(1) A split three-characteristic method combined with linear DDM is introduced for the investigation of stress wave propagation in the rock mass with double-scale discontinuities. The effect of micro-defects and macro-joints on stress wave propagation is evaluated by studying the transmission coefficients of MD and MJ. The effect of MJ and MD-MJ on transmission stress waves is further compared. It is found that considering the propagation of stress waves in the double-scale discontinuous rock mass is of significance to studying the real dynamic mechanical properties of the rock mass. The following conclusions can be drawn based on the analytical results.

The proposed characteristic method can be utilized for calculating the propagation of stress waves in double-scale discontinuous rock mass considering the effect of micro-defect and macro-joint.

The mechanism of MD and MJ on stress wave propagation is different. The effect of MJ on stress wave propagation is related to the macro-joint stiffness, while the effect of MD on stress wave propagation is related to the viscous coefficient and propagation distance.

When the propagation distance of the stress wave is small enough, the MJ can replace the MD-MJ to obtain an approximate solution. However, when the propagation distance of the stress wave is large enough, the effect of micro-defect on the stress wave attenuation cannot be ignored.

For the transmission coefficient, the micro-defect in MD-MJ has a more obvious influence in the medium-frequency region than in other frequency regions. For the effective velocity, the micro-defect in MD-MJ has a more obvious influence in the small-frequency region than in other frequency regions.

In addition, when the macro-joint stiffness is sufficiently large, the micro-defect in MD-MJ has a significant effect on both of transmission coefficient and effective velocity.

It is summarized that the present study can be used to investigate the effect of double-scale discontinuities, which is different from previous studies. Moreover, the present method can be efficiently used in the numerical simulation of its different formats.

(2) A method of combining the split three-characteristic lines with the piecewise linearly DDM is introduced to study the transmission properties of a double-scale discontinuous rock mass with a nonlinearly deformational macro-joint and compared with the transmission properties of a double-scale discontinuous rock mass with a linearly deformational macro-joint. The effect of incident stress wave amplitude and the length of representative elementary volume (REV) on the transmitted stress wave is analyzed. Moreover, by comparing the effect of nonlinearly and linearly deformational macro-joint in the double-scale discontinuous rock mass on stress wave propagation, the difference in the transmission coefficients is further studied.

The wave transmitted through a nonlinearly deformational macro-joint in the double-scale discontinuous rock mass is related to the amplitude. The transmitted wave through micro-defects in the double-scale discontinuous rock mass is related to the length of representative elementary volume (REV).

A linearly deformational macro-joint can be used to replace a nonlinearly deformational macro-joint when the incident stress wave amplitude is sufficiently small. However, the effect of the nonlinearly deformational behavior of macro-joint on the stress wave propagation cannot be neglected when the incident stress wave amplitude is large. In addition, the nonlinearly deformational macro-joint can be simplified as a linearly deformational macro-joint to study large amplitude stress wave propagation when the frequency is within the small-frequency or large-frequency region.

The effect of the nonlinearly deformational macro-joint on the stress wave propagation can be neglected when the incident stress wave amplitude and initial macro-joint stiffness are sufficiently large, and the effect of micro-defect in the double-scale discontinuous rock mass must be considered. In addition, the influence degree of micro-defect on the stress wave propagation through the double-scale discontinuous rock mass increases with the increase of incident stress wave frequency.

(3) The present section extends traditional stress waves with a single pulse propagation to a stress wave with multiple pulse propagation. A modified split three-characteristic method considering the different mechanical behaviors of macro-joint under multiple pulses is proposed, which is used to investigate the multiple pulse propagation in the rock mass with micro-defect and macro-joint efficiently.

It is concluded that the study of multiple pulse propagation in the rock mass with micro-defect and macro-joint can be used to predict the wave transmission properties in practical engineering more accurately. For the stress waves with multiple pulse propagation, the transmission coefficient, effective velocity and energy transmission coefficient of subsequent pulses are larger than those of the former pulses, and the subsequent pulses cannot be ignored.

For the transmission coefficient and energy transmission coefficient, the first pulse, the second pulse and the third pulse have more obvious differences in the medium-frequency region than other frequency regions. For the effective velocity, the first pulse, second pulse and third pulse have more obvious differences in the small-frequency region than other frequency regions.

When the incident pulse amplitude is small, the transmission coefficient, the effective velocity and energy transmission coefficient of the first pulse, the second pulse and the third pulse are significantly different. When the amplitude of the incident pulses is large, the difference becomes smaller and the rock mass with micro-defect and macro-joint can be regarded as the rock mass with micro-defect, from which the influence degree of micro-defect on the stress wave propagation can be evaluated.

In addition, the frequency-dependence of the effective velocity decreases as the number of pulses increases, which has similar effect rules for the amplitude-dependence of the transmission coefficient, the effective velocity and the energy transmission coefficient. It can be predicted that when the number of pulses is sufficiently large, the frequency-dependence of the effective velocity can be ignored, and the amplitude-dependence of the transmission coefficient, the effective velocity and energy transmission coefficient can also be ignored.

(4) The stress wave propagation through a double-scale discontinuous rock mass considering unloading behavior is theoretically investigated, in which the different loading and unloading behaviors of macro-joint are considered. The effect of incident wave waveform, incident wave frequency, incident wave amplitude and propagation distance on the energy transmission coefficient and amplitude transmission coefficient is analyzed, respectively. The results of the present study are compared with those of traditional study I considering unloading behavior, and traditional study II without considering unloading behavior.

It is concluded that the present study can effectively study the effect of unloading behavior on stress wave propagation in the double-scale discontinuous rock mass. The effect of unloading behavior of the macro-joint and the micro-defect on stress wave propagation cannot be ignored. The unloading behaviors of macro-joint cause a decrease in the energy transmission coefficient, while not affecting the amplitude transmission coefficient. The micro-defect results in decreases in the energy and amplitude transmission coefficients.

The energy and amplitude transmission coefficients for rectangular waves, half-sinusoidal waves and triangular waves successively decrease. The energy and amplitude transmission coefficients decrease as the incident wave frequency increases, and increases as the incident wave amplitude increases. In addition, the energy and amplitude transmission coefficients of the present study decrease as the propagation distance increases.

The energy and amplitude transmission coefficients of the present study are always smaller than those of the two traditional studies. However, when the propagation distance is small, the traditional study I can be used as a special case of the present study to investigate the energy and amplitude transmission coefficients. In addition, the traditional study II can also be used as a special case of the present study to investigate the amplitude transmission coefficients.

(5) The wave propagation through double-scale discontinuous rock mass under combined static-dynamic loads is investigated by the split three-characteristic method considering the static prestress. The variation in the amplitude and energy transmission coefficient with static prestress is investigated. The influence of incident wave amplitude, incident wave frequency and propagation distance on amplitude and energy transmission coefficients is analyzed under different static prestresses. The results of the present study are compared with those of traditional studies without considering micro-defects. The differences in the amplitude and energy transmission coefficients are discussed. The effect of micro-defects on wave propagation in a rock mass is evaluated.

It is demonstrated that the properties of wave transmission are highly dependent on the static prestress. The amplitude and energy transmission coefficients increase and tend to a constant value as the static prestress increases. However, when the incident wave amplitude is large, the effect of static prestress on the amplitude transmission coefficient is negligible, while its effect on the energy transmission coefficient is not negligible. When the incident wave frequency is extremely small or extremely large, the effect of the static prestress on the amplitude and energy transmission coefficients is negligible.

For an arbitrary static prestress, the amplitude and energy transmission coefficients considering micro-defects are smaller than those without considering micro-defects, which is caused by the effect of micro-defects on stress wave propagation. The effect of micro-defects on wave propagation is slight for small static prestress. However, the effect of micro-defects on wave propagation is relatively significant for large static prestress.

In addition, when the static prestress and propagation distance are small or when the incident wave frequency is small, the difference in the amplitude transmission coefficient and the difference in the energy transmission coefficient are close to zero. In this case, the effect of micro-defects in the rock mass on wave propagation can be disregarded. The traditional study without considering micro-defects can be considered a special case of the present study.

References

1. Chen X, Li JC, Cai MF, et al. A further study on wave propagation across a single joint with different roughness. Rock Mech Rock Eng. 2016;49(7):2701–9.
2. Feng P, Dai F, Liu Y, et al. Coupled effects of static-dynamic strain rates on the mechanical and fracturing behaviors of rock-like specimens containing two unparallel fissures. Eng Fract Mech. 2019;207:237–53.
3. Niu LL, Zhu WC, Li SH, et al. Determining the viscosity coefficient for viscoelastic wave propagation in rock bars. Rock Mech Rock Eng. 2018;51(5):1347–59.
4. Zhao J, Cai JG. Transmission of elastic p-waves across single fractures with a nonlinear normal deformational behavior. Rock Mech Rock Eng. 2001;34(1):3–22.
5. Zhao J, Cai JG, Zhao XB, et al. Dynamic model of fracture normal behaviour and application to prediction of stress wave attenuation across fractures. Rock Mech Rock Eng. 2008;41(5):671–93.
6. Pyrak-Nolte LJ, Myer LR, Cook NGW. Transmission of seismic waves across single natural fractures. J Geophys Res. 1990;95(B6):8617–38.
7. Wang LL, Labibes K, Azari Z, et al. Generalization of split Hopkinson bar technique to use viscoelastic bars. Int J Impact Eng. 1994;15(5):669–86.
8. Deng XF, Zhu JB, Chen SG, et al. Some fundamental issues and verification of 3DEC in modeling wave propagation in jointed rock masses. Rock Mech Rock Eng. 2012;45(5):943–51.
9. Ren F, Fan L, Ma G. Simulation of viscoelastic behavior of defected rock by using numerical manifold method. Front Archit Civ Eng China. 2011;5(2):199–207.
10. Li J, Ma G, Huang X. Analysis of wave propagation through a filled rock joint. Rock Mech Rock Eng. 2009;43(6):789–98.
11. Li H, Liu T, Liu Y, et al. Numerical modeling of wave transmission across rock masses with nonlinear joints. Rock Mech Rock Eng. 2015;49(3):1115–21.
12. Fakhimi AA, Fairhurst C. A model for the time-dependent behavior of rock. Int J Rock Mech Min Sci Geomech Abstr. 1994;31(2):117–26.
13. Niu LL, Zhu WC, Li S, et al. Spalling of a one-dimensional viscoelastic bar induced by stress wave propagation. Int J Rock Mech Min. 2020;131: 104317.

14. Bandis SC, Lumsden AC, Barton NR. Fundamentals of rock joint deformation. Int J Rock Mech Min. 1983;20(6):249–68.
15. Barton N, Bandis S, Bakhtar K. Strength, deformation and conductivity coupling of rock joints. Int J Rock Mech Min Sci Geomech Abstr. 1985;22(3):121–40.
16. Kulhawy FH. Stress deformation properties of rock and rock discontinuities. Eng Geol. 1975;9(4):327–50.
17. Zhao XB, Zhao J, Cai JG. P-wave transmission across fractures with nonlinear deformational behaviour. Int J Numer Anal Met. 2006;30(11):1097–112.
18. Zhao XB, Zhu JB, Zhao J, et al. Study of wave attenuation across parallel fractures using propagator matrix method. Int J Numer Anal Met. 2012;36(10):1264–79.
19. Zhu J, Hu SS, Wang LL. An analysis of stress uniformity for concrete-like specimens during SHPB tests. Int J Impact Eng. 2009;36(1):61–72.
20. Fan LF, Wang M, Wu ZJ. Effect of nonlinear deformational macrojoint on stress wave propagation through a double-scale discontinuous rock mass. Rock Mech Rock Eng. 2020;54(3):1077–90.
21. Ma GW, An XM. Numerical simulation of blasting-induced rock fractures. Int J Rock Mech Min. 2008;45(6):966–75.
22. Zhang QB, Zhao J. A review of dynamic experimental techniques and mechanical behaviour of rock materials. Rock Mech Rock Eng. 2013;47(4):1411–78.
23. Cai JG, Zhao J. Effects of multiple parallel fractures on apparent attenuation of stress waves in rock masses. Int J Rock Mech Min. 2000;37(4):661–82.
24. Cook NGW. Natural joints in rock—mechanical, hydraulic and seismic behavior and properties under normal stress. Int J Rock Mech Min Sci Geomech Abstr. 1992; 29(3):198–223.
25. Fan LF, Wong LNY. Stress wave transmission across a filled joint with different loading/unloading behavior. Int J Rock Mech Min. 2013;60:227–34.
26. Fan LF, Wang M, Wu ZJ. A split three-characteristics method for stress wave propagation through a rock mass with double-scale discontinuities. Rock Mech Rock Eng. 2020;53(12):5767–79.
27. Mindlin RD. Waves and vibrations in isotropic elastic plates. New York: Pergamon Press; 1960.
28. Miller RK. An approximate method of analysis of the transmission of elastic waves through a frictional boundary. J Appl Mech-T ASME. 1977;44(4):652–6.
29. Ma GW, Fan LF, Li JC. Evaluation of equivalent medium methods for stress wave propagation in jointed rock mass. Int J Numer Anal Met. 2013;37(7):701–15.
30. Schoenberg M. Elastic wave behavior across linear slip interfaces. J Acoust Soc Am. 1980;68(5):1516–21.
31. Ichikawa Y, Kawamura K, Uesugi K, et al. Micro- and macrobehavior of granitic rock: Observations and viscoelastic homogenization analysis. Comput Method Appl Mech Eng. 2001;191(1–2):47–72.
32. Huang XL, Qi SW, Guo SF, et al. Experimental study of ultrasonic waves propagating through a rock mass with a single joint and multiple parallel joints. Rock Mech Rock Eng. 2014;47(2):549–59.
33. Zheng Y, Chen CX, Liu TT, et al. Numerical study of p-waves propagating across deep rock masses based on the hoek–brown model. Int J Geomech. 2020;20(2):04019152.
34. Li JC, Ma GW. Experimental study of stress wave propagation across a filled rock joint. Int J Rock Mech Min. 2009;46(3):471–8.
35. Li ZL, Li JC, Li HB, et al. Effects of a set of parallel joints with unequal close-open behavior on stress wave energy attenuation. Wave Random Complex. 2020;2:1–25.
36. Wu W, Li HB, Zhao J. Dynamic responses of non-welded and welded rock fractures and implications for p-wave attenuation in a rock mass. Int J Rock Mech Min. 2015;77:174–81.
37. Han ZY, Li DY, Zhou T, et al. Experimental study of stress wave propagation and energy characteristics across rock specimens containing cemented mortar joint with various thicknesses. Int J Rock Mech Min. 2020;131: 104352.
38. Avanthi Isaka BL, Ranjith PG, Rathnaweera TD, et al. Quantification of thermally-induced microcracks in granite using x-ray ct imaging and analysis. Geothermics. 2019;81:152–67.

39. Li ZL, Li JC, Li HB, et al. Effects of a set of parallel joints with unequal close-open behavior on stress wave energy attenuation[J]. Wave Random Complex. 2021;31(6):2427–51.
40. Tao M, Ma A, Cao WZ, et al. Dynamic response of pre-stressed rock with a circular cavity subject to transient loading. Int J Rock Mech Min. 2017;99:1–8.
41. Hao XJ, Du WS, Zhao YX, et al. Dynamic tensile behaviour and crack propagation of coal under coupled static-dynamic loading. Int J Min Sci Technol. 2020;30(5):659–68.
42. Chen X, Xu ZY. The ultrasonic p-wave velocity-stress relationship of rocks and its application. B Eng Geol Environ. 2016;76(2):661–9.

Chapter 9
Dual-Mesh Three-Characteristic Method for Stress Wave Propagation Through the Micro-Defected Rock Mass with a Thin Layer-Filled Macro-Joint

Investigation of the effect of discontinuities in the rock mass on stress wave attenuation has always been an important concern in rock mechanics. The discontinuities in the rock mass include macro-joint and micro-defect, which affect the propagation of stress waves. In addition, it should be noted that macro-joints are not only widely distributed in natural rock mass, but also often filled with various materials, such as sand and clay [1, 2]. The transmission properties of stress waves in macro-joint are mostly controlled by the filling material. Therefore, it is difficult to exactly predict the wave attenuation due to the complexity of the multiple factors.

The purpose of this section is to investigate the stress wave propagation through the micro-defected rock mass with the thin layer-filled macro-joint by the proposed dual-mesh three-characteristic method. In the study, the filled macro-joint is equivalent to a thin viscoelastic layer. The micro-defected rock mass is equivalent to a viscoelastic medium. In order to analyze the effect of the micro-defect and the filled macro-joint on the stress wave propagation, the results of the present study are compared with the existing analytical results without considering the micro-defect, and without considering the micro-defect and the macro-joint thickness. Finally, the parametric studies are carried out with respect to the viscosity coefficient in the equivalent viscoelastic model of the micro-defected rock mass, the viscosity coefficient in the equivalent viscoelastic model of the filled macro-joint and the thickness of the filled macro-joint.

9.1 The Dual-Mesh Three-Characteristic Method

The micro-defect and macro-joint are widely distributed in the natural rock mass [3, 4]. The representative elementary volume (REV) of the rock mass with micro-defect and the filled macro-joint is shown in Fig. 9.1, which can be described by a sandwich model. For the given length of the representative elementary volume (REV)

L. Fan et al., *The Method of Characteristics for Stress Wave Propagation in the Rock Mass*, https://doi.org/10.1007/978-981-97-1440-7_9

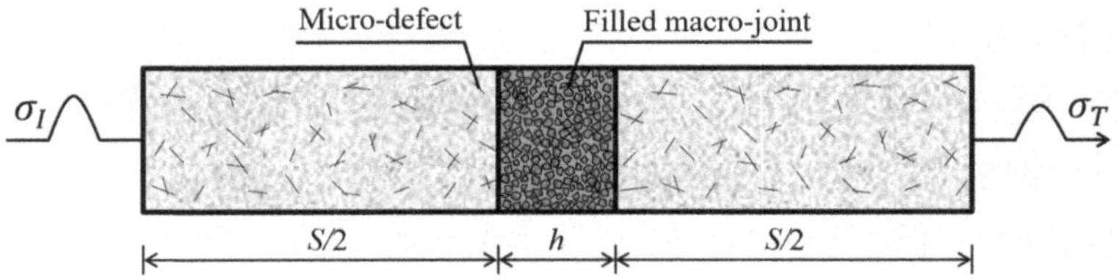

Fig. 9.1 The representative elementary volume (REV) of the rock mass with micro-defect and the thin layer-filled macro-joint

in Fig. 9.1, the effect of micro-defect on the propagation of stress waves is mainly determined by the density of micro-defect in the rock mass. The effect of the macro-joint on the propagation of stress waves is mainly determined by the stiffness of the macro-joint. The density of micro-defects and the stiffness of the filled macro-joint can be reflected by the equivalent viscosity of the viscoelastic medium. Therefore, the micro-defected rock mass beside the macro-joint is equivalently regarded as a viscoelastic medium in Fig. 9.1. It should be noted that when the incident wave frequency is small, that is, the wavelength is larger than the maximum geometric dimension of the micro-defect, the equivalent viscoelastic medium of the micro-defected rock mass is effective to study the stress wave propagation. In addition, the filling material is also equivalent to a viscoelastic medium different from the adjacent rock mass. The filled macro-joint is welded to the adjacent rock mass.

Figure 9.2 shows the viscoelastic sandwich model representing the rock mass with micro-defect and the filled macro-joint. The three-element viscoelastic model of the micro-defected rock mass and filling material is composed of a Maxwell element and a spring in parallel as shown in Fig. 9.2, which can express the rate-dependent effect of the rock mass under stress waves. When the loading rate is small enough, the Maxwell element term is neglected and the model is degraded to a spring model containing only the static elastic modulus of the rock mass. When the loading rate is large enough, the Maxwell element term plays an important role in the viscous behavior of rock mass. Considering that the traditional three-characteristic method can only calculate the stress wave propagation through the single-scale discontinuities [5], a dual-mesh three-characteristic method is proposed as shown in Fig. 9.3, which can be used to calculate the stress wave propagation through the double-scale discontinuities with micro-defect and the filled macro-joint. The three characteristic lines at the location of the filled macro-joint are different from those at the location of the micro-defected rock mass. Figure 9.3 schematically shows that the position for the interface of the left and right sides of the filled macro-joint are at x_1 and x_2, respectively.

Figure 9.4 shows in detail the five basic elements that compose the dual-mesh three-characteristic method. Figure 9.4a b shows the triangle element and diamond element at the location of the micro-defected rock mass, similar to the traditional three-characteristic method, which is used to solve the wave propagation at the left boundary points $P_1(0, j)$ and interior points $P_2(i, j)$ of a rock mass, respectively. In order to further solve the problem of wave propagation on the left side of the filled

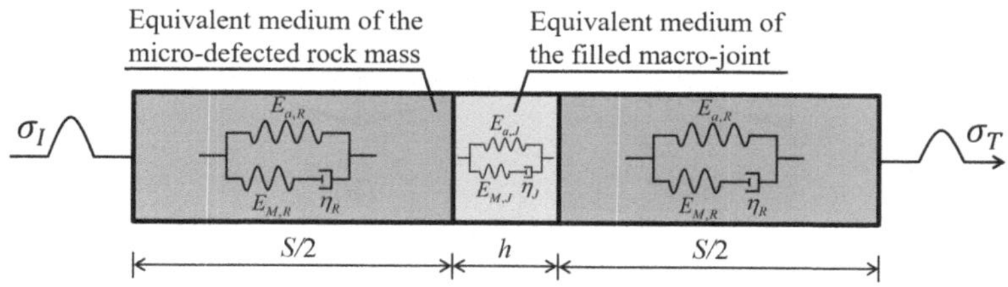

Fig. 9.2 The viscoelastic sandwich model of the rock mass with micro-defect and the filled macro-joint

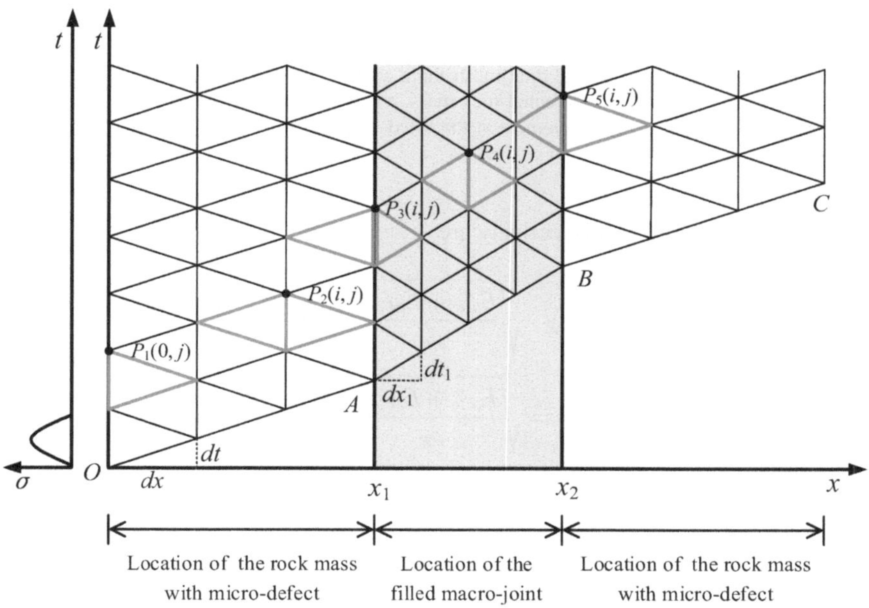

Fig. 9.3 Dual-mesh three-characteristic method

macro-joint at points $P_3(i, j)$, the inside of the filled macro-joint at points $P_4(i, j)$ and the right side of the filled macro-joint at points $P_5(i, j)$. The double triangle element at the left side of the filled macro-joint, the diamond element at the filled macro-joint and the double triangle element at the right side of the filled macro-joint are introduced as can be seen from Fig. 9.4c to Fig. 9.4e.

In the dual-mesh three-characteristic method, the equations of three sets of characteristic lines at the location of the micro-defected rock mass and the corresponding three sets of characteristic compatibility relationship can be obtained, according to the constitutive equation of the three-element viscoelastic model of the micro-defected rock mass, the motion equation and the continuity equation [6]. The equations of the left-running, right-running and up-running characteristic lines at the location of the

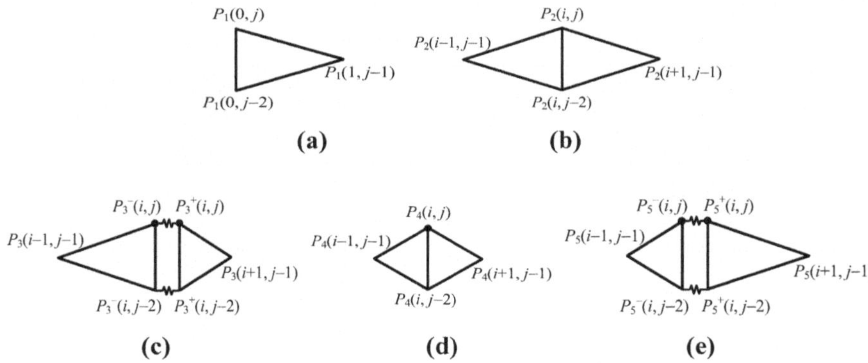

Fig. 9.4 Five basic elements of the dual-mesh three-characteristic method. **a** Triangle element. **b** Diamond element at the micro-defected rock mass. **c** Double triangle element at the left side of the filled macro-joint. **d** Diamond element at the filled macro-joint. **e** Double triangle element at the right side of the filled macro-joint

micro-defected rock mass respectively are

$$\frac{dx}{dt} = -\sqrt{\frac{E_{a,R} + E_{M,R}}{\rho_R}} = -C_R \tag{9.1}$$

$$\frac{dx}{dt} = \sqrt{\frac{E_{a,R} + E_{M,R}}{\rho_R}} = C_R \tag{9.2}$$

and

$$dx = 0 \tag{9.3}$$

The compatibility conditions corresponding to the left-running, right-running and up-running characteristic lines at the location of the micro-defected rock mass respectively are

$$dv = -\frac{1}{\rho_R C_R} d\sigma + \left[\frac{(\sigma - E_{a,R}\varepsilon) E_{M,R}}{(E_{a,R} + E_{M,R}) \eta_R} \right] dx \tag{9.4}$$

$$dv = \frac{1}{\rho_R C_R} d\sigma + \left[\frac{(\sigma - E_{a,R}\varepsilon) E_{M,R}}{(E_{a,R} + E_{M,R}) \eta_R} \right] dx \tag{9.5}$$

and

$$d\varepsilon - \frac{d\sigma}{E_{a,R} + E_{M,R}} - \frac{(\sigma - E_{a,R}\varepsilon) E_{M,R}}{(E_{a,R} + E_{M,R}) \eta_R} dt = 0 \tag{9.6}$$

By a similar derivation, the equations of three sets of characteristic lines at the location of the filled macro-joint and the corresponding three sets of characteristic compatibility relationships can be obtained. The equations of the left-running, right-running and up-running characteristic lines at the location of the filled macro-joint respectively are

$$\frac{dx}{dt} = -\sqrt{\frac{E_{a,J} + E_{M,J}}{\rho_J}} = -C_J \tag{9.7}$$

$$\frac{dx}{dt} = \sqrt{\frac{E_{a,J} + E_{M,J}}{\rho_J}} = C_J \tag{9.8}$$

and

$$dx = 0 \tag{9.9}$$

The compatibility conditions corresponding to the left-running, right-running and up-running characteristic lines at the location of the filled macro-joint respectively are

$$dv = -\frac{1}{\rho_J C_J} d\sigma + \left[\frac{(\sigma - E_{a,J}\varepsilon) E_{M,J}}{(E_{a,J} + E_{M,J})\eta_J} \right] dx \tag{9.10}$$

$$dv = \frac{1}{\rho_J C_J} d\sigma + \left[\frac{(\sigma - E_{a,J}\varepsilon) E_{M,J}}{(E_{a,J} + E_{M,J})\eta_J} \right] dx \tag{9.11}$$

and

$$d\varepsilon - \frac{d\sigma}{E_{a,J} + E_{M,J}} - \frac{(\sigma - E_{a,J}\varepsilon) E_{M,J}}{(E_{a,J} + E_{M,J})\eta_J} dt = 0 \tag{9.12}$$

where x and t are Cartesian coordinate and time, respectively. σ, ε and v are the particle stress, particle strain and particle velocity, respectively. In addition, ρ is the density. C is the wave velocity. η is the viscous coefficient in the three-element viscoelastic model. E_M and E_a are the elastic modulus of spring in the three-element viscoelastic model, respectively. The subscripts "R" and "J" in the above parameters refer to the mechanical parameters of the micro-defected rock mass and the filled macro-joint, respectively.

The six sets of characteristic lines in Fig. 9.3 can be determined according to Eqs. (9.1) to (9.3) and Eqs. (9.7) to (9.9). When the initial and boundary conditions of the rock mass are given, by using the six sets of compatibility conditions along the six sets of characteristic lines, the stress $\sigma(i, j)$, velocity $v(i, j)$ and strain $\varepsilon(i, j)$ of stress wave propagation through the rock mass with micro-defect and the filled macro-joint can be obtained. According to the Hugoniot relations and Eq. (9.5) of compatibility conditions along characteristic line OA, the three unknowns, $\sigma(x, t)$, $v(x, t)$ and $\varepsilon(x, t)$, along characteristic line OA can be obtained firstly as

$$\sigma(x,t) = \sigma(0,0)\exp\left[-\frac{\rho_R C_R x}{2\eta_R\left(1 + E_{a,R}/E_{M,R}\right)^2}\right] \tag{9.13}$$

$$v(x,t) = -\frac{\sigma(0,0)}{\rho_R C_R}\exp\left[-\frac{\rho_R C_R x}{2\eta_R\left(1 + E_{a,R}/E_{M,R}\right)^2}\right] \tag{9.14}$$

and

$$\varepsilon(x,t) = \frac{\sigma(0,0)}{E_{a,R} + E_{M,R}}\exp\left[-\frac{\rho_R C_R x}{2\eta_R\left(1 + E_{a,R}/E_{M,R}\right)^2}\right] \tag{9.15}$$

In addition, the velocity field and stress field on two sides of the welded interface at x_1 are regarded to be continuous. Therefore, the initial stress $\sigma(x_1, t)$ along the characteristic line AB is known. By combining the initial stress $\sigma(x_1, t)$ with the Hugoniot relations and Eq. (9.11) of compatibility conditions along characteristic line AB, the three unknowns, $\sigma(x, t)$, $v(x, t)$ and $\varepsilon(x, t)$, along characteristic line AB can be obtained:

$$\sigma(x,t) = \sigma(0,0)\exp\left[-\frac{\rho_R C_R x_1}{2\eta_R\left(1 + E_{a,R}/E_{M,R}\right)^2} + \frac{\rho_J C_J (x_1 - x)}{2\eta_J\left(1 + E_{a,J}/E_{M,J}\right)^2}\right] \tag{9.16}$$

$$v(x,t) = -\frac{\sigma(0,0)}{\rho_J C_J}\exp\left[-\frac{\rho_R C_R x_1}{2\eta_R\left(1 + E_{a,R}/E_{M,R}\right)^2} + \frac{\rho_J C_J (x_1 - x)}{2\eta_J\left(1 + E_{a,J}/E_{M,J}\right)^2}\right] \tag{9.17}$$

and

$$\varepsilon(x,t) = \frac{\sigma(0,0)}{E_{a,J} + E_{M,J}}\exp\left[-\frac{\rho_R C_R x_1}{2\eta_R\left(1 + E_{a,R}/E_{M,R}\right)^2} + \frac{\rho_J C_J (x_1 - x)}{2\eta_J\left(1 + E_{a,J}/E_{M,J}\right)^2}\right] \tag{9.18}$$

By derivation similar to the above, the three unknowns, $\sigma(x, t)$, $\varepsilon(x, t)$ and $v(x, t)$, along characteristic line BC are expressed as

$$\sigma(x,t) = \sigma(0,0)\exp\left[-\frac{\rho_R C_R (x_1 - x_2 + x)}{2\eta_R\left(1 + E_{a,R}/E_{M,R}\right)^2} + \frac{\rho_J C_J (x_1 - x_2)}{2\eta_J\left(1 + E_{a,J}/E_{M,J}\right)^2}\right] \tag{9.19}$$

$$v(x,t) = -\frac{\sigma(0,0)}{\rho_R C_R}\exp\left[-\frac{\rho_R C_R(x_1 - x_2 + x)}{2\eta_R\left(1 + E_{a,R}/E_{M,R}\right)^2} + \frac{\rho_J C_J(x_1 - x_2)}{2\eta_J\left(1 + E_{a,J}/E_{M,J}\right)^2}\right] \tag{9.20}$$

and

$$\varepsilon(x,t) = \frac{\sigma(0,0)}{E_{a,R} + E_{M,R}}\exp\left[-\frac{\rho_R C_R(x_1 - x_2 + x)}{2\eta_R\left(1 + E_{a,R}/E_{M,R}\right)^2} + \frac{\rho_J C_J(x_1 - x_2)}{2\eta_J\left(1 + E_{a,J}/E_{M,J}\right)^2}\right] \tag{9.21}$$

When the unknowns, $\sigma(i,j)$, $\varepsilon(i,j)$ and $v(i,j)$, along characteristic lines *OA*, *AB* and *BC* are obtained, the remaining solutions inside the dual-mesh three characteristic lines can be obtained according to the five basic elements in Fig. 9.4. Similar to the traditional three-characteristic method, according to the boundary conditions and the compatibility conditions corresponding to characteristic lines $P_1(0,j)$-$P_1(0,j–2)$ and $P_1(0,j)$-$P_1(1,j–1)$ in triangle elements of Fig. 9.4a, the three unknowns at the left boundary points $P_1(0,j)$ can be obtained. In addition, according to the compatibility conditions corresponding to characteristic lines $P_2(i,j)$-$P_2(i–1,j–1)$, $P_2(i,j)$-$P_2(i+1,j–1)$ and $P_2(i,j)$-$P_2(i,j–2)$ in diamond elements at the micro-defected rock mass of Fig. 9.4b, the three unknowns at the interior points $P_2(i,j)$ can be obtained. Similarly, according to the compatibility conditions corresponding to characteristic lines $P_4(i,j)$-$P_4(i–1,j–1)$, $P_4(i,j)$-$P_4(i+1,j–1)$ and $P_4(i,j)$-$P_4(i,j–2)$ in diamond elements at the filled macro-joint of Fig. 9.4d, the three unknowns at the interior points $P_4(i,j)$ can be obtained. The detailed calculation process can be found in the references [7].

Different from the traditional three-characteristic method, the double triangle elements at the left and right sides of the filled macro-joint are proposed to solve the stress wave propagation through the interface at positions x_1 and x_2. The interface at positions x_1 and x_2 is considered welded; the velocity field and stress field on two sides of the welded interface are regarded to be continuous

$$v^-(i,j) = v^+(i,j) \tag{9.22}$$

and

$$\sigma^-(i,j) = \sigma^+(i,j) \tag{9.23}$$

where the superscripts " − " refer to the left side of the position at x_1 and x_2, and the superscripts " + " refer to the right side of the positions at x_1 and x_2.

Substituting Eqs. (9.22) and (9.23) into the compatibility conditions corresponding to characteristic lines in double triangle elements at the left side of the filled macro-joint of Fig. 9.4c, the unknowns, $\sigma(i,j)$, $v(i,j)$, $\varepsilon^-(i,j)$ and $\varepsilon^-(i,j)$, of the points $P_3(i,j)$ on the interface at position x_1 can be obtained. The compatibility conditions corresponding to characteristic lines $P_3^-(i,j)$-$P_3(i–1,j–1)$ is

$$
\begin{aligned}
v^{-}(i,j)-v(i-1,j-1) &= \frac{1}{\rho_R C_R}\left[\sigma^{-}(i,j)-\sigma(i-1,j-1)\right]\\
&+\frac{\left[\sigma(i-1,j-1)-E_{a,R}\varepsilon(i-1,j-1)\right]E_{M,R}}{\left(E_{a,R}+E_{M,R}\right)\eta_R}\left[x(i,j)-x(i-1,j-1)\right]
\end{aligned}
\tag{9.24}
$$

The compatibility conditions corresponding to characteristic line $P_3^-(i,j)$-$P_3(i+1,j-1)$ is

$$
\begin{aligned}
v^{+}(i,j)-v(i+1,j-1) &= -\frac{1}{\rho_J C_J}\left[\sigma^{+}(i,j)-\sigma(i+1,j-1)\right]\\
&+\frac{\left[\sigma(i+1,j-1)-E_{a,J}\varepsilon(i+1,j-1)\right]E_{M,J}}{\left(E_{a,J}+E_{M,J}\right)\eta_J}\left[x(i,j)-x(i+1,j-1)\right]
\end{aligned}
\tag{9.25}
$$

The compatibility conditions corresponding to characteristic line $P_3^-(i,j)$-$P_3^-(i,j-2)$ is

$$
\begin{aligned}
\varepsilon^{-}(i,j)-\varepsilon(i,j-2) &= \frac{1}{E_{a,R}+E_{M,R}}\left[\sigma^{-}(i,j)-\sigma(i,j-2)\right]\\
&+\frac{\left[\sigma(i,j-2)-E_{a,R}\varepsilon(i,j-2)\right]E_{M,R}}{\left(E_{a,R}+E_{M,R}\right)\eta_R}\left[t(i,j)-t(i,j-2)\right]
\end{aligned}
\tag{9.26}
$$

The compatibility conditions corresponding to characteristic line $P_3^-(i,j)$-$P_3^-(i,j-2)$ is

$$
\begin{aligned}
\varepsilon^{+}(i,j)-\varepsilon(i,j-2) &= \frac{1}{E_{a,J}+E_{M,J}}\left[\sigma^{+}(i,j)-\sigma(i,j-2)\right]\\
&+\frac{\left[\sigma(i,j-2)-E_{a,J}\varepsilon(i,j-2)\right]E_{M,J}}{\left(E_{a,J}+E_{M,J}\right)\eta_J}\left[t(i,j)-t(i,j-2)\right]
\end{aligned}
\tag{9.27}
$$

Substituting Eqs. (9.22) and (9.23) into Eqs. (9.24) and (9.25), the $\sigma(i,j)$ at x_1 can be obtained:

$$
\sigma(i,j) = -\frac{\rho_J C_J \rho_R C_R}{\rho_J C_J+\rho_R C_R}
\begin{bmatrix}
\frac{\left[\sigma(i-1,j-1)-E_{a,R}\varepsilon(i-1,j-1)\right]E_{M,R}}{\left(E_{a,R}+E_{M,R}\right)\eta_R}dx\\
+v(i-1,j-1)-\frac{1}{\rho_R C_R}\sigma(i-1,j-1)\\
+\frac{\left[\sigma(i+1,j-1)-E_{a,J}\varepsilon(i+1,j-1)\right]E_{M,J}}{\left(E_{a,J}+E_{M,J}\right)\eta_J}dx_1\\
-v(i+1,j-1)-\frac{1}{\rho_J C_J}\sigma(i+1,j-1)
\end{bmatrix}
\tag{9.28}
$$

Substituting Eq. (9.28) into Eqs. (9.24) and (9.25), the $v(i,j)$ at x_1 can be obtained. In addition, substituting Eq. (9.28) into Eqs. (9.26) and (9.27), the $\varepsilon^-(i,j)$ and $\varepsilon^-(i,j)$ at left and right sides of the position at x_1 can be obtained, respectively.

Furthermore, substituting Eqs. (9.22) and (9.23) into the compatibility conditions corresponding to characteristic lines in double triangle elements at the right side of the filled macro-joint of Fig. 9.4e, the unknowns, $\sigma(i, j)$, $v(i, j)$, $\varepsilon^-(i, j)$ and $\varepsilon^-(i, j)$ of the points $P_5(i, j)$ on the interface at position x_2 can be obtained. The compatibility condition corresponding to characteristic line $P_5^-(i, j)$-$P_5(i–1, j–1)$ is

$$\begin{aligned} v^-(i,j) - v(i-1,j-1) &= \frac{1}{\rho_J C_J}\left[\sigma^-(i,j) - \sigma(i-1,j-1)\right] \\ &+ \frac{\left[\sigma(i-1,j-1) - E_{a,J}\varepsilon(i-1,j-1)\right]E_{M,J}}{\left(E_{a,J}+E_{M,J}\right)\eta_J}\left[x(i,j) - x(i-1,j-1)\right] \end{aligned} \tag{9.29}$$

The compatibility condition corresponding to characteristic line $P_5^-(i, j)$-$P_5(i + 1, j–1)$ is

$$\begin{aligned} v^+(i,j) - v(i+1,j-1) &= -\frac{1}{\rho_R C_R}\left[\sigma^+(i,j) - \sigma(i+1,j-1)\right] \\ &+ \frac{\left[\sigma(i+1,j-1) - E_{a,R}\varepsilon(i+1,j-1)\right]E_{M,R}}{\left(E_{a,R}+E_{M,R}\right)\eta_R}\left[x(i,j) - x(i+1,j-1)\right] \end{aligned} \tag{9.30}$$

The compatibility condition corresponding to characteristic line $P_5^-(i, j)$-$P_5^-(i, j–2)$ is

$$\begin{aligned} \varepsilon^-(i,j) - \varepsilon(i,j-2) &= \frac{1}{E_{a,J}+E_{M,J}}\left[\sigma^-(i,j) - \sigma(i,j-2)\right] \\ &+ \frac{\left[\sigma(i,j-2) - E_{a,J}\varepsilon(i,j-2)\right]E_{M,J}}{\left(E_{a,J}+E_{M,J}\right)\eta_J}\left[t(i,j) - t(i,j-2)\right] \end{aligned} \tag{9.31}$$

The compatibility conditions corresponding to characteristic line $P_5^-(i, j)$-$P_5^-(i, j–2)$ is

$$\begin{aligned} \varepsilon^+(i,j) - \varepsilon(i,j-2) &= \frac{1}{E_{a,R}+E_{M,R}}\left[\sigma^+(i,j) - \sigma(i,j-2)\right] \\ &+ \frac{\left[\sigma(i,j-2) - E_{a,R}\varepsilon(i,j-2)\right]E_{M,R}}{\left(E_{a,R}+E_{M,R}\right)\eta_R}\left[t(i,j) - t(i,j-2)\right] \end{aligned} \tag{9.32}$$

Substituting Eqs. (9.22) and (9.23) into Eqs. (9.29) and (9.30), the $\sigma(i, j)$ at x_2 can be obtained:

$$\sigma(i,j) = -\frac{\rho_J C_J \rho_R C_R}{\rho_J C_J + \rho_R C_R}\left[\begin{array}{l} \frac{\left[\sigma(i-1,j-1)-E_{a,J}\varepsilon(i-1,j-1)\right]E_{M,J}}{\left(E_{a,J}+E_{M,J}\right)\eta_J}dx_1 \\ -\frac{1}{\rho_J C_J}\sigma(i-1,j-1) + v(i-1,j-1) \\ +\frac{\left[\sigma(i+1,j-1)-E_{a,R}\varepsilon(i+1,j-1)\right]E_{M,R}}{\left(E_{a,R}+E_{M,R}\right)\eta_R}dx \\ -\frac{1}{\rho_R C_R}\sigma(i+1,j-1) - v(i+1,j-1) \end{array}\right] \tag{9.33}$$

Substituting Eqs. (9.33) into Eqs. (9.29) and (9.30), the $v(i,j)$ at x_2 can be obtained. In addition, substituting Eqs. (9.33) into Eqs. (9.31) or (9.32), the $\varepsilon^-(i, j)$ and $\varepsilon^-(i, j)$ at left and right sides of the position at x_2 can be obtained, respectively.

According to the previous calculation, the stress $\sigma(i, j)$, strain $\varepsilon(i, j)$ and velocity $v(i, j)$ of stress wave propagation through the rock mass with micro-defect and the filled macro-joint can be obtained.

9.2 Stress Wave Transmission Based on Dual-Mesh Three-Characteristic Method

In this section, the wave propagation through the rock mass with micro-defect and the thin layer-filled macro-joint is investigated based on the dual-mesh three-characteristic method. The results are compared with the existing analytical results without considering micro-defect, and without considering micro-defect and macro-joint thickness. The results without considering the micro-defect are obtained from analytical studies of the wave propagation through the filled macro-joint, in which the filled macro-joint is modeled as a thin-layer interface model [91]. The filled macro-joint and the rock mass on both sides are regarded as an elastic medium. The results without considering the micro-defect and the macro-joint thickness are based on the DDM to study wave propagation through a linearly macro-jointed rock mass, in which the macro-joint is considered as a zero-thickness interface, and the rock mass on both sides are regarded as an elastic medium.

The parameters of the filled macro-joint and the micro-defected rock mass in the present study are shown in Table 9.1. When the micro-defect effect is not considered, the parameters of the filled macro-joint and the rock mass are shown in.

Moreover, the macro-joint stiffness k_n is defined as the ratio of the elastic modulus of the filled macro-joint to the thickness of the filled macro-joint, i.e., $k_n = E_J/h$. The incident stress wave is assumed to be half-sinusoidal as Table 9.2

$$\sigma(0, t) = \begin{cases} A_0 sin(2\pi f_0 t) & ,\text{when} \quad 0 \le t \le 1/(2f_0) \\ 0 & \quad\quad\quad \text{others} \end{cases} \tag{9.34}$$

Table 9.1 Parameters of the filled macro-joint and the micro-defected rock mass of the present study

Parameters of the filled macro-joint	Value	Parameters of micro-defected rock mass	Value
Elastic modulus $E_{a,J}$	1.50 GPa	Elastic modulus $E_{a,R}$	38 GPa
Elastic modulus $E_{M,J}$	0.95 GPa	Elastic modulus $E_{M,R}$	12 GPa
Viscous coefficient η_J	1.20 MPa·s	Viscous coefficient η_R	25 MPa·s
Density ρ_J	1700 kg/m^3	Density ρ_R	2500 kg/m^3

Table 9.2 Parameters of the filled macro-joint and the rock mass without considering micro-defect

Parameters of the filled macro-joint	Value	Parameters of the micro-defected rock mass	Value
Elastic modulus E_J	2.45 GPa	Elastic modulus E_R	50 GPa
Density ρ_J	1700 kg/m^3	Density ρ_R	2500 kg/m^3

where A_0 is the amplitude and f_0 is the frequency of the incident stress wave.

In the calculation, the amplitude and frequency of the incident wave are $A_0 = 0.02$ MPa and $f_0 = 1000$ Hz, respectively. The length of the rock mass is selected as $S = 10.0$ m, and the thickness of the filled macro-joint is $h = 0.04$ m.

Figure 9.5a shows the transmitted waves of the present study that stress wave transmission through the micro-defected rock mass with the thin layer-filled macro-joints, and compares it with transmitted waves when micro-defects are not considered. Figure 9.5b shows the transmitted waves of the present study, and compares it with transmitted waves when micro-defect and the macro-joint thickness are not considered. In addition, the symbol T in Fig. 9.5 represents the transmission coefficient, which is the ratio of the amplitude of the transmitted stress wave to that of the incident stress wave. It can be observed from Fig. 9.5a that the amplitude of the transmitted stress wave obtained from the present study and the study of micro-defect not considered is attenuated. By comparison, it is observed from Fig. 9.5a that the amplitude of the transmitted wave obtained from the study of micro-defects not considered is larger than that obtained from the present study. In addition, it can be observed from Fig. 9.5b that the amplitude of the transmitted stress wave obtained from the study of micro-defect and the macro-joint thickness not considered is attenuated, and the amplitude of transmitted wave obtained from the study of micro-defect and the macro-joint thickness not considered is larger than that obtained from the present study. It can be observed from Fig. 9.5 that the present study can effectively consider the effect of micro-defects and the thin layer-filled macro-joint on wave propagation.

It also can be observed from Figs. 9.5a and b that the transmission coefficient obtained from the study of micro-defect not considered is $T = 0.8868$, which is the largest compared with the transmission coefficient obtained from the study of micro-defect and the macro-joint thickness not considered and the present study. The transmission coefficient obtained from the study of micro-defect and the macro-joint thickness not considered is $T = 0.8790$, which is slightly smaller than that from the study of micro-defect not considered. The transmission coefficient obtained from the present study is the smallest, which is $T = 0.7730$.

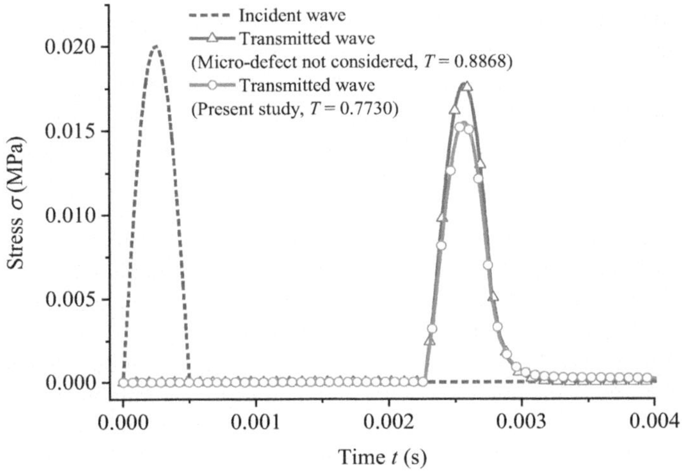

(a)

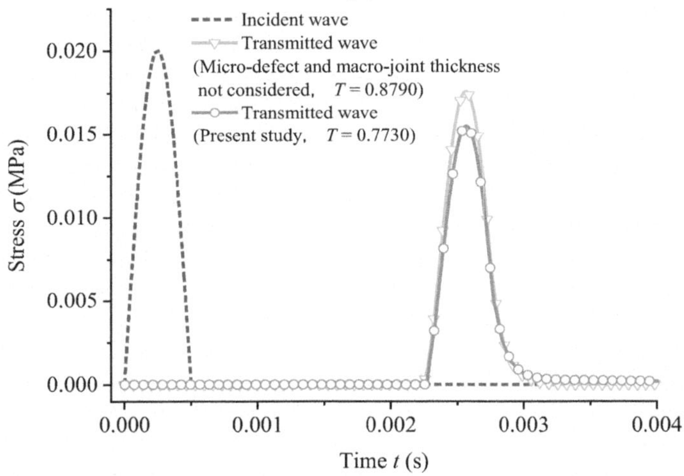

(b)

Fig. 9.5 The transmitted waves of the present study compared with **a** the study of micro-defect not considered and **b** the study of micro-defect and the macro-joint thickness not considered

9.3 Transmission Coefficient of Stress Wave

9.3.1 *Effect of the Viscosity Coefficient of the Micro-Defected Rock Mass on Transmission Coefficient*

Figure 9.6a shows that the effect of the viscosity coefficient in the equivalent viscoelastic model of the micro-defected rock mass on the transmission coefficient obtained from the present study and the study of micro-defect not considered. Figure 9.6b shows that the effect of the viscosity coefficient in the equivalent viscoelastic model of the micro-defected rock mass on transmission coefficient obtained from the present study and the study of micro-defect and the macro-joint thickness not considered. The thickness of the filled macro-joint is selected as $h = 0.02$, 0.05 and 0.08 m, respectively. The incident stress wave frequency is $f_0 = 1000$ Hz, and the viscosity coefficient in the equivalent viscoelastic model of the filled macro-joint is $\eta_J = \infty$.

It is observed from Fig. 9.6a that the transmission coefficient obtained from the present study decreases as the viscosity coefficient in the equivalent viscoelastic model of the micro-defected rock mass increases. After achieving a minimum value, it increases as the viscosity coefficient in the equivalent viscoelastic model of the micro-defected rock mass further increases. The transmission coefficient obtained from the study of micro-defect not considered is not affected by the viscosity coefficient in the equivalent viscoelastic model of the micro-defected rock mass, which remains constant as the increase of the viscosity coefficient in the equivalent viscoelastic model of the micro-defected rock mass as shown in Fig. 9.6a. It is also observed from Fig. 9.6a that the larger thickness of the filled macro-joint results in the small transmission coefficient obtained from the present study and the study of micro-defect not considered for a given viscosity coefficient in the equivalent viscoelastic model of the micro-defected rock mass. By comparison, it is observed from Fig. 9.6a that the transmission coefficient obtained from the study of micro-defects not considered is larger than that obtained from the present study, and the difference first increases, then decreases as the viscosity coefficient in the equivalent viscoelastic model of the micro-defected rock mass increases. When the viscosity coefficient in the equivalent viscoelastic model of the micro-defected rock mass is enough large, the difference can be ignored.

It is observed from Fig. 9.6b that as the viscosity coefficient in the equivalent viscoelastic model of the micro-defected rock mass increases, the transmission coefficient obtained from the present study first decreases and then increases, which is the same as that seen in Fig. 9.6a. The transmission coefficient obtained from the study of micro-defect and the macro-joint thickness not considered remains constant as the increase of viscosity coefficient in the equivalent viscoelastic model of the micro-defected rock mass. It is also observed from Fig. 9.6b that the cases with a larger thickness of the filled macro-joint result in the small transmission coefficient obtained from the study of micro-defect and the macro-joint thickness not considered, which is the same as that of the present study. In addition, Fig. 9.6b also shows that,

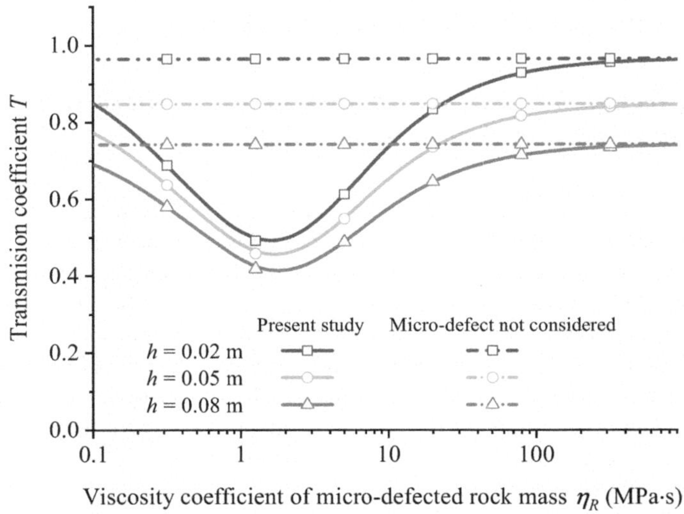

(a)

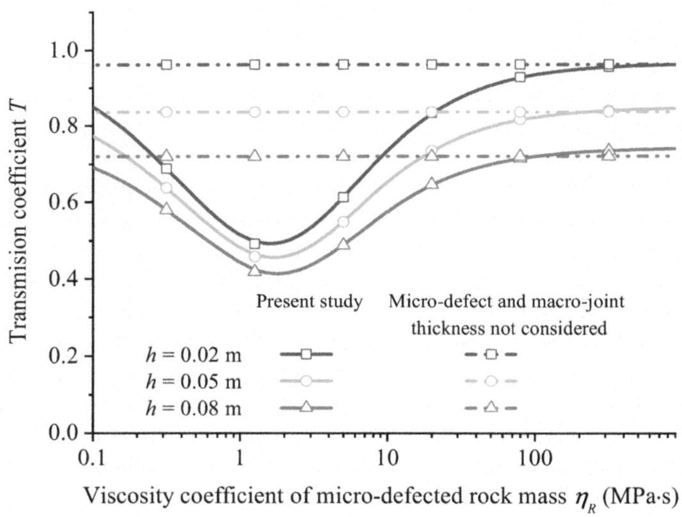

(b)

Fig. 9.6 Effect of the viscosity coefficient of the micro-defected rock mass on transmission coefficient. **a** The transmission coefficient obtained from the present study and the study of micro-defect not considered. **b** The transmission coefficient obtained from the present study and the study of micro-defect and the macro-joint thickness not considered for a small thickness of the filled macro-joint, the transmission coefficients obtained from the study of micro-defect and the macro-joint thickness not considered are larger than those obtained from the present study; the difference first increases, then decreases and gradually approaches zero as the viscosity coefficient in the equivalent viscoelastic model of the micro-defected rock mass increases. For a large thickness of the filled macro-joint, the transmission coefficients obtained from the study of micro-defect and the macro-joint thickness not considered are larger than those obtained from the present study when the viscosity coefficient in the equivalent viscoelastic model of the micro-defected rock mass is small. However, the transmission coefficients obtained from the study of micro-defect and the macro-joint thickness not considered are smaller than those obtained from the present study, when the viscosity coefficient in the equivalent viscoelastic model of the micro-defected rock mass is sufficiently large.

9.3.2 Effect of the Viscosity Coefficient of the Filled Macro-Joint on Transmission Coefficient

Figure 9.7a shows that the effect of the viscosity coefficient in the equivalent viscoelastic model of the filled macro-joint on the transmission coefficient obtained from the present study and the study of micro-defect not considered. Figure 9.7b shows that the effect of the viscosity coefficient in the equivalent viscoelastic model of the filled macro-joint on transmission coefficient obtained from the present study and the study of micro-defect and the macro-joint thickness not considered. The thickness of the filled macro-joint is $h = 0.02$, 0.05 and 0.08 m, respectively. The incident stress wave frequency is $f_0 = 1000$ Hz, and the viscosity coefficient in the equivalent viscoelastic model of the micro-defected rock mass is $\eta_R = \infty$.

It is observed from Fig. 9.7a that the transmission coefficient obtained from the present study increases, and then tends to be flat as the viscosity coefficient in the equivalent viscoelastic model of the filled macro-joint increases. It is also observed from Fig. 9.7a that the transmission coefficient obtained from the study of micro-defect not considered remains constant as the increase of the viscosity coefficient in the equivalent viscoelastic model of the filled macro-joint. Similar to the results obtained from the previous sections, the larger thickness of the filled macro-joint results in the small transmission coefficient obtained from the present study and the study of micro-defects not considered for a given viscosity coefficient in the equivalent viscoelastic model of the filled macro-joint. Moreover, Fig. 9.7a also shows that the transmission coefficients obtained from the study of micro-defects not considered are larger than those obtained from the present study, the difference decreases and gradually decreases to zero as the viscosity coefficient in the equivalent viscoelastic model of the filled macro-joint increases.

It is observed from Fig. 9.7b that the transmission coefficient obtained from the present study increases and then tends to be flat as the viscosity coefficient in the

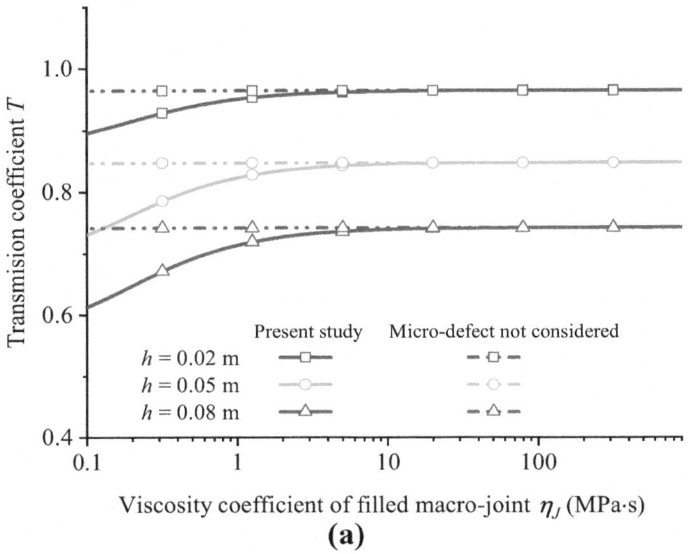

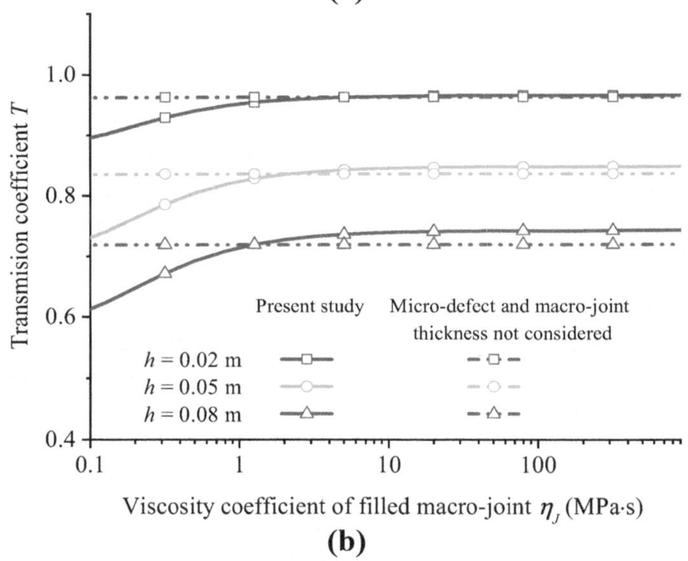

Fig. 9.7 Effect of the viscosity coefficient of the filled macro-joint on transmission coefficient. **a** The transmission coefficient obtained from the present study and the study of micro-defect not considered. **b** The transmission coefficient obtained from the present study and the study of micro-defect and the macro-joint thickness not considered

equivalent viscoelastic model of the filled macro-joint increases, which is the same as that seen in Fig. 9.7a. The transmission coefficient obtained from the study of micro-defect and the macro-joint thickness not considered remains constant as the increase of viscosity coefficient in the equivalent viscoelastic model of the filled macro-joint. Figure 9.7b also shows that the larger thickness of the filled macro-joint results in the small transmission coefficient obtained from the present study and the study of micro-defect and the macro-joint thickness not considered; similar results have been drawn in Fig. 9.7b. In addition, for a small thickness of the filled macro-joint, the transmission coefficients obtained from the study of micro-defect and the macro-joint thickness not considered are larger than those obtained from the present study, the difference decreases and gradually decreases to zero as the viscosity coefficient in the equivalent viscoelastic model of the filled macro-joint increases. For a large thickness of the filled macro-joint, the transmission coefficients obtained from the study of micro-defect and the macro-joint thickness not considered are larger than those obtained from the present study, when the viscosity coefficient in the equivalent viscoelastic model of the filled macro-joint is small. However, the transmission coefficients obtained from the study of micro-defect and the macro-joint thickness not considered are smaller than those obtained from the present study, when the viscosity coefficient in the equivalent viscoelastic model of the filled macro-joint is sufficiently large.

9.3.3 Effect of the Thickness of Filled Macro-Joint on Transmission Coefficient

Figure 9.8a shows the effect of the thickness of the filled macro-joint on the transmission coefficient obtained from the present study and the study of micro-defects not considered. Figure 9.8b shows the effect of the thickness of the filled macro-joint on the transmission coefficient obtained from the present study and the study of micro-defect and the macro-joint thickness not considered. The incident stress wave frequency is $f_0 = 300$, 500 and 700 Hz, respectively. The viscosity coefficient in the equivalent viscoelastic model of the filled macro-joint is $\eta_J = 1.2$ MPa·s. The viscosity coefficient in the equivalent viscoelastic model of the micro-defected rock mass is $\eta_R = \infty$.

It is observed from Fig. 9.8a that the transmission coefficient obtained from the present study and the study of micro-defects not considered decreases as the thickness of the filled macro-joint increases. Figure 9.8a also shows that, for a given thickness of the filled macro-joint, the smaller frequency results in larger transmission coefficients obtained from the present study and the study of micro-defect not considered. In addition, the transmission coefficient obtained from the study of micro-defects not considered is always larger than that obtained from the present study, and the difference increases as the increase of the thickness of the filled macro-joint.

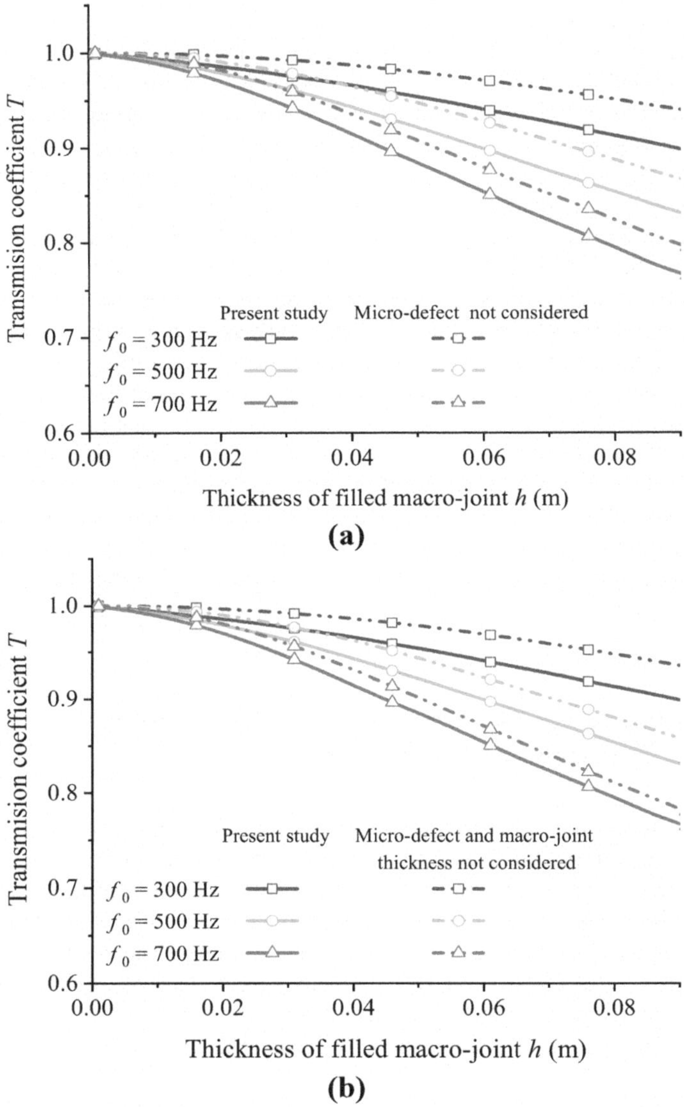

Fig. 9.8 Effect of the thickness of the filled macro-joint on transmission coefficient. **a** The transmission coefficient obtained from the present study and the study of micro-defect not considered. **b** The transmission coefficient obtained from the present study and the study of micro-defect and the macro-joint thickness not considered

It is observed from Fig. 9.8b that the transmission coefficient obtained from the study of micro-defect and the macro-joint thickness not considered decreases as the thickness of the filled macro-joint increases, which is the same as that of the present study. Figure 9.8b also shows that, for a given thickness of the filled macro-joint, the smaller frequency results in larger transmission coefficients obtained from the study of micro-defects not considered, which is also the same as that of the present study. In addition, it can be also observed from Fig. 9.8a that the transmission coefficient obtained from the study of micro-defect and the macro-joint thickness not considered is larger than that obtained from the present study, and the difference increases as the increase of the thickness of the filled macro-joint.

9.4 Conclusions

A dual-mesh three-characteristic method is proposed to investigate the stress wave propagation through the rock mass with micro-defect and the thin layer-filled macro-joint. By comparison with the existing method without considering micro-defect, and without considering micro-defect and the macro-joint thickness, the dual-mesh three-characteristic method is verified to effectively analyze stress wave propagation through the rock mass with micro-defect and the thin layer-filled macro-joint.

It is concluded that the viscosity coefficient in the equivalent viscoelastic model of the micro-defected rock mass, the viscosity coefficient in the equivalent viscoelastic model of the filled macro-joint and the thickness of the filled macro-joint have a distinct effect on the wave attenuation. The transmission coefficient decreases initially and then increases as the viscosity coefficient in the equivalent viscoelastic model of the micro-defected rock mass increases. The transmission coefficient increases as the viscosity coefficient in the equivalent viscoelastic model of the filled macro-joint increases. In addition, the transmission coefficient decreases as the thickness of the filled macro-joint increases.

The present study is compared with the existing study that without considering micro-defect, and without considering micro-defect and the macro-joint thickness. The results show that the effect of micro-defect can be ignored, when the viscosity coefficient in the equivalent viscoelastic model of the micro-defected rock mass and the filled macro-joint are sufficiently large. In addition, the effect of the micro-defect and the macro-joint thickness can be ignored, when the viscosity coefficient in the equivalent viscoelastic model of the micro-defected rock mass and the filled macro-joint are sufficiently large, and the thickness of the filled macro-joint is sufficiently small. Therefore, the present study can be reduced to the existing study without considering micro-defect, and without considering micro-defect and macro-joint thickness.

This section studies the stress wave propagation through the double-scale discontinuous rock mass by the dual-mesh three-characteristic method. Both the properties of the micro-defected rock mass and thin layer-filled macro-joint can be combinate considered in the present study, which is different from previous studies of wave propagation through the jointed rock mass. In addition, the present study focuses on the analysis of the transmission coefficient of stress wave propagation through the micro-defected rock mass with a thin layer-filled macro-joint. In future research, the energy dissipation of stress waves caused by micro-defects and the filled macro-joint can be comprehensively analyzed to extend the present investigation.

References

1. Huang XL, Qi SW, Williams A, et al. Numerical simulation of stress wave propagating through filled joints by particle model. Int J Solids Struct. 2015;69(70):23–33.
2. Zhu JB, Perino A, Zhao GF, et al. Seismic response of a single and a set of filled joints of viscoelastic deformational behaviour. Geophys J Int. 2011;186(3):1315–30.
3. Zhao J, Zhao XB, Cai JG. A further study of p-wave attenuation across parallel fractures with linear deformational behaviour. Int J Rock Mech Min. 2006;43(5):776–88.
4. Niu LL, Zhu WC, Li SH, et al. Determining the viscosity coefficient for viscoelastic wave propagation in rock bars. Rock Mech Rock Eng. 2018;51(5):1347–59.
5. Zhu J, Hu SS, Wang LL. An analysis of stress uniformity for concrete-like specimens during shpb tests. Int J Impact Eng. 2009;36(1):61–72.
6. Wang LL, Labibes K, Azari Z, et al. Generalization of split hopkinson bar technique to use viscoelastic bars. Int J Impact Eng, 1994; 15(5):669.686.
7. Fan LF, Wang M, Wu ZJ. A split three-characteristics method for stress wave propagation through a rock mass with double-scale discontinuities. Rock Mech Rock Eng. 2020;53(12):5767–79.
8. Li JC, Rong LF, Li HB, et al. An shpb test study on stress wave energy attenuation in jointed rock masses. Rock Mech Rock Eng. 2018;52(2):403–20.
9. Zou Y, Li JC, Zhao J. A novel experimental method to investigate the seismic response of rock joints under obliquely incident wave. Rock Mech Rock Eng. 2019;52(9):3459–66.
10. Feng P, Wei MD, Dai F, et al. Dem investigation on the mechanical behaviors of flawed specimens subjected to coupled static-dynamic loads. Soil Dyn Earthq Eng. 2020;135: 106220.
11. Zhu JB, Zhao XB, Wu W, et al. Wave propagation across rock joints filled with viscoelastic medium using modified recursive method. J Appl Geophys, 2012; 86(Null).
12. Huang XL, Qi SW, Liu YS, et al. Stress wave propagation through viscous-elastic jointed rock masses using propagator matrix method (pmm). Geophys J Int. 2015;200(1):452–70.

The manufacturer's authorised representative in the EU is Springer Nature Customer Service Centre GmbH, Europaplatz 3, 69115 Heidelberg, Germany. If you have any concerns regarding our products, please contact ProductSafety@springernature.com

Printed and bound by CPI Group (UK) Ltd, Croydon, CR0 4YY
01/10/2026
02225849-0008